Nutritional Neurosciences

Series Editor

Mohamed Essa, Sultan Qaboos University, Qaboos, Oman

This book series aims to publish volumes focusing on both basic and clinical research in the field of nutritional neuroscience with a focus on delineating the effect of nutrition on brain function and behavior. The books will examine the role of different nutrients, food agents and supplements (both macro and micro) on brain health, neurodevelopment, neurochemistry, and behaviour. The books will examine the influence of diet, including phytochemicals, antioxidants, dietary supplements, food additives, and other nutrients on the physiology and metabolism of neurons, neurotransmitters and their receptors, cognition, behavior, and hormonal regulations.

The books will also cover the influence of nutrients and dietary supplements on the management of neurological disorders. It details the mechanism of action of phytonutrients on signaling pathways linked with protein folding, aggregation, and neuroinflammation. The books published in the series will be useful for neuroscientists, nutritionists, neurologists, psychiatrists, and those interested in preventive medicine.

Amit Kumar Tripathi • Malini Kotak
Editors

Gut Microbiome in Neurological Health and Disorders

Springer

Editors
Amit Kumar Tripathi
Clinical Research Division, School of Basic and Applied Science
Galgotias University
Greater Noida, Uttar Pradesh, India

Malini Kotak
Personal Care Division – Health and Hygiene
ITC Life Sciences and Technology Centre
Bangalore, Karnataka, India

ISSN 2730-6712 ISSN 2730-6720 (electronic)
Nutritional Neurosciences
ISBN 978-981-19-4529-8 ISBN 978-981-19-4530-4 (eBook)
https://doi.org/10.1007/978-981-19-4530-4

This Springer imprint is published by the registered company Springer Nature Singapore Pte Ltd.
The registered company address is: 152 Beach Road, #21-01/04 Gateway East, Singapore 189721, Singapore

Preface

In 2006, neuroscientist Jane Foster and her team were working with two groups of mice: one with a healthy gut microbiome, and the other that lacked a microbiome. They observed that the mice without gut bacteria seemed less anxious than their healthy equivalents. When these mice were placed in a maze with some open paths and some walled-in ones, they preferred the exposed paths. They concluded that the gut bacteria influence the brain and behaviour of these mice. When she tried to publish this finding, it was rejected multiple times as people thought that this was an artefact. Eventually, after 3 years and 7 submissions, her findings were published.

In recent years, however, the gut-brain axis has been established and numerous publications over the past decade have revealed that the gut microbiome could have profound effects on the brain and might be involved in neurological disorders. Animal and preliminary human studies suggest that microbiome can trigger or alter the course of neurological conditions like Parkinson's disease and Autism. Several therapies that modulate the gut microbiome have been reported to have shown promising results and few of these therapies are already being tested in human clinical trials.

The interaction between the gastrointestinal tract (GIT), gut microbiome, and central nervous system (CNS) is complex and bidirectional, for example, a healthy gut environment with diverse microorganisms ensures normal functioning of brain, whereas the CNS modulates most aspects of GIT functioning and physiology. As a result, the term microbiota-gut-brain axis (MGBA) was coined to stress the importance of these interactions. The interactions between MGBA occur by means of neural, endocrine, immune, and humoral links. This book provides an overview of bidirectional communication between gut-microbiome-brain, various pathways, nutrients, metabolites, etc. involved in MGBA interactions, relevance of this axis in context of neurological disorders and potential therapeutic interventions, involving gut microbiome or pro/pre-biotic, which can ameliorate neurological disorders.

Briefly, we have introduced microbiota-gut-brain axis, its interaction with immunity and various metabolites. We have also discussed neurological disorders with focus on neurodegenerative diseases and cerebrovascular diseases. We also provide an overview of potential microbiome-based therapeutics interventions for neurological disorders. Lastly, considering that microbiome is an evolving subject and with an aim of continuous education process of students and researchers, this book provides a detailed understanding of methodologies which can be instrumental in studying microbiome. The text reflects the current state of evidence available in the field of gut microbiome in context of neural health and disease and presents some novel findings including reports on SARS-CoV-2 and COVID-19.

Greater Noida, India Amit Kumar Tripathi
Bangalore, India Malini Kotak

Contents

1 **Gut Microbiome Brain Axis: An Introduction** 1
Santosh Anand, Sunil Kumar Mishra, and Jayam Samlin

2 **Cross Talk Between Gut Microbiota and Host Immune Cells** 7
Ankit Verma, Awakash Soni, and Priya Gupta

3 **Microbiota–Gut–Brain Axis and Neurodegenerative Disorder** 27
Awakash Soni, Ankit Verma, and Priya Gupta

4 **Gut Microbiota Regulation of Cerebral Stroke** 47
Kaushlendra Kumar, Hema Kumari, and Amit Kumar Tripathi

5 **Aging: Impact of Gut Microbiota** 71
Santosh Anand, Ranoji Rao Narasinga Rao Lakshmikanth, Kannasandra Ramaiah Manjula, Doddhakathanahalli Ramanath Jayashree, and Tekupalli Ravikiran

6 **Gut Microbiome Regulation of Appetite and Role in Neurological Disorders** 83
Ankita Singh, Om Prakash Verma, and Rajavashisth Tripathi

7 **Human Diets, Gut Microbiome, and Neuroinflammation** 107
Jyoti Singh, Zoya Khan, and Tripathi Rajavashisth

8 **Dietary Fatty Acids, Gut Microbiome, and Gut–Brain Communication: A Current Perspective** 121
Santosh Anand, S. K. Sukrutha, B. R. Shilpa, and A. Nagarathna

9 **Role of Short-Chain Fatty Acids from Gut Microbiota in Neuroendocrine Pathogenesis Management** 139
Neha Sahu, Prabhat Upadhyay, and Sunil Kumar Mishra

10 Potential Role of Probiotics on Gut Microbiota in Neurological Disease . . . 153
Jovel Varghese Jose and S. Aliya

11 Reversal of Metabolic Disorder Through the Restoration of Gut Microbiota . . . 179
Prabhat Upadhyay, Diya Kalra, Sarika Gupta, and Sunil Kumar Mishra

12 Gut Microbiome and Diet: Promising Approach for Treatment of Cognitive Impairment . . . 195
Awakash Soni, Priya Gupta, and Ankit Verma

13 Nanoplastics, Gut Microbiota, and Neurodegeneration . . . 211
Ananya Rai

14 Gut Microbiome, COVID-19, and Neurological Impairment . . . 235
Richa Das, Riya Singh, and Amit Kumar Tripathi

15 Tools to Study Gut Microbiome . . . 253
K. S. Sreevatshan, Veena G. Nair, C. S. Srinandan, and Ganesh Babu Malli Mohan

16 Germ-free Mice Technology: Opportunity for Future Research . . . 271
Ashish Jain and Anand Maurya

17 Gut Microbiome and Neurodegeneration: A Bioinformatics Approach . . . 297
Swetanshu and Pratichi Singh

About the Editors

Amit Kumar Tripathi is working as Assistant Professor, Clinical Research Division, School of Basic and Applied Science, Galgotias University, Greater Noida, India. He has done his Doctoral Research at CSIR-Central Drug Research Institute and was awarded by Jamia Hamdard, New Delhi. His doctoral and post-doctoral research was focused on the molecular mechanism to understand cerebral stroke and cancer. He has several years of research experience on animal models of various neurological disorders. He has published his research work in various peer-reviewed international journals, books, and chapters. He is involved in teaching various courses such as Recombinant DNA Technology, Nanotechnology, Molecular Therapeutics, Microbial Technology, Animal Biotechnology, Yellow Biotechnology at Bachelor, and Master of Science in School of Basic and Applied Science, Galgotias University. He has been awarded several national and international awards for his contribution to research in health and disease.

Malini Kotak is an Associate Scientist in the Health and Hygiene segment at ITC Ltd., India's leading fast moving consumer goods company. During her doctoral training at the University of Texas at Arlington and post-doctoral training at ITC Ltd., she has extensively worked on gut, hand, skin, and vaginal microbiome and its association with human health and disease. She has more than 10 years of work experience in industry and academia and has done cutting-edge research in fields like biotherapeutics, microbiome, precision medicine, genomics, and cancer biology. She has published several research papers in reputed journals, book chapter, and is a recipient of several research fellowships in India and abroad. Her current research involves the development of novel products which are of relevance in health and hygiene industry.

Chapter 1
Gut Microbiome Brain Axis: An Introduction

Santosh Anand, Sunil Kumar Mishra, and Jayam Samlin

Abstract Human gut is inhabited by diversified microorganisms that play a pivotal role in the maintenance of health and development. The commensal microbes that dwell in the intestine predominantly control host physiology at both local and systemic levels. Human–microbiome connection can be regarded as a step toward integration in the evolutionary process. Numerous emerging diseases are now linked to the loss of one or more of the gut microbiome population. Variations of the gut microbiome in relation to its microbial constitution and function have been intricated in various conditions and diseases. Modulation of this microbiome by various strategies including fortification of pre-, pro- and synbiotics, fatty acids, vitamins, etc. can be employed as a plausible treatment for the management of gut dysbiosis. In this chapter, systematic information has been provided with regard to the impact of gut microbiota on the maintenance of human health.

Keywords Gut microbiota · Brain · Immune system · Aging · Fatty acids · Pre/probiotics

1.1 Introduction

The human gastrointestinal tract is colonized by an array of microbes referred to as microbiota. The gut microbial constitution varies between individuals and evolves all through the host's lifespan and influenced by intrinsic and extrinsic factors

S. Anand (✉)
Department of Biotechnology, School of Applied Sciences, REVA University, Bengaluru, India
e-mail: santosh.anand@reva.edu.in

S. K. Mishra
Department of Pharmaceutical Engineering and Technology, Indian Institute of Technology (Banaras Hindu University), Varanasi, India
e-mail: skmishra.phe@iitbhu.ac.in

J. Samlin
Clinical Research Team, Dozee Health Care, Bengaluru, India

A. K. Tripathi, M. Kotak (eds.), *Gut Microbiome in Neurological Health and Disorders*, Nutritional Neurosciences,
https://doi.org/10.1007/978-981-19-4530-4_1

(Catalkaya et al. 2020). Even though microbes in the gut have been investigated to a greater extent, explorations of the role of these entities in the human gut are limited. Understanding the relatedness of the gut microbiome axis in the maintenance of homeostasis and to the progression of ailments has become the recent focus of the scientific community (Ghaisas et al. 2016). Owing to the multifaceted role of gut microbiota axis and its association with several positive and negative health outcomes, an attempt has been made to collect the information from various studies to emphasize the impact, that these microorganisms have on human health.

1.2 Gut Microbiota and Brain

A growing body of literature revealed the influence of dynamic modulations in the gut microbiota on brain physiology and behavior. It is now evident that several unrelated factors to the nervous system, including the gut microbiota residing in the gastrointestinal tract, play an important part in the regulation of cognitive anomalies leading to neurodegenerative diseases (Gonzalez-Santana and Diaz Heijtz 2020). Although the knowledge of the role and molecular machinery in establishing the association between gut microbial community and the central nervous system is at its infancy, clinical approaches have correlated the variation in the gut microbiota constitution between stroke subjects and healthy individuals and found a close link between the symbiotic gut microbes and stroke regulation. Also, dietary phytochemicals have a role in reversing stroke-induced neuronal cell death by reversing the gut microbiota (Mukherjee et al. 2019; Zhao et al. 2022; Tripathi et al. 2022).

1.3 Gut Microbiota and Immune System

The gut microbiota has surfaced as a key regulatory component of the host immune system by restoring intestinal homeostasis and suppressing inflammation (Dhanesha et al. 2020). The diversified microbial population in the gut metabolizes complex Biomolecules into metabolic products that orchestrate cross-talk between the gut epithelia and cells of the immune system. Epithelial cells in the gut, as a part of immunodefense mechanism, secrete a mucosal barrier to separate the host immune cells and the gut microbes thereby lowering intestinal permeability (Zhang and Frenette 2019). Any discrepancy in the interaction between intestinal microbiota and the host immune cells has been found to enhance the vulnerability toward infections (Yoo et al. 2020).

1.4 Gut Microbiota and Aging

The global population of older adults is rapidly mounting and the health status of this elderly population is therefore posing an alarming issue to be addressed. Aging is a multifaceted process accompanied with several physiological changes affecting metabolic, genomic, and immunological functions (Nagpal et al. 2018; Lewis et al. 2020). These physiological alterations lead to an augmented susceptibility to various infections and diseases thereby increasing mortality. Despite the fact that, the etiologies of age-associated ailments are relatively divergent, significant reports suggest chronic, low-grade inflammation (inflammaging) among the most reliable physiological alteration in age-linked diseases (Buford 2017; Varricchi et al. 2020). The gut microbiota (GM) is modulated during the aging process in relation to composition and functionality (Clements and Carding 2018; Ahmadi et al. 2020). Several findings support the importance of a favorable GM behind healthy aging and any alteration in physiological functions during the aging process has the ability to influence the composition and functions of microbial species inhabiting GM (Tiihonen et al. 2010).

1.5 Gut Microbiota and Diseases

The cross talk between the gut microbiota and the host has gained considerable interest because of its participation in diverse ailments including renal, neurodegenerative, and cardiovascular diseases. The constant communication of the gut microbes and vital organs of the host contributes to restore the host's health and homeostasis (Kumar et al. 2016; Yang et al. 2018). In the recent decade, prevalence of chronic kidney disease is rapidly mounting owing to the aging population. It is primarily caused by glomerulonephritis, hypertension, and diabetes mellitus, leading to end-stage renal disease with high mortality rate linked to cardiovascular diseases, irrespective of active medical strategies. The microbiota has also been documented to influence the cognitive and physiological functions of the nervous system, although the underlying mechanisms have not been established yet. The metabolites synthesized by the gut microbiota have been found to be prognostic markers of arterial thrombosis. Evidences from the preclinical studies have revealed the potency of these metabolites in augmenting the thrombotic potential and in the induction of platelet hyperreactivity (Kim and Song 2020).

1.6 Gut Microbiota and Fatty Acids

Intake of a healthy diet is the major imperative factor influencing host nutrition and metabolism. A well-balanced diet is a prerequisite to maintain healthy aging. Diet enriched with high fat alters the homeostasis, metabolism, and gut commensal microflora. Evidences suggest that, high fat diet (HFD) may lead to the development of a wide array of health impediments such as obesity (Jansen et al. 2021), inflammatory diseases such as colitis (Black et al. 2013), gut dysbiosis (Ananthakrishnan et al. 2014), and intestinal gut barrier impairment (Agus et al. 2016). The present chapter highlights the impact of long- and short-chain fatty acids on gut microbiota in ameliorating various gastrointestinal complications/gut dysbiosis. Long-chain fatty acids, omega fatty acids, also known as polyunsaturated fatty acids and essential fatty acids through fish oil have been documented to have a wide spectrum of health benefits such as in cardiovascular disease (Mozaffarian and Wu 2011; Rohr et al. 2020; Tripathi et al. 2021) neurological disorders (Schunck et al. 2018), pregnancy (Shrestha et al. 2020; Stavrinou et al. 2020), infant growth (Carlson et al. 2013; Derbyshire 2018), aging (Schunck et al. 2018), and inflammatory diseases (Simopoulos 2002; Lauritzen et al. 2016; Innes and Calder 2018).

1.7 Gut Microbiota and Pre/Probiotics

The role of probiotics, prebiotics, and synbiotics influence the healthy microbiota. There is a strong experimental evidence to support this hypothesis. Probiotics comprise lactic acid bacteria and bifidobacteria and prebiotics comprises nondigestible oligosaccharides. It is observed that a combination of probiotics and prebiotics reduced inflammation by reducing salt hydrolase activity and increasing taurine abundance thereby enhancing lifespan by reducing adiposity and leaky gut (Calder 2017; Chander et al. 2018; Kerry et al. 2018; Sanders et al. 2019).

References

Agus A, Denizot J, Thevenot J, Martinez-Medina M, Massier S, Sauvanet P et al (2016) Western diet induces a shift in microbiota composition enhancing susceptibility to adherent-invasive *E. coli* infection and intestinal inflammation. Sci Rep 6:19032

Ahmadi S, Wang S, Nagpal R, Wang B, Jain S, Razazan A et al (2020) A human-origin probiotic cocktail ameliorates aging-related leaky gut and inflammation via modulating the microbiota/taurine/tight junction axis. JCI Insight 5(9):e132055

Ananthakrishnan AN, Khalili H, Konijeti GG, Higuchi LM, Silva PD, Fuchs CS et al (2014) Long-term intake of dietary fat and risk of ulcerative colitis and Crohn's disease. Gut 63(5):776–784

Black MH, Watanabe RM, Trigo E, Takayanagi M, Lawrence JM, Buchanan TA et al (2013) High-fat diet is associated with obesity-mediated insulin resistance and β-cell dysfunction in Mexican Americans. J Nutr 143(4):479–485

Buford TW (2017) (Dis) Trust your gut: the gut microbiome in age-related inflammation, health, and disease. Microbiome 5(80):1–11

Calder PC (2017) Omega-3 fatty acids and inflammatory processes: from molecules to man. Biochem Soc Trans 45:1105–1115

Carlson SE, Colombo J, Gajewski BJ, Gustafson KM, Mundy D, Yeast J et al (2013) DHA supplementation and pregnancy outcomes. Am J Clin Nutr 97(4):808–815

Catalkaya G, Venema K, Lucini L, Rocchetti G, Delmas D, Daglia M, De Filippis A, Xiao H, Quiles JL, Xiao J, Capanoglu E (2020) Interaction of dietary polyphenols and gut microbiota: microbial metabolism of polyphenols, influence on the gut microbiota, and implications on host health. Food Front 1:109–133

Chander AM, Yadav H, Jain S, Bhadada SK, Dhawan DK (2018) Cross-talk between gluten, intestinal microbiota and intestinal mucosa in celiac disease: recent advances and basis of autoimmunity. Front Microbiol 9:2597

Clements SJ, Carding SR (2018) Diet, the intestinal microbiota, and immune health in aging. Crit Rev Food Sci Nutr 58(4):651–661

Derbyshire E (2018) Brain health across the lifespan: a systematic review on the role of omega-3 fatty acid supplements. Nutrients 10(8):1094

Ghaisas S, Maher J, Kanthasamy A (2016) Gut microbiome in health and disease: linking the microbiome-gut-brain axis and environmental factors in the pathogenesis of systemic and neurodegenerative diseases. Pharmacol Ther 158:52–62

Gonzalez-Santana A, Diaz Heijtz R (2020) Bacterial peptidoglycans from microbiota in neurodevelopment and behavior. Trends Mol Med 26(8):729–743

Innes JK, Calder PC (2018) Omega-6 fatty acids and inflammation. Prostaglandins Leukot Essent Fatty Acids 132:41–48

Jansen VL, Gerdes VE, Middeldorp S, van Mens TE (2021) Gut microbiota and their metabolites in cardiovascular disease. Best Pract Res Clin Endocrinol Metab 35(3):101492

Kerry RG, Patra JK, Gouda S, Park Y, Shin HS, Das G (2018) Benefaction of probiotics for human health: a review. J Food Drug Anal 26(3):927–939

Kim SM, Song IH (2020) The clinical impact of gut microbiota in chronic kidney disease. Korean J Intern Med 35(6):1305–1316

Kumar M, Babaei P, Ji B, Nielsen J (2016) Human gut microbiota and healthy aging: recent developments and future prospective. Nutr Healthy Aging 4(1):3–16

Lauritzen L, Brambilla P, Mazzocchi A, Harslof LB, Ciappolino V, Agostoni C (2016) DHA effects in brain development and function. Nutrients 8(1):6

Lewis MC, Childs CE, Perez-Cano FJ (2020) Editorial: sustained effects of early nutrition on immune development and microbiome-immune crosstalk. Front Immunol 11:1687

Mozaffarian D, Wu JH (2011) Omega-3 fatty acids and cardiovascular disease. J Am Coll Cardiol 58(20):2047–2067

Mukherjee S, Tripathi AK, Kumar G, Patnaik R, Dhanesha N, Mishra D (2019) Neuroprotective potential of small molecule phytochemicals in stroke therapy. In: Patnaik R, Tripathi A, Dwivedi A (eds) Advancement in the pathophysiology of cerebral stroke. Springer, Singapore. https://doi.org/10.1007/978-981-13-1453-7-12

Nagpal R, Mainali R, Ahmadi S, Wang S, Singh R, Kavanagh K et al (2018) Gut microbiome and aging: physiological and mechanistic insights. Nutr Healthy Aging 4(4):267–285

Rohr MW, Narasimhulu CA, Rudeski-Rohr TA, Parthasarathy S (2020) Negative effects of a high-fat diet on intestinal permeability: a review. Adv Nutr 11(1):77–91

Sanders ME, Merenstein DJ, Reid G, Gibson GR, Rastall RA (2019) Probiotics and prebiotics in intestinal health and disease: from biology to the clinic. Nat Rev Gastroenterol Hepatol 16(10): 605–616

Schunck WH, Konkel A, Fischer R, Weylandt KH (2018) Therapeutic potential of omega-3 fatty acid-derived epoxyeicosanoids in cardiovascular and inflammatory diseases. Pharmacol Ther 183:177–204

Shrestha N, Sleep SL, Cuffe JSM, Holland OJ, Perkins AV, Yau SY et al (2020) Role of omega-6 and omega-3 fatty acids in fetal programming. Clin Exp Pharmacol Physiol 47(5):907–915

Simopoulos AP (2002) Omega-3 fatty acids in inflammation and autoimmune diseases. J Am Coll Nutr 21(6):495–505

Stavrinou PS, Andreou E, Aphamis G, Pantzaris M, Ioannou M, Patrikios IS et al (2020) The effects of a 6-month high dose omega-3 and omega-6 polyunsaturated fatty acids and antioxidant vitamins supplementation on cognitive function and functional capacity in older adults with mild cognitive impairment. Nutrients 12(2):325

Tiihonen K, Ouwehand AC, Rautonen N (2010) Human intestinal microbiota and healthy ageing. Ageing Res Rev 9(2):107–116

Tripathi AK, Singh RS, Soni A, Tripathi R, Patnaik R (2021) Rodent stroke model guidelines: an update. In: Tripathi AK, Singh AK (eds) Models and techniques in stroke biology. Springer, Singapore. https://doi.org/10.1007/978-981-33-6679-4_1

Tripathi AK, Ray AK, Mishra SK (2022) Molecular and pharmacological aspects of piperine as a potential molecule for disease prevention and management: evidence from clinical trials. Beni-Suef Univ J Basic Appl Sci 11(1):1–24

Varricchi G, Bencivenga L, Poto R, Pecoraro A, Shamji MH, Rengo G (2020) The emerging role of T follicular helper (TFH) cells in aging: influence on the immune frailty. Ageing Res Rev 61: 101071

Yang T, Richards EM, Pepine CJ, Raizada MK (2018) The gut microbiota and the brain-gut-kidney axis in hypertension and chronic kidney disease. Nat Rev Nephrol 14(7):442–456

Yoo JY, Groer M, Dutra SVO, Sarkar A, McSkimming DI (2020) Gut microbiota and immune system interactions. Microorganisms 8(10):1587

Zhang D, Frenette PS (2019) Cross talk between neutrophils and the microbiota. Blood 133(20): 2168–2177

Zhao L, Yang L, Guo Y, Xiao J, Zhang J, Xu S (2022) New insights into stroke prevention and treatment: gut microbiome. Cell Mol Neurobiol 42(2):455–472

Chapter 2
Cross Talk Between Gut Microbiota and Host Immune Cells

Ankit Verma, Awakash Soni, and Priya Gupta

Abstract It has been extensively recognized that large numbers of microorganisms inhabit in our gastrointestinal tract. In recent years gut microbiota has attracted considerable and significant attention due to its diverse and extensive impacts on various aspects of the host's pathophysiology. Numerous studies have approved that vital communication between the intestinal microbiota and immune cells, performed an essential role in the induction, development, training as well as the function of the host's immune response.

The gut microbial communities are essential and support to establish a mutualistic relationship on various levels. This host–microbial relationship influences the host immune system's development and homeostasis. This combination of gut microbiota and the immune system allows protective responses to pathogens and is also linked to the maintenance of regulatory pathways that offer tolerance to mild antigens.

Here, in this chapter, we review the latest research on the role of the interactions between gut microbiota and immune system cross talk as well as the implications of these findings on human health. We also confer microbiome modulation strategies and ongoing studies as well as potential research areas that merit concentrated research efforts.

A. Verma (✉)
Department of Life Sciences and the National Institute for Biotechnology in the Negev, Ben Gurion University of Negev, Beer Sheva, Israel
e-mail: verma@post.bgu.ac.il

A. Soni
Department of Microbiology and Molecular Genetics, The Kuvin Center for the Study of Infectious and Tropical Diseases, The Hebrew University-Hadassah Medical School, Jerusalem, Israel

P. Gupta
Division of Pharmacology, Council of Scientific and Industrial Research-Central Drug Research Institute, Lucknow, India

A. K. Tripathi, M. Kotak (eds.), *Gut Microbiome in Neurological Health and Disorders*, Nutritional Neurosciences,
https://doi.org/10.1007/978-981-19-4530-4_2

Keywords Gut microbiota · Immune cells · Immune system and homeostasis

Abbreviations

CD4 + T cells	Cluster of differentiation 4 of T helper cells
DC	Dendritic cells
GF mice	Germ-free mice
GIT	Gastrointestinal tract
IFN	Interferon
IgA	Immunoglobulin A
IL	Interleukin
ILCs	Innate lymphoid cells
ILCs	Lymphoid cells
LPS	Lipopolysaccharides
M cells	Microfold cells
MLN	Mesenteric lymph node
PSA	Polysaccharide A
SCFAs	Short-chain fatty acids
Th	T helper cells
TLR	Toll-like receptors
TNF-α	Tumor necrosis factor
Treg	Cells-Regulatory T cells

2.1 Introduction

A wide variety of commensal bacteria collectively called the microbiota populates the human body. These microorganisms continue to live on within the human system and exhibit symbiosis with the host for mutual benefits. The human gastrointestinal tract (GI) system is one of the foremost alliances among the host, microbes, environmental factors as well as antigens in the human body (Hooper et al. 2012; Lynch and Pedersen 2016). Although the last two decades, the researcher has the focus of intense studies on gut microbiota. Now, it's well reported that microbiota, has a close correlation with human health and the progression of diseases (Wang et al. 2017; Durack and Lynch 2019; Zheng et al. 2020). The latest developments in microbiota research have also revealed that gut microbes are not only a passive onlooker but also actively participate in multiple host functions, such as dietary responses, pathogen defense, immune functions, and regulating metabolic activities (Zheng et al. 2020). Across all tissues and organs, the human immune system includes a diverse network of innate and adaptive immune systems as well as plays an active role in defending the host against various potentially harmful foreign agents and endogenous homeostasis disruptions (Cho and Blaser 2012). Further, it was recently revealed that the influence of the microbiota is not restricted to the GI tract, but it also plays an important and vital role in the bidirectional communication between the host immune response and GI tract. The mammalian intestine acts as a suitable habitat for host–microbiota interactions. From an ecological point of view, mammals and their intestinal microbial ecosystem have evolved together toward mutualism and hemostasis. This type of an intimate relationship includes the proper functioning of the host's immunity and the prevention of overexploitation of the host's resources, by commensals and the preservation of immune tolerance to harmless stimuli (Hooper et al. 2012; Zheng et al. 2020). In the GI tract, gut microbiota plays an effective immune response stimulator. Microbiota acts as a vital role in the preparation, growth, and work of the host immune system's major components. Respectively, immune signals stimulated by the intestinal microbiota, in effects, serve as potential tools for modulating the intestinal commensals and protecting from an invasion of pathogens (Mao et al. 2018). It is important to know the reciprocal and confounding contact communication between the intestinal microbiome and the host immune system. This cross talk acts as a vital role on the initiation and development of regulatory T cells (Treg cells), T helper (Th) cells that account for the majority of effect on the intestine and innate lymphoid cells (ILCs) as well as B cells that produce immunoglobulin A (IgA). On the other hand, gut microbial metabolites like short-chain fatty acids, polysaccharides, and other molecules play an important role in the host's immune response.

Here, we discuss aspects of existing knowledge and recent evidence suggesting that a beneficial partnership of the gut microbiome with the development and functions of the immune system. We do need to better understand how our microbial allies, as well as the nonbacterial members, contribute or defend against disease. The ultimate goal of translating these findings into clinical applications is a key focus of recent initiatives.

2.2 Gut Microbiota and Immune System Interaction During Development

The gut microbiota, as we introduced in the preceding chapter, is the microbial population of commensal bacteria that populates our gastrointestinal tract (GIT). Recent studies and ongoing development of new tools in the field strongly describe the link between gut microbiota and health and disease. Now question is that, why gut microbiota is essential for a healthy immune system? Various approaches have been used to validate that the cascade of signals generated by gut microbiota plays important role in the professional development of the host's immune system, whereas the immune system coordinates and maintains the key characteristics host–microbe symbiotic relationship. The immune system is essential during both health and diseases because it's responsible for the recognition, response as well as adaptation to countless self and foreign molecules.

However, it is classically assumed that the immune system has evolved as a shield against microbial infection, but animals peacefully coexist with a large and complex microbiota that interacts extensively with the immune system. It is generally recognized that the starting thousands of days from pregnancy and after birth are the most important timescales for intervention and whatever modulation performed at this time has an enormous capacity to enhance the development and growth of the child (Gordon et al. 2012). In normal conditions, the fetal GIT is considered sterile, when the immune system is first exposed to commensals that arise during transit via the birth canal. Such early exposures are deemed to set the long-term tone of the systemic and mucosal immune system. It is still not well understood the mechanism whereby neonate tissues respond to the difficult task of microbial colonization.

Although certain factors in breast milk are believed to describe some of those early responses to commensals. Besides, this factor is associated with the development of the GIT microbiota and the immune system (Belkaid and Hand 2014). Breast milk and colostrum comprise live bacteria, metabolites, IgA, immune cells, and cytokines, as they play a protective role in infants. These factors are synergistic in the formation of the infant microbiota (fed with breast milk) and in the response of the host to these microbes, which regulates the maturation of the newborn intestine and the composition of microbial community (Field 2005). Furthermore, the ability to tolerate the microbes can also be clarified by the relative immateriality of the infant immune system at birth and by the tolerated surroundings conditions that define the early existence of mammals. The evolving immune structure is also considered for the production of inflammatory cytokine output and deform development of T and B cells in favor of regulating the response (PrabhuDas et al. 2011; Siegrist 2001). Recent studies indicated that a given population of erythroid cells enriched in neonates helps sustain this immunoregulatory environment and limits inflammation of the mucosa after colonization by microbiota (Papaioannou et al. 2019). Initial commensals exposure to the host can also suppress cells like invariant natural killer T cells that are involved in the activation of inflammatory reactions, which have long-term implications for the host's ability to precede for inflammatory diseases (Olszak et al. 2012; An et al. 2014).

The recognition of microbial associated molecular patterns contain numerous bacterial antigens (e.g., capsular polysaccharides, LPS, muramic acid, flagellin, and unmethylated bacterial DNA)] is one of the key means of cross talk between the host and gut microbiota (Francino 2014). The innate neonate immune system combines such signals as an exclusive tool to facilitate a healthy microbial population. Such as, in the small intestine, epithelial and lymphoid cell membrane toll-like receptors (TLRs) are involved in this differential recognition, which is responsible for the normal development and growth of the mucosal immune system (Belkaid and Hand 2014). Furthermore, the response of molecular signals by microbial antigens and metabolites induced TLR stimulation. Even though, signaling of TLR expression by neonate innate cells and their sense to microbial ligands has appeared similar to adults. But it's diverse from mature cells with a prominent deficiency in the synthesis of inflammatory moderators such as oxygen radicals as well as the development of regulatory cytokines namely IL-10 (Chassin et al. 2010; Kollmann et al. 2012).

Moreover, the activation of the immune response to various stimuli by NFκB is a key signaling channel pathway and its representation as a significant "conversation node" between microbes and intestinal epithelial cells. NFκB activates various genes that code for cytokines, chemokines as well as other effectors of the humoral immune response (Belkaid and Hand 2014; Thomas and Versalovic 2010). The innate immune system participates in microbial signals, the function of which remains unclear. However, some research findings revealed that epithelial cell expression of enzymes that modify the epigenome may be necessary for the regulation of commensal-dependent gut homeostasis (Thomas and Versalovic 2010; Zhang et al. 2020). The study conducted in GF animals has shown that the gut microbiota performed a significant role in the production and proliferation of secondary and lymphoid structures (Kamada and Núñez 2013; Zhang et al. 2016). These effects are especially evident in the GIT with a smaller patch size of Peyer and decreased number of CD4 + T as well as plasma cells generating IgA (Zhang et al. 2020; Smith et al. 2007). Intestinal tertiary lymphoid structures, such as cytopathic or isolated lymphoid follicles, are caused by commensal exposure after birth (Ohnmacht et al. 2011; Bouskra et al. 2008). Furthermore, commensals can also help to improve the intestinal barrier through several mechanisms, such as promoting the maturation of epithelial cells and angiogenesis (Belkaid and Hand 2014).

The highly regulated process of the newborn's immune system and the function of commensals in the growth and training of this system, contribute to establishing a long-lasting as well as host's homeostatic and commensal relationship. Such important interactions between the gut microbiota and the host's immune organization have significant and long-term human health consequences.

2.3 Translocation of Microbes in the Gastrointestinal Tract

The movement of viable microbes (bacteria) from GIT to other intestinal locations, [mesenteric lymph node (MLN) complex, bloodstream, and remote organs and tissues] is recognized as bacterial translocation. Three key methods facilitate the migration of microbes from the GIT in animal models are recognized: (a) intestinal bacterial overgrowth followed by the distraction of the GI microecology, (b) increased the gut mucosal permeability, and (c) immune deficiencies and immunosuppression (Berg 1985). There has been a very low rate of continuous translocation of many microbial communities through the gut mucosa in healthy adults. Also, host immune defense normally kills these spontaneously translocating bacteria (Belkaid and Hand 2014; Takiishi et al. 2017). Therefore, it is very unusual for viable indigenous gut bacteria to be cultured from the MLN or other extraintestinal sites of a normal adult host with a healthy gut barrier and a proficient immune system. The key path of the passage of native GIT bacteria into a host with the gut barrier is intracellular via the intestinal epithelial cells (IECs) (Takiishi et al. 2017; Tannock 2012). IEC can act as nonprofessional phagocytes that readily absorb any microbes or particles that come into close contact with each other.

Moreover, translocating bacteria move through the IECs lining in the GIT and are transferred through the lymphatics to the MLN. The migrating bacteria may then move from the MLN to other extraintestinal sites (Tannock 2012). Intestinal epithelium tight junction protein (TJ) regulates the passage of micro- and macromolecules from the intestinal lumen to the host. Various studies have shown that luminal bacteria are tested by dendritic cells (DC) which are anchored between epithelial cells via TJ receptors. It has been shown that dendrite protrusions can pass through epithelial junctions to "catch" bacteria from the lumen (Farache et al. 2013; Rescigno et al. 2001; Niess et al. 2005). Following capture by DC in the subepithelial dome region, IgA response activation is activated locally and at sites distant from the mucosa, which may be the desired outcome. It was further suggested that due to the low degree of physiological losses in the epithelial barrier, DC can translocate bacteria that invade the lamina propria (Corthésy et al. 2007).

Another way to pass the epithelium depends on bacterial adhesion to the microfold cells (M cells) that cover the patches of Peyer. Under stable conditions, M cells, macrophages, and DCs extend lumps that sample luminal content and allow access to certain commensals as well as pathogens beyond the epithelial layer without inducing damage (Corthésy et al. 2007; Mabbott et al. 2013). Such types of machinery are essential for the normal stimulation of tolerance to T cells and IgA, as well as for bacterial clearance. For example, live commensal *Enterobacter cloacae* are absorbed by M cells and transported from DC to MLN for IgA responses (Hooper et al. 2012). Furthermore, commensals with lymphoid tissues improve tolerance by inducing IL-10 and IL-22 from DC and innate type 3 lymphoid cells, respectively, which simultaneously inhibit the response of Th17 and encourage bacterial colonization (Fung et al. 2016). In the presence of intact MLNs as well as in the absence of host virulence or deficiencies, such commensals are prevented from further spreading, partially due to effective killing by macrophages (Macpherson and Uhr 2004). On the other side, pathogens have built machinery to manipulate such type of sampling cells and allow them to invade. The paracellular route is attained by pathogens like *Toxoplasma gondii*, *Entamoeba histolytica*, and *Streptococcus* (group-A), which disrupt tight and adherens to increase permeability and assist their translocation. *Salmonella typhimurium* and *Shigella flexneri* target M cells, leading to interruption of the intestinal epithelium and subsequently prompting inflammation and cell death that leads to additional bacterial translocation (Ribet and Cossart 2015; Cywes and Wessels 2001; Hawkins and Byrne 2015).

Therefore, it is evidenced that indigenous bacterial translocation is a crucial early stage in the pathogenesis of opportunistic infections.

2.4 Communication Between the Host's Immune Cells and the Intestinal Microbiome

The immune system is the host's vital defense structure, consisting of a complex network of innate and adaptive machinery in all tissues, which protect against several destructive external agents. It is well recognized that the human GIT is constantly in contact with antigenic load in the form of commensal microbes, metabolites, and dietary antigens. These antigenic agents influence peripheral immune cell populations and thereby modulate the immune system. Cross talk between immune cells of host and the microbiota plays a vital role in maintaining homeostasis in immune response to self and foreign antigen. Now, it is not surprising that interplay between some gut bacteria and immune response has been connected to autoimmune diseases. In this segment, we will try to understand, the relationship between gut microbes and immune machinery (Fig. 2.1).

2.4.1 Dendritic Cell Relationship with Gut Bacteria

Dendritic cells (DC) are professional and frontline antigen-presenting cells which are likely to be fundamental in the balance between tolerance and active immunity against commensal microorganisms (Stagg et al. 2003). Lymphoid and intestinal tissues are the core of a large network of innate immune cells with their role, including macrophages, traditional DCs such as CD11c + DCs and plasmacytoid DCs (Coombes and Powrie 2008). DCs are present throughout the gut-associated lymphoid tissue (GALT) structures of the intestinal immune system, such as the

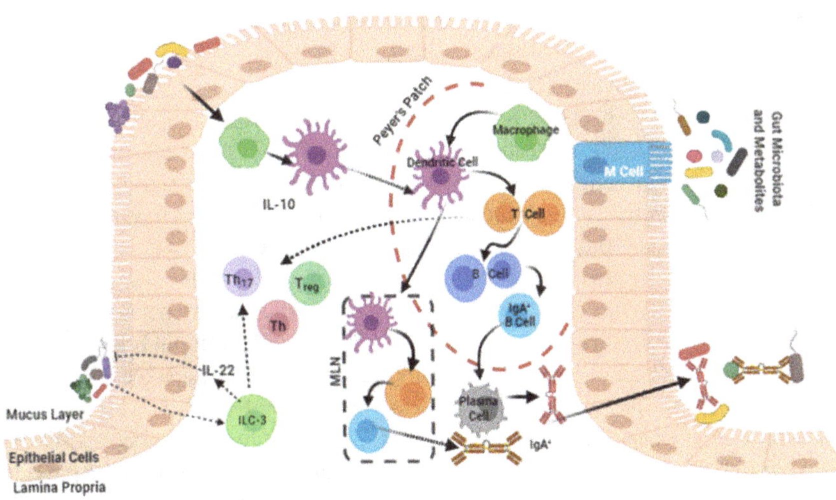

Fig. 2.1 Interaction between the intestinal microbiota and the host immune response

Peyer patches and MLNs, and in the small bowel and the colonic lamina propria (Varol et al. 2010). DCs interact with bacteria that have gained access to the GALT via transcytosis across specialized enterocytes called microfold or M cells, and sense these microbes through pattern recognition receptors such as TLRs (Mabbott et al. 2013). DCs coordinate the optimum balance between different types of effector and regulatory Th cells in their development. DC can also contribute to immune responses by generating interleukin (IL)-12, IL-18, and IL-23. Such cytokines create a microenvironment that favors the production of nonresponsive IgA and T cells to feed antigens (Macpherson and Uhr 2004). Various studies proved that cytokine production by DCs depends on the nature of the environmental stimulus including microbial metabolites. For instance, *Bifidobacterium longum* cell walls stimulate the production of predominately IL-10 while LPS induced IL-12 and slightly IL-10 production (Stagg et al. 2003; Goldsmith et al. 2009).

Further details on how communications between immune cells and microbial metabolites assist in the regulation of immune response, along with the possibility that commensal organisms can be responsible for controlling gut immune homeostasis, as outlined in the following section.

2.4.2 IgA and Gut Microbiome

Immunoglobulin A (IgA) is the most frequently occurring antibody isotype present in mucosal secretions. Secreted IgA target to antigen, thereby preventing them from direct interaction with the host's epithelial cells. The variety of IgA functions on the intestinal surface is a useful starting point when considering the role of IgA in relation to the intestinal microbiota. Though, unlike IgG and IgA in serum, human and murine intestines, IgA is expressed as polymeric IgA (pIgA). pIgA interacts with the polymeric Ig receptor (pIgR) and interaction is essential for active transport as well as secretion of IgA across mucosal surfaces (Pickard et al. 2017; Johansen and Kaetzel 2011). Usually, secreted IgA (sIgA) is generated in a T cell-dependent manner. In particular, bacterial antigen promotes the proliferation of IgA^+ B cells from Peyer's patches to the intestinal stromal layer where IgA^+ B cells develop IgA and secrete it into the gut lumen (Honda and Littman 2016; Matsuo et al. 2018a). Intestinal plasma cells can also produce IgA via an independent mechanism of T cells. For example, the endoplasmic reticulum stress of IEC induces the proliferation and activation of peritoneal B1b cells independently of T cells and microbiota, leading to an intense barrier-protective IgA response. Mucosae-associated epithelial chemokine plays an indirect role by promoting IgA secretion through inducing the homing of IgA antibody-secreting cells (Grootjans et al. 2019; Matsuo et al. 2018b).

Furthermore, in the GF mice, a significant reduction of gut IgA-secreting cells and failure to produce IgA has been reported. Interestingly, this defect was easily recovered after bacterial colonization, suggesting that gut symbiotic bacteria provide crucial stimulatory signals for local sIgA biosynthesis (Hapfelmeier et al. 2010). *Bacteroides fragilis* is an important human gut commensal, that widely modulates its surface components to facilitate the IgA binding in mice, as well as IgA increased mucosal colonization of multiple *B. fragilis* strains. It has also been reported that this commensal has favorable effects that ameliorate inflammatory and behavioral symptoms in experimental animal models (Donaldson et al. 2018). Studies have shown that IgA has raised the adhesion of *Escherichia coli*, *Bifidobacterium lactis*, and *Lactobacillus rhamnosus* to epithelial cells, indicating that all these microbes could also advantage from IgA to establish a mucosa-associated bacterial community (Donaldson et al. 2018; Randal Bollinger et al. 2003; Mathias et al. 2010). Another group of microbes like, prototypical, segmented filamentous bacteria are considered to occupy an unusual niche near the ileal epithelium, where they stimulate the production of IgA and CD4 + Th17 cells proliferation (Bunker and Bendelac 2018).

In addition to IgA antibodies, numerous studies have shown that IgM and IgG antibodies are also capable of reacting to microbiota. There are practically no detectable IgM or IgG-expressing plasma cells in the intestine of murine, although, the IgM^+ and IgG^+ plasma cells are easily detected in the human intestine (Benckert et al. 2011; Rios et al. 2016). While various aspect remains unclear regarding gut's IgG^+ and IgM^+ plasma cells, their specificity similar to IgAs indicates that these cells can be derived from similar precursors (Bunker and Bendelac 2018; Magri et al. 2017). However, IgA helps to prevent possible harmful irritation of the mucosal immune system through the coating and aggregating of the luminal antigens thus firming the intestinal physical barrier (Zhang et al. 2019). Further, during health, IgA fosters colonization of the microbiota in the mucosa with beneficial effects, while pathological states can induce by IgA responses to pathogens that interrupt healthy microbiome equilibrium (Donaldson et al. 2018).

Though IgA stimulates health by controlling the function and architecture of intestinal microbiota, but again the molecular characteristics for this function of homeostatic IgA remain unknown (Nakajima et al. 2018), thus needing more studies in this prominent research area.

2.4.3 Microbiota-Mediated Regulation of Treg Cells, Th17 Cells, and Th1 Cells

T lymphocyte (T cell) is a type of leukocyte, which grows in the thymus gland and acts as a key function in the immune system. As we know that T cells are clusters of subsets based on their functions. In this connection Treg, Th17, and Th1 cells are important subsets of T cells that have vital roles in numerous immune response processes, including inflammation along with disease progression (Zhu and Paul 2010). The dynamic population of Tregs and Th17 cells have been found in abundance in gut mucosa which is also known to possess the largest and most diverse microbiota. Treg modulates the immune response by maintaining tolerance to self and foreign antigens and thus helps in preventing autoimmune disease (Zheng et al. 2020).

FOXP3 (forkhead box P3) is a leading regulator of the regulatory pathway in the progression and function of Treg cells. A prevalent human commensal, *Bacteroides fragilis* in intestinal lamina propria, guides the development of Foxp3+ Treg. Foxp3 + Treg, peripherally separated Treg (pTreg) cells which provide immunological tolerance for nonpathogenic antigens (Round et al. 2011; Fournier and Parkos 2012). Adaptive immunity grows moderately in both antibiotic-treated mice and GF mice. This is characterized by a lack of intestinal Treg cells and by susceptibility to autoimmune responses mediated by Th1, Th2, or Th17 (Ohnmacht et al. 2011; Campbell et al. 2018). Furthermore, IL-10 produce through a self-modulation pathway of Th1 cells which transform pro-inflammatory Th1 cells into regulatory T cells (Rutz et al. 2008).

The microbiome-derived short-chain fatty acids-induced signaling cascade that potentially modulates the Th1 response that may provide novel possibilities for therapeutic approaches in Th1-driven immune diseases (Sun et al. 2018). So, it is evidenced that the gut microbiome has a positive regulatory impact on the initiation, differentiation, and function of Treg cells (Ohnmacht et al. 2011; Fournier and Parkos 2012).

2.4.4 The Gut Microbiome and Innate Lymphoid Cells

Innate lymphoid cells (ILCs) are newly defined innate immune cell that mirrors the features and function of CD4+ T cell subsets. This is found in almost every tissue but specifically enriched at mucosal surfaces, where they are thought to play a key role in preserving epithelial barrier integrity and regulating innate immune responses. ILCs lack antigen-specific receptors and assist as an arm of the innate immune system. Moreover, it's triggered through stress signals, microbial components, and the cytokine microenvironment of the neighboring tissue, rather than via antigen. ILCs are three subtypes ILC1s, ILC2s, and ILC3s, in which ILC3s play a key role in communication between immune cells and gut microbiome as well as to form a network that assists and manages gut homeostasis. ILC3, modulate local processes by which mononuclear phagocytes capture commensal associated antigens in the lamina propria (Pokrovskii et al. 2018; Emgård et al. 2018; Gury-BenAri et al. 2016).

Furthermore, it is well established that ILC3s act as pivotal regulators of adaptive immunity. For instance, it is reported that disruptive adaptive immune responses to the bacterial population were observed in the lack of ILC3 in conventional mice (Hepworth et al. 2013). ILCs maintain intestinal homeostasis through interactions dependent on the major histocompatibility complex II with CD4+ T cells that restrict the adaptive pathological responses of immune cells to commensal bacteria (Sano et al. 2015). However, macrophage-derived IL-1β through microbial sensing, triggered the secretion of granulocyte macrophage colony-stimulating factor (GMCSF) derived from ILC3. This factor facilitates the development of Treg and immune tolerance through the generation of IL-10 by DC and monocytes (Mortha et al. 2014; Shi et al. 2006).

On the other hand, segmented filamentous bacteria induce secretion of IL22 from ILC3, this cytokine further induces epithelial serum amyloid-A (SAA) generation in a Stat3-dependent manner. IEC-derived SAA stimulates Th17 cell differentiation in the microenvironment (Sano et al. 2015). Studies also indicate that the microbiome and ILC3 association often include glial cells (non-neuronal cells in the CNS) in the lamina propria for a healthy gut. The glia cell's neuroregulatory receptor recognizes pathogenic bacteria and produces neurotrophic factors that induce IL-22 secretion derived from ILC3. This provides a significant description of the connection between innate immune modulation and intestinal neurons (Zhang et al. 2019; Ibiza et al. 2016; Bogunovic 2016). ILC3s are widely explored in the ILC family, due to their dynamic roles in regulating the balance between immune tolerance to nonpathogenic antigens and immune response to pathogenic stimuli (Zhang et al. 2019). Thus, the gut microbiome creates fine-tune cross talk with the host.

Numerous immune mechanisms work in parallel to the gut microbiota response and contribute to intestinal homeostasis. Gut microbiota and metabolic factors constitute the main antigen load of GIT. Macrophages and dendritic cells look over intestinal luminal contents. The intestinal and metabolic factors affect the activation and migration of DCs to MLN, where naive T cells are activated, and subsequently, B cells are transformed into IgA-secreting plasma cells and inducing the production of IgA antibodies. DCs prompt induction of Treg, which is liable for the creation of IL-10. What's more, the antimicrobial proteins discharged by the host cells can balance the synthesis of the microbiota. ILC3 stimulates Th17 cell differentiation by generating IL-22.

2.5 Microbial Metabolite-Mediated Modulation of Host Immunity

The relatively slight consideration has been given to an additional cross talk between the immune system and commensal bacteria, viz. the immune function effects of microbial metabolites. Additionally, the intermediates of essential microbial metabolism, the structural components of the commensal microbes such as the secondary metabolites can be involved within the communication between microbiota and the host immune system. All these metabolites influence the immune cell metabolism and stimulate immune cell subsets signaling which can trigger immunomodulatory effects. The investigation of this additional communication has taken our interpretation of the host–microbiota interactions to a new mechanistic stage.

Short-chain fatty acids (SCFAs) are aliphatic carboxylic acids (including acetate, propionate, and butyrate) and fermentation end products of dietary fibers by the anaerobic gut microbiota. Numerous favorable effects of SCFAs have been reported on mammalian energy metabolism. Certain intestinal anaerobic bacteria, such as

Roseburia intestinalis, Faecalibacterium prausnitzii, and *Anaerostipes butyraticus* are effective in converting nondigestible carbohydrates into fermentation products such as SCFA (Zhang et al. 2019). Studies have shown that SCFA derived from the gut microbiota facilitated the regulation of Treg (Arpaia et al. 2013). The predicted SCFA activity resulting from the modulation of the several cells participating in the activation of regulatory reactions and therefore the impact of SCFA on both dendritic and T cells has been related to this process (Mao et al. 2018). As for innate immune functions, SCFA induces the release of prostaglandins E2 and anti-inflammatory cytokine IL-10 expression via PTX-sensitive G-protein-coupled receptors, thus inhibiting inflammatory reactions in human monocytes (Cox et al. 2009). There is increasing evidence that in addition to SCFA, the microbial fermentation of dietary fibers (in particular polysaccharides and oligosaccharides) may also produce succinate, which acts as a precursor to propionate in microbial metabolism and as an intermediate in the TCA cycle (Zhang et al. 2019; Koh et al. 2016). Furthermore, succinate has drawn new attention due to its key role in immune modulation. For instance, in the experimental mice model succinate raised by longan polysaccharide administration, which can enhance the host's immune function in the sense of stress, that may be credited to favorable in the gut immune key like IgA, IFN-Δ, IL-6, and transforming growth factor-β (Zheng et al. 2020; Zhang et al. 2017).

Moreover, the prominent gut commensal *Bacteroides fragilis* produced metabolites like polysaccharide A (PSA) that act as a model symbiosis factor. Modulated CD4 + T cell homeostasis and PSA-dependent development of cytokines have been reported in GF mice colonized with *B. fragilis*, thus ameliorating intestinal inflammation (Mazmanian et al. 2008; Round and Mazmanian 2010). The intestinal recognition of a bacterial polysaccharide can also affect systemic immune regulation, such as *B. fragilis* colonization enhances CNS inflammation and neurodegeneration (Lee et al. 2011; Hsiao et al. 2013). The study of other commensal carbohydrates for their related immunoregulatory effects on T cells has not yet been studied.

Additionally, lactic acid (lactate), is the key component of milk which is a diet-derived microbial metabolite, provides a various variety of metabolic and immune activities like acts as a significant source of energy for cell regeneration and in signalling molecules (Koh et al. 2016). Oral administration of lactate from *L. helveticus*-derived induces DCs protrusion through GPR31 in intestinal CX3CR1+ cells that take up antigens of gut lumen, which activates an intense antigen-specific immune response to *Salmonella* infection (Morita et al. 2019). In the same way, neonatal mice infected with the intestinal pathogen *Vibrio cholerae* treated by *L. lactis* which reduces the infection of the pathogen and increases survival (Mao et al. 2018). These findings have demonstrated that intestinal lactate-mediated immunomodulation is beneficial for the decolonization of intestinal pathogens. Furthermore, integration with lactate-producing microbes like *Bifidobacterium* and *Lactobacillus*, which have increased lactate levels, facilitates the proliferation of intestinal stem cells through Wnt/β-catenin signals of Paneth and gut stromal cells. In particular, mice lacking GPR81 show reduced ISC-mediated epithelial development, suggesting that lactate favorably regulates the role of the intestinal barrier in a GPR81-dependent manner (Lee et al. 2018). Such analyses have revealed that the harmful or beneficial impacts of lactate possibly depend on dosage, experimental model, and host's immunocompetence. Several other microbial-mediated metabolites facilitated immunomodulation, such as desaminotyrosine (DAT), derived from flavonoids produced by the intestinal commensal *C. orbiscindens*. Its metabolite is adequate for improving the development of type I IFN and to protect against influenza infection (Kuss et al. 2011; Lobel and Garrett 2017). Additionally, certain groups of rare bacteria like *Bacteroidetes* produced flowing metabolites such as mevalonate and dimethylglycine. Such metabolites have been shown to promote systemic proliferation of CD8 + T cells, thus fighting intracellular pathogens and improving the immune checkpoint inhibitor-mediated antitumor immunity in melanoma mice (Tanoue et al. 2019; Reticker-Flynn and Engleman 2019). These innovative findings have shown that members of unusual microbiomes potentially show significant effects on host immunity. But in contrast, some microbial metabolites appear to adversely affect host immunity. For instance, microbiota-derived 1,2-propanediol enhances the expression of the virulence factor in pathogens, stimulates gut colonization, and spread of pathogens like *C. rubentium* (Zhang et al. 2019; Connolly et al. 2018).

In addition to all these microbial metabolites, several new molecules produced by the intestinal microbiota act as messengers for the transmission of microbial signals to different parts of the host. Transmission of these signals leads to vital communication between the gut microbiota and the host and exerts functional or adverse impacts on the outcomes of a variety of diseases, like intestinal inflammation, autoimmune, metabolic diseases, and tumors. Thus, more systematic investigations are required to uncover the functions and mechanisms of these additional metabolites.

2.6 Probiotics: An Immune Modulator

The beneficial effect of probiotics on human health and its immunomodulatory effects are well established. With this, probiotics have the potential to prevent certain pathologies. The deeper and better understanding of the immunomodulatory effects of probiotics may give some new-gen to design an effective approach to prevent inflammatory or allergic diseases (Sharma and Im 2018; Isolauri et al. 2001; Klaenhammer et al. 2012).

It has been evidenced that host's immune system is influenced by different probiotic bacterial strains to wield their probiotic effects. Thus, it is clear that probiotics can modulate the host immune response. Probiotic bacteria are most often described as "live microorganisms which, if ingested or locally applied in sufficient quantities, contribute health benefits to the host." Probiotics like *Lactobacillus* spp. and *Bifidobacterium* is commonly present in several dietary supplements products or drugs (Valdes et al. 2018). Following human intestines, probiotic microorganisms act symbiotically and stimulate, modulate, and regulate several functions, such as digestion, metabolism, epithelial innate immunity, brain–gut communication, and competitive exclusion of pathogens. In the past few years, probiotics have been widely explored for their humoral, cellular, and nonspecific immune modulations, as well as for supporting the immunological barrier. Orally administered probiotics could interact with GI mucosa and gut-associated lymphoid tissue (GALT), where immune cells are localized. Probiotics also interact with IECs, DCs, and macrophages diversely in the GI. DCS in lamina propria could contact with probiotics in the gut lumen through the dendrites and then import them into the lumen (Taverniti and Guglielmetti 2011). Studies demonstrated that administration of a rationally selected mixture of probiotics (IRT5: *B. bifidum, L. casei, L. acidophilus, L. reuteri*, and *Streptococcus thermophiles*) could upregulate iTregs (CD41Foxp31) population through the generation of regulatory DCs in MLN (Kang and Im 2015). Cell surface molecules of *Lactobacillus* strains have also been reported to exhibit TNF-α inducing activity in macrophages through TLR2 signalling (Mikelsaar et al. 2011).

Instead of these probiotics, bacteria may influence the gut microbiota by promoting the growth of symbiotic bacteria such as bifidobacteria, or by inhibiting the in vitro adherence of certain pathogens such as *H. pylori* (Myllyluoma et al. 2008). Another research reported by the inactivation process, the target or model immune system, may have a major effect on the experimental results. For instance, the heat-inactivated strain of *Lactobacillus casei Shirota* was slightly less efficient in inducing certain pro-inflammatory cytokines, like IL-12 and TNF-α, compared to viable cells, while the induction of IL-10 was the same among the inactivated and viable cells (Cross et al. 2004). Various studies are indicating that probiotics interact with IECs and DCs in the GI tract and exert immune modulation. Numerous *Lactobacillus* spp. have been shown to change the phenotype of DCs and their cytokine forms. Moreover, it also observed that probiotics induce TReg cells, upregulate IL-10, and transforming growth factor-β as well as an increase in local IgA production (Campbell et al. 2018; Klaenhammer et al. 2012; Ding et al. 2017). Additionally, it is also reported that regulation of immune responses by efficient probiotic bacteria also depends on a complex communication between the host immune system and various bacterial compounds, such as chromosomal DNA, cell wall components, and soluble metabolites (Tannock 2012; Jijon et al. 2004).

Thus, altogether these findings highlight the impacts of probiotics on the host's immune system which is now already has been moved into a new and captivating era of research. The growing probiotic research can offer a novel and useful way of modulating host immunity to protect or treat a wide range of human and animal diseases.

2.7 Conclusion and Perspectives

The gut immune microenvironment is characterized by reciprocal correlations between the immune structure and intestinal microbiota. To attain such a state, host immune cells must learn to exhibit a tolerogenic behavior toward most microbial antigens. The regulation of miscellaneous mechanisms by the commensal microbiota reflects the strong symbiotic relationship with its host. The gut microorganism communicates with the immune system at several levels and may be involved in the pathophysiological functions of the host. This can also be used to develop treatments for many immunological disorders, such as autoimmune and inflammatory diseases, opportunistic infections, allergies, dysbiosis, and cancer. Diversified gut microbiota is vital for a healthy life. Recognition of gut microbiota and their specific occupations are essential and can provide the basis for the development of safer and better strategies to establish host defense and block the pathogens. The gut flora and their metabolites are well established for their immunomodulatory properties in the host.

Further, the immunomodulatory effects of gut microbiota/microbiota-based molecules can be exploited as therapeutic interventions to rescue the host from different disease conditions. Besides, the mechanistic details of microbial metabolites may assist in the development of novel combating strategies for drug-resistant pathogens. Also, gut flora can be further exploited for the identification of potential biomarkers for different diseases which help in developing new screening tools that may further support in prevention, identification, and treatment of several diseases.

References

An D, Oh SF, Olszak T, Neves JF, Avci FY, Erturk-Hasdemir D, Lu X, Zeissig S, Blumberg RS, Kasper DL (2014) Sphingolipids from a symbiotic microbe regulate homeostasis of host intestinal natural killer T cells. Cel 156(1–2):123–133

Arpaia N, Campbell C, Fan X, Dikiy S, van der Veeken J, Deroos P, Liu H, Cross JR, Pfeffer K, Coffer PJ, Rudensky AY (2013) Metabolites produced by commensal bacteria promote peripheral regulatory T-cell generation. Nature 504(7480):451–455

Belkaid Y, Hand TW (2014) Role of the microbiota in immunity and inflammation. Cell 157(1): 121–141

Benckert J, Schmolka N, Kreschel C, Zoller MJ, Sturm A, Wiedenmann B, Wardemann H (2011) The majority of intestinal IgA+ and IgG+ plasmablasts in the human gut are antigen-specific. J Clin Investig 121(5):1946–1955

Berg RD (1985) Bacterial translocation from the intestines. Exp Anim 34(1):1–6

Bogunovic M (2016) ILC3s and the willow tree of voices. Immunity 45(2):238–239

Bouskra D, Brézillon C, Bérard M, Werts C, Varona R, Boneca IG, Eberl G (2008) Lymphoid tissue genesis induced by commensals through NOD1 regulates intestinal homeostasis. Nature 456(7221):507–510

Bunker JJ, Bendelac A (2018) IgA Responses to Microbiota. Immunity 49(2):211–224

Campbell C, Dikiy S, Bhattarai SK, Chinen T, Matheis F, Calafiore M, Hoyos B, Hanash A, Mucida D, Bucci V, Rudensky AY (2018) Extrathymically generated regulatory T cells establish a niche for intestinal border-dwelling bacteria and affect physiologic metabolite balance. Immunity 48(6):1245–1257

Chassin C, Kocur M, Pott J, Duerr CU, Gütle D, Lotz M, Hornef MW (2010) miR-146a mediates protective innate immune tolerance in the neonate intestine. Cell Host Microbe 8(4):358–368

Cho I, Blaser MJ (2012) The human microbiome: at the interface of health and disease. Nat Rev Genet 13(4):260–270

Connolly JP, Slater SL, O'Boyle N, Goldstone RJ, Crepin VF, Ruano-Gallego D, Herzyk P, Smith DG, Douce GR, Frankel G, Roe AJ (2018) Host-associated niche metabolism controls enteric infection through fine-tuning the regulation of type 3 secretion. Nat Commun 9(1):1–4

Coombes JL, Powrie F (2008) Dendritic cells in intestinal immune regulation. Nat Rev Immunol 8(6):435–446

Corthésy B, Gaskins HR, Mercenier A (2007) Cross-talk between probiotic bacteria and the host immune system. J Nutr 137(3):781S–790S

Cox MA, Jackson J, Stanton M, Rojas-Triana A, Bober L, Laverty M, Yang X, Zhu F, Liu J, Wang S, Monsma F (2009) Short-chain fatty acids act as antiinflammatory mediators by regulating prostaglandin E2 and cytokines. World J Gastroenterol 15(44):5549

Cross ML, Ganner A, Teilab D, Fray LM (2004) Patterns of cytokine induction by gram-positive and gram-negative probiotic bacteria. FEMS Microbiol Immuno 42(2):173–180

Cywes C, Wessels MR (2001) Group A Streptococcus tissue invasion by CD44-mediated cell signalling. Nature 414(6864):648–652

Ding YH, Qian LY, Pang J, Lin JY, Xu Q, Wang LH, Huang DS, Zou H (2017) The regulation of immune cells by Lactobacilli: a potential therapeutic target for anti-atherosclerosis therapy. Oncotarget 8(35):59915

Donaldson GP, Ladinsky MS, Yu KB, Sanders JG, Yoo BB, Chou WC, Conner ME, Earl AM, Knight R, Bjorkman PJ, Mazmanian SK (2018) Gut microbiota utilize immunoglobulin A for mucosal colonization. Science 360(6390):795–800

Durack J, Lynch SV (2019) The gut microbiome: relationships with disease and opportunities for therapy. J Exp Med 216(1):20–40

Emgård J, Kammoun H, García-Cassani B, Chesné J, Parigi SM, Jacob JM, Cheng HW, Evren E, Das S, Czarnewski P, Sleiers N (2018) Oxysterol sensing through the receptor GPR183 promotes the lymphoid-tissue-inducing function of innate lymphoid cells and colonic inflammation. Immunity 48(1):120–132

Farache J, Koren I, Milo I, Gurevich I, Kim KW, Zigmond E, Furtado GC, Lira SA, Shakhar G (2013) Luminal bacteria recruit CD103+ dendritic cells into the intestinal epithelium to sample bacterial antigens for presentation. Immunity 38(3):581–595

Field CJ (2005) The immunological components of human milk and their effect on immune development in infants. J Nutr 135(1):1–4

Fournier BM, Parkos CA (2012) The role of neutrophils during intestinal inflammation. Mucosal Immunol 5(4):354–366

Francino MP (2014) Early development of the gut microbiota and immune health. Pathogens 3(3): 769–790

Fung TC, Bessman NJ, Hepworth MR, Kumar N, Shibata N, Kobuley D, Wang K, Ziegler CG, Goc J, Shima T, Umesaki Y (2016) Lymphoid-tissue-resident commensal bacteria promote members of the IL-10 cytokine family to establish mutualism. Immunity 44(3):634–646

Goldsmith M, Avni D, Ernst O, Glucksam Y, Levy-Rimler G, Meijler MM, Zor T (2009) Synergistic IL-10 induction by LPS and the ceramide-1-phosphate analog PCERA-1 is mediated by the cAMP and p38 MAP kinase pathways. Mol Immunol 46(10):1979–1987

Gordon JI, Dewey KG, Mills DA, Medzhitov RM (2012) The human gut microbiota and undernutrition. Sci Transl Med 4(137):137ps12

Grootjans J, Krupka N, Hosomi S, Matute JD, Hanley T, Saveljeva S, Gensollen T, Heijmans J, Li H, Limenitakis JP, Ganal-Vonarburg SC (2019) Epithelial endoplasmic reticulum stress orchestrates a protective IgA response. Science 363(6430):993–998

Gury-BenAri M, Thaiss CA, Serafini N, Winter DR, Giladi A, Lara-Astiaso D, Levy M, Salame TM, Weiner A, David E, Shapiro H (2016) The spectrum and regulatory landscape of intestinal innate lymphoid cells are shaped by the microbiome. Cell 166(5):1231–1246

Hapfelmeier S, Lawson MA, Slack E, Kirundi JK, Stoel M, Heikenwalder M, Cahenzli J, Velykoredko Y, Balmer ML, Endt K, Geuking MB (2010) Reversible microbial colonization of germ-free mice reveals the dynamics of IgA immune responses. Science 328(5986): 1705–1709

Hawkins RD, Byrne JH (2015) Associative learning in invertebrates. Cold Spring Harb Perspect Biol 7(5):a021709

Hepworth MR, Monticelli LA, Fung TC, Ziegler CG, Grunberg S, Sinha R, Mantegazza AR, Ma HL, Crawford A, Angelosanto JM, Wherry EJ (2013) Innate lymphoid cells regulate CD4+ T-cell responses to intestinal commensal bacteria. Nature 498(7452):113–117

Honda K, Littman DR (2016) The microbiota in adaptive immune homeostasis and disease. Nature 535(7610):75–84

Hooper LV, Littman DR, Macpherson AJ (2012) Interactions between the microbiota and the immune system. Science 336(6086):1268–1273

Hsiao EY, McBride SW, Hsien S, Sharon G, Hyde ER, McCue T, Codelli JA, Chow J, Reisman SE, Petrosino JF, Patterson PH (2013) Microbiota modulate behavioral and physiological abnormalities associated with neurodevelopmental disorders. Cell 155(7):1451–1463

Ibiza S, García-Cassani B, Ribeiro H, Carvalho T, Almeida L, Marques R, Misic AM, Bartow-McKenney C, Larson DM, Pavan WJ, Eberl G (2016) Glial-cell-derived neuroregulators control type 3 innate lymphoid cells and gut defence. Nature 535(7612):440–443

Isolauri E, Sütas Y, Kankaanpää P, Arvilommi H, Salminen S (2001) Probiotics: effects on immunity. Am J Clin Nutr 73(2):444s–450s

Jijon H, Backer J, Diaz H, Yeung H, Thiel D, McKaigney C, De Simone C, Madsen K (2004) DNA from probiotic bacteria modulates murine and human epithelial and immune function. Gastroenterology 126(5):1358–1373

Johansen FE, Kaetzel CS (2011) Regulation of the polymeric immunoglobulin receptor and IgA transport: new advances in environmental factors that stimulate pIgR expression and its role in mucosal immunity. Mucosal Immunol 4(6):598–602

Kamada N, Núñez G (2013) Role of the gut microbiota in the development and function of lymphoid cells. J Immunol 190(4):1389–1395

Kang HJ, Im SH (2015) Probiotics as an immune modulator. J Nutr Sci Vitamino 61(Supplement): S103–S105

Klaenhammer TR, Kleerebezem M, Kopp MV, Rescigno M (2012) The impact of probiotics and prebiotics on the immune system. Nat Rev Immunol 12(10):728–734

Koh A, De Vadder F, Kovatcheva-Datchary P, Bäckhed F (2016) From dietary fiber to host physiology: short-chain fatty acids as key bacterial metabolites. Cell 165(6):1332–1345

Kollmann TR, Levy O, Montgomery RR, Goriely S (2012) Innate immune function by Toll-like receptors: distinct responses in newborns and the elderly. Immunity 37(5):771–783

Kuss SK, Best GT, Etheredge CA, Pruijssers AJ, Frierson JM, Hooper LV, Dermody TS, Pfeiffer JK (2011) Intestinal microbiota promote enteric virus replication and systemic pathogenesis. Science 334(6053):249–252

Lee YK, Menezes JS, Umesaki Y, Mazmanian SK (2011) Proinflammatory T-cell responses to gut microbiota promote experimental autoimmune encephalomyelitis. Proc Natl Acad Sci 108 (Supplement 1):4615–4622

Lee YS, Kim TY, Kim Y, Lee SH, Kim S, Kang SW, Yang JY, Baek IJ, Sung YH, Park YY, Hwang SW (2018) Microbiota-derived lactate accelerates intestinal stem-cell-mediated epithelial development. Cell Host Microbe 24(6):833–846

Lobel L, Garrett WS (2017) Take DAT, Flu! Immunity 47(3):400–402

Lynch SV, Pedersen O (2016) The human intestinal microbiome in health and disease. N Engl J Med 375(24):2369–2379

Mabbott NA, Donaldson DS, Ohno H, Williams IR, Mahajan A (2013) Microfold (M) cells: important immunosurveillance posts in the intestinal epithelium. Mucosal Immunol 6(4): 666–677

Macpherson AJ, Uhr T (2004) Induction of protective IgA by intestinal dendritic cells carrying commensal bacteria. Science 303(5664):1662–1665

Magri G, Comerma L, Pybus M, Sintes J, Lligé D, Segura-Garzón D, Bascones S, Yeste A, Grasset EK, Gutzeit C, Uzzan M (2017) Human secretory IgM emerges from plasma cells clonally related to gut memory B cells and targets highly diverse commensals. Immunity 47(1):118–134

Mao K, Baptista AP, Tamoutounour S, Zhuang L, Bouladoux N, Martins AJ, Huang Y, Gerner MY, Belkaid Y, Germain RN (2018) Innate and adaptive lymphocytes sequentially shape the gut microbiota and lipid metabolism. Nature 554(7691):255–259

Mathias A, Duc M, Favre L, Benyacoub J, Blum S, Corthésy B (2010) Potentiation of polarized intestinal Caco-2 cell responsiveness to probiotics complexed with secretory IgA. J Biol Chem 285(44):33906–33913

Matsuo K, Nagakubo D, Yamamoto S, Shigeta A, Tomida S, Fujita M, Hirata T, Tsunoda I, Nakayama T, Yoshie O (2018a) CCL28-deficient mice have reduced IgA antibody–secreting cells and an altered microbiota in the colon. J Immunol 200(2):800–809

Matsuo K, Nagakubo D, Yamamoto S, Shigeta A, Tomida S, Fujita M et al (2018b) CCL28-deficient mice have reduced IgA antibody–secreting cells and an altered microbiota in the colon. J Immunol 200(2):800–809

Mazmanian SK, Round JL, Kasper DL (2008) A microbial symbiosis factor prevents intestinal inflammatory disease. Nature 453(7195):620–625

Mikelsaar M, Lazar V, Onderdonk A, Donelli G (2011) Do probiotic preparations for humans really have efficacy? Microb Ecol Health Dis 22(1):10128

Morita N, Umemoto E, Fujita S, Hayashi A, Kikuta J, Kimura I, Haneda T, Imai T, Inoue A, Mimuro H, Maeda Y (2019) GPR31-dependent dendrite protrusion of intestinal CX3CR1+ cells by bacterial metabolites. Nature 566(7742):110–114

Mortha A, Chudnovskiy A, Hashimoto D, Bogunovic M, Spencer SP, Belkaid Y, Merad M (2014) Microbiota-dependent crosstalk between macrophages and ILC3 promotes intestinal homeostasis. Science 343(6178):1249288

Myllyluoma E, Ahonen AM, Korpela R, Vapaatalo H, Kankuri E (2008) Effects of multispecies probiotic combination on Helicobacter pylori infection in vitro. Clin Vaccine Immunol 15(9): 1472–1482

Nakajima A, Vogelzang A, Maruya M, Miyajima M, Murata M, Son A, Kuwahara T, Tsuruyama T, Yamada S, Matsuura M, Nakase H (2018) IgA regulates the composition and metabolic function of gut microbiota by promoting symbiosis between bacteria. J Exp Med 215(8):2019–2034

Niess JH, Brand S, Gu X, Landsman L, Jung S, McCormick BA, Vyas JM, Boes M, Ploegh HL, Fox JG, Littman DR (2005) CX3CR1-mediated dendritic cell access to the intestinal lumen and bacterial clearance. Science 307(5707):254–258

Ohnmacht C, Marques R, Presley L, Sawa S, Lochner M, Eberl G (2011) Intestinal microbiota, evolution of the immune system and the bad reputation of pro-inflammatory immunity. Cell Microbiol 13(5):653–659

Olszak T, An D, Zeissig S, Vera MP, Richter J, Franke A, Glickman JN, Siebert R, Baron RM, Kasper DL, Blumberg RS (2012) Microbial exposure during early life has persistent effects on natural killer T cell function. Science 336(6080):489–493

Papaioannou NE, Pasztoi M, Schraml BU (2019) Understanding the functional properties of neonatal dendritic cells: a doorway to enhance vaccine effectiveness? Front Immunol 10(9): 3123

Pickard JM, Zeng MY, Caruso R, Núñez G (2017) Gut microbiota: role in pathogen colonization, immune responses, and inflammatory disease. Immunol Rev 279(1):70–89

Pokrovskii M, Hall JA, Ochayon DE, Yi R, Chaimowitz NS, Seelamneni H, Carriero N, Watters A, Waggoner SN, Littman DR, Bonneau R (2018) Transcriptional regulatory networks that

promote and restrict identities and functions of intestinal innate lymphoid cells. Available at SSRN 3291328
PrabhuDas M, Adkins B, Gans H, King C, Levy O, Ramilo O, Siegrist CA (2011) Challenges in infant immunity: implications for responses to infection and vaccines. Nat Commun 12(3): 189–194
Randal Bollinger R, Everett ML, Palestrant D, Love SD, Lin SS, Parker W (2003) Human secretory immunoglobulin a may contribute to biofilm formation in the gut. Immunology 109(4):580–587
Rescigno M, Urbano M, Valzasina B, Francolini M, Rotta G, Bonasio R, Granucci F, Kraehenbuhl JP, Ricciardi-Castagnoli P (2001) Dendritic cells express tight junction proteins and penetrate gut epithelial monolayers to sample bacteria. Nat Commun 2(4):361–367
Reticker-Flynn NE, Engleman EG (2019) A gut punch fights cancer and infection. Nature 565:573–574
Ribet D, Cossart P (2015) How bacterial pathogens colonize their hosts and invade deeper tissues. Microbes Infect 17(3):173–183
Rios D, Wood MB, Li J, Chassaing B, Gewirtz AA, Williams IR (2016) Antigen sampling by intestinal M cells is the principal pathway initiating mucosal IgA production to commensal enteric bacteria. Mucosal Immunol 9(4):907–916
Round JL, Mazmanian SK (2010) Inducible Foxp3+ regulatory T-cell development by a commensal bacterium of the intestinal microbiota. Proc Natl Acad Sci 107(27):12204–12209
Round JL, Lee SM, Li J, Tran G, Jabri B, Chatila TA, Mazmanian SK (2011) The Toll-like receptor 2 pathway establishes colonization by a commensal of the human microbiota. Science 332(6032):974–977
Rutz S, Janke M, Kassner N, Hohnstein T, Krueger M, Scheffold A (2008) Notch regulates IL-10 production by T helper 1 cells. Proc Natl Acad Sci 105(9):3497–3502
Sano T, Huang W, Hall JA, Yang Y, Chen A, Gavzy SJ, Lee JY, Ziel JW, Miraldi ER, Domingos AI, Bonneau R (2015) An IL-23R/IL-22 circuit regulates epithelial serum amyloid A to promote local effector Th17 responses. Cell 163(2):381–393
Sharma G, Im SH (2018) Probiotics as a potential immunomodulating pharmabiotics in allergic diseases: current status and future prospects. Allergy Asthma Immunol Res 10(6):575–590
Shi Y, Liu CH, Roberts AI, Das J, Xu G, Ren G, Zhang Y, Zhang L, Yuan ZR, Tan HS, Das G (2006) Granulocyte-macrophage colony-stimulating factor (GM-CSF) and T-cell responses: what we do and don't know. Cell Res 16(2):126–133
Siegrist CA (2001) Neonatal and early life vaccinology. Vaccine 19(25–26):3331–3346
Smith K, McCoy KD, Macpherson AJ (2007) Use of axenic animals in studying the adaptation of mammals to their commensal intestinal microbiota. InSeminars in immunology. Academic Press 19(2):59–69
Stagg AJ, Hart AL, Knight SC, Kamm MA (2003) The dendritic cell: its role in intestinal inflammation and relationship with gut bacteria. Gut 52(10):1522–1559
Sun M, Wu W, Chen L, Yang W, Huang X, Ma C, Chen F, Xiao Y, Zhao Y, Ma C, Yao S (2018) Microbiota-derived short-chain fatty acids promote Th1 cell IL-10 production to maintain intestinal homeostasis. Nat Commun 9(1):1–5
Takiishi T, Fenero CI, Câmara NO (2017) Intestinal barrier and gut microbiota: shaping our immune responses throughout life. Tissue Barriers 5(4):e1373208
Tannock GW (ed) (2012) Medical importance of the normal microflora. Springer Science & Business Media, Cham
Tanoue T, Morita S, Plichta DR, Skelly AN, Suda W, Sugiura Y, Narushima S, Vlamakis H, Motoo I, Sugita K, Shiota A (2019) A defined commensal consortium elicits CD8 T cells and anti-cancer immunity. Nature 565(7741):600–605
Taverniti V, Guglielmetti S (2011) The immunomodulatory properties of probiotic microorganisms beyond their viability (ghost probiotics: proposal of paraprobiotic concept). Genes Nutr 6(3): 261
Thomas CM, Versalovic J (2010) Probiotics-host communication: modulation of signaling pathways in the intestine. Gut Microbes 1(3):148–163

Valdes AM, Walter J, Segal E, Spector TD (2018) Role of the gut microbiota in nutrition and health. BMJ 13(361):k2179

Varol C, Zigmond E, Jung S (2010) Securing the immune tightrope: mononuclear phagocytes in the intestinal lamina propria. Nat Rev Immunol 10(6):415–426

Wang B, Yao M, Lv L, Ling Z, Li L (2017) The human microbiota in health and disease. Engineering 3(1):71–82

Zhang Z, Li J, Zheng W, Zhao G, Zhang H, Wang X, Guo Y, Qin C, Shi Y (2016) Peripheral lymphoid volume expansion and maintenance are controlled by gut microbiota via RALDH+ dendritic cells. Immunity 44(2):330–342

Zhang J, Yang G, Wen Y, Liu S, Li C, Yang R, Li W (2017) Intestinal microbiota are involved in the immunomodulatory activities of longan polysaccharide. Mol Nutr Food Res 61(11): 1700466

Zhang Z, Tang H, Chen P, Xie H, Tao Y (2019) Demystifying the manipulation of host immunity, metabolism, and extraintestinal tumors by the gut microbiome. Signal Transduct Target Ther 4(1):1–34

Zhang PP, Li LL, Han X, Li QW, Zhang XH, Liu JJ, Wang Y (2020) Fecal microbiota transplantation improves metabolism and gut microbiome composition in db/db mice. Acta Pharmacol Sin 14:1–8

Zheng D, Liwinski T, Elinav E (2020) Interaction between microbiota and immunity in health and disease. Cell Res 20:1–5

Zhu J, Paul WE (2010) Peripheral CD4+ T-cell differentiation regulated by networks of cytokines and transcription factors. Immunol Rev 238(1):247–262

Chapter 3
Microbiota–Gut–Brain Axis and Neurodegenerative Disorder

Awakash Soni, Ankit Verma, and Priya Gupta

Abstract Diverse and apt gut microbiota is a vital part of the human body and obligatory to maintain a healthy life. Significant involvement of gut flora in the central nervous system (CNS) development and its function has been revealed. The communication of gut microbiome and CNS is bidirectional and thus a person's mental state can also be influenced by gut microbiome composition. Increased intestinal permeability and impaired intestinal barrier are associated with gut dysbiosis which in turn influence the structure and functions of CNS. Also, there are resilient connections between the microbiome and neurological diseases. In this chapter, we discussed the biological association of gut microbiome in the CNS function and development including blood–brain barrier, myelination, and neurogenesis. We also discussed the factors that influence the gut microbiome and CNS including specific and nonspecific factors, age, probiotic, prebiotic, vegetarian diet, Western diet, and antibiotics. We also tried to cast some light on the connection between gut dysbiosis and neurodegenerative diseases/disorders with clinical evidence and underlying mechanisms. New insights into factors influencing the gut microbiota and CNS connections may be exploited as capable tools to tackle neurological diseases.

Keywords Gut microbiota · Microbiome · CNS · Blood–brain barrier · Neurological disorders

---

A. Soni (✉)
Department of Microbiology and Molecular Genetics, The Kuvin Center for the Study of Infectious and Tropical Diseases, Hebrew University-Hadassah Medical School, Jerusalem, Israel

A. Verma
Department of Life Sciences and the National Institute for Biotechnology in the Negev, Ben Gurion University of Negev, Beer Sheva, Israel

P. Gupta
Department of Developmental Biology and Cancer Research, Hebrew University-Hadassah Medical School, Jerusalem, Israel

A. K. Tripathi, M. Kotak (eds.), *Gut Microbiome in Neurological Health and Disorders*, Nutritional Neurosciences,
https://doi.org/10.1007/978-981-19-4530-4_3

Abbreviations

AD	Alzheimer's disease
AMPs	Antimicrobial peptides
ANS	Autonomic nervous system
ASD	Autism spectrum disorder
BBB	Blood–brain barrier
CNS	Central nervous system
ENS	Enteric nervous system
GF	Germ-free mice
GI	Gastrointestinal tract
IEC	Intestinal epithelial cells
IgA	Immunoglobulin A
IL	Interleukin
NFκB	Nuclear factor-kappa B
PD	Parkinson's disease
TLR	Toll-like receptor
TNF-α	Tumor necrosis factor

3.1 Introduction

Microbial communities including bacteria, archaea, fungi, and viruses coexist with humans and other animals in the gastrointestinal tract known as microbiota (Ferreiro et al. 2018). The beneficial effects of gut flora have been known for three decades and have implications in both health and diseases (Maynard et al. 2012). Past decade studies suggest a multifaceted and bidirectional communication between the gastrointestinal (GI) tract and the brain, which is known as the "Gut–Brain Axis" (GBA). GBA works as a coordinator between the gut and the brain for different physiological communications relayed from the immunological, endocrine, nutritional, and neuronal systems. The autonomic nervous system (ANS), the enteric nervous system (ENS), the immune system, as well as microbial metabolites, circulating hormones, and other neuromodulatory molecules are vital to establish a connection between the gut microbiome and the brain. This bidirectional cross talk plays a significant role in maintaining gut homeostasis and influences the key processes of central nervous system such as neurotransmission and behavior, including its cognitive functions (Ferreiro et al. 2018; Maynard et al. 2012). Moreover, gut microbiome plays a significant role in postnatal development and maturation of key systems that may influence the programming of CNS with the immune and endocrine systems. Also, many microbial metabolites, which are present in the peripheral blood, have profound effects on the function of CNS. The neuroactive metabolites, synthesized by gut microbes were found to be implicated with neurodegenerative disorders including Parkinson's disease (PD), Alzheimer's disease (AD), multiple sclerosis, and amyotrophic lateral sclerosis (Bäckhed et al. 2005). Additionally, immunologic and neuropsychiatric conditions such as schizophrenia, autism, and depression are also found to be associated with dysbiosis (maladaptation of gut microbiome), thus gut microbes are considered a key health regulator. Preclinical findings also support the involvement of microbiota in behavioral abnormalities observed in different neurological conditions (Bäckhed et al. 2005). Additionally, understanding the mechanisms underlying the influence of gut microbes and its influencing capabilities on the brain may help to design a new strategy to treat different neurological disorders.

3.2 The Gut Microbiome and CNS Connection

Gut microbiota has coevolved with the host over thousands of years to form a complex and mutually beneficial relationship (Neish 2009). It is roughly estimated that ~60 tons of food passes through the human GI tract in the average lifetime of a person, which contains plenty of microorganisms from the surrounding which impose a huge threat on native gut microbiota (Bäckhed et al. 2005; Bengmark 1998). Recently, the estimated weight of bacterial content in the body is approximately 0.2 kg, and the ratio of human: bacterial cells are closer to 1:1 (Sender et al. 2016), but earlier studies cited it as 1:10 (Gill et al. 2006). There are six major bacterial phyla present in the gut including *Firmicutes*, *Bacteroidetes*, *Proteobacteria*, *Actinomycetes*, *Verrucomicrobia*, *and Fusobacteria* (Eckburg et al. 2005). These gut bacteria offer several benefits to the host, including strengthening gut integrity or shaping the intestinal epithelium (Natividad and Verdu 2013), harvesting energy from indigestible food (den Besten et al. 2013), protection from pathogen (Bäumler and Sperandio 2016) and regulation of host immune response (Gensollen et al. 2016). However, an imbalance in gut microbiota composition has a potential risk to these vital processes.

There are clear connections between CNS function and gut microbiome, including brain circuitry, neurophysiology, and behavior. Signals from gut microbes support CNS functions (Heijtz et al. 2011; Sharon et al. 2016). Also, multiple CNS diseases have a crucial connection with the gut–brain axis (Sharon et al. 2016; Hsiao et al. 2013; Kundu et al. 2017). The effect of gut microbiota on the biosynthesis of host serotonin is known to affect multiple aspects of host behavior both in mice and humans (Bravo et al. 2011; Desbonnet et al. 2015; Fröhlich et al. 2016). It has also been shown that gut microbiota regulates the permeability of the blood–brain barrier (BBB) (Braniste et al. 2014; Hoyles et al. 2018). Furthermore, a lifelong close correlation between the brain metabolome and the intestinal flora has been recently recognized (Chen et al. 2018). Excessive stress response of germ-free (GF) mice and association of gut dysbiosis in AD and autism spectrum disorder (ASD) in animal models have also been proved (Heijtz et al. 2011; Sharon et al. 2016; Hsiao et al. 2013; Kundu et al. 2017; Bercik et al. 2011a). There are many factors generated by gut microbiota that significantly affect the CNS functions. Thus, these studies support the hypothesis of gut and brain connection.

3.3 Development and Influence of Gut Microbiome

The development of gut microbiota seems to be initiated before birth, as certain commensal bacteria (including *Firmicutes*, *Tenericutes*, *Proteobacteria*, *Bacteroidetes*, and *Fusobacteria* phyla) are detected in the placenta of the mother. But the general opinion is that gut microbiota establishment in infants begins at birth, and different phyla of bacteria quickly colonize the child's gut, and then continue to grow and establish residency in the intestine (Aagaard et al. 2014; Rodríguez et al. 2015). Different events of life for instance illness, antibiotic treatment, and diet changes are significantly associated with the change in the microbiota (Rodríguez et al. 2015; Koenig et al. 2011). It has also been shown that the mode of delivery of infants affects the composition of gut microflora. A high abundance of lactobacilli has been found in the vaginally delivered newborn in the early days which may be due to an abundance of lactobacilli in the vaginal flora (Rutayisire et al. 2016). While infants delivered by C-section are inhabited by facultative anaerobes *Clostridium* species that are later substituted by *Bacteroides* genus. It is noteworthy that there is ~72% similarity in the fecal microbiota of vaginally delivered infants with and their mothers, while in C-section delivered infants and their mothers only ~41% such similarity was observed (Martin et al. 2016; Milani et al. 2017). Thus, it is evinced that the mode of delivery of infants also affects the infant gut microbiota during early life. The diversity of gut microbiota is low in the early stages of development, as only two main phyla *Actinobacteria* and *Proteobacteria* were reported to be dominant (Senghor et al. 2018). During development, microbial diversity increases, and around the age of 2.5 years, a child's gut microbiota resembles adult microbiota (Rodríguez et al. 2015). Reports indicate that gut microbiota diversity decreases in the elderly population (Odamaki et al. 2016; Ogawa et al. 2018).

Many specific and nonspecific factors produced by the host influence the composition of microbiota. Intestinal epithelial cells (IEC) secrete mucus, antimicrobial peptides (AMPs), and immunoglobulin A (IgA), which promote the growth of gut native microbial species and inhibit others. IEC outer layer mucus occupied by O-glycans serves as a nutrient source and a binding site for gut microbiota (Podolsky et al. 1993; Artis et al. 2004). The presence of miRNA in feces was reported and the main sources are IEC and Hopx-positive cells and considered as the specific host factors which control gut microbiota. Many studies have indicated that miRNAs may serve as potential markers for intestinal malignancy, and have the potential ability to affect the gut microbiota composition. For example, miRNA-515-5P can stimulate *Fusobacterium nucleatum* growth and miRNA-1226-5p induces *E. coli* growth (Ahmed et al. 2009; Link et al. 2012; Liu et al. 2016).

Hippocrates, more than 2000 years ago said "Let food be thy medicine and medicine be thy food." This quote seems still germane, since there is a strong connection between diet with healthy life and mental fitness (Gómez-Pinilla 2008; Reza et al. 2019; Spencer et al. 2017; Beilharz et al. 2016). Structure and diversity of the gut microflora are significantly shaped by diet. In fact, it has been observed that the first effect on gut flora is caused by infant diet (Kumbhare et al. 2019).

Lactobacillus and *Bifidobacterium* species are dominant in breastfeeding infants, while *Enterococcus*, *Enterobacteria*, *Bacteroides*, *Clostridia*, and *Streptococcus* phyla have been observed in infants fed the formula (Guaraldi and Salvatori 2012; Groer et al. 2014; Yoshioka et al. 1983; Stark et al. 1982). These bacteria are also connected with child immunity (Grönlund et al. 2000). Vegetarian diets and Western diets are also reported as key contributors to the diversity of gut microflora. Healthy and diverse gut microbiota is a characteristic feature of vegetarian diets, dominated by *Ruminococcus*, *Roseburia*, and *Eubacterium* species, which are capable of metabolizing insoluble carbohydrates (Walker et al. 2011; Tang et al. 2013), while nonvegetarian diet has been linked with less diverse gut flora, an increase in *Bacteroides* number and a decrease in *Firmicutes* (David et al. 2014). Plant-based diet encourages growth of fiber fermenting microbes and their metabolites with health benefits (O'Grady et al. 2019). Polyphenols consumed with foods are extensively metabolized in the gastrointestinal tract and were shown to affect gut flora (Neish 2009; Bowden and Ross 1965; Clemente et al. 2012). Polyphenols, present in plants, increase *Bifidobacterium* and *Lactobacillus* species and have proven to have anti-pathogenic and anti-inflammatory effects as well as to provide cardiovascular protection (Tomova et al. 2019). Also, antioxidant, anticarcinogenic, antiadipogenic, antidiabetic, and neuroprotective properties of dietary polyphenols suggest that they may play a role in disease prevention (Cardona et al. 2013). Studies have also shown that the route of administration of flavonoids determines their biological activity. Oral route administration of flavonoids showed anxiolytic effect, while this effect was absent in intraperitoneally administered animals (Vissiennon et al. 2012).

Antibiotics are magic molecules for the treatment of pathogenic infections. Yet, they kill not only pathological agents but also beneficial microbes in the gut, resulting in dysbiosis (Klingensmith and Coopersmith 2016). The antibiotics treatment causes a diminution of gut microbiota and disturbs secondary bile acid and serotonin metabolism in the colon, thus causing a delay in intestinal motility (Ge et al. 2017). Antibiotics clindamycin, clarithromycin, metronidazole, and ciprofloxacin were reported to affect gut flora structure for an extended period of time (Jernberg et al. 2007; Jakobsson et al. 2010; Dethlefsen and Relman 2011).

Probiotics have a beneficial effect on health by contributing to favorable gut microbe's growth (Sánchez et al. 2017). *Lactobacillus*, *Bifidobacteria*, and yeasts, such as *Saccharomyces boulardii* are the most commonly used probiotic species (Dinleyici et al. 2012). There are considerable shreds of evidence showing positive effects of probiotics, which include preventing harmful bacterial species to adhere to gut epithelial cells (Servin 2004), producing antimicrobial compounds (Cleusix et al. 2007), inducing host immune responses, and reducing inflammation (Belkaid and Hand 2014) and total cholesterol (Wu et al. 2017). A mucin-degrading bacteria *Akkermansia muciniphila* restores the gut barrier and has probiotic properties (Everard et al. 2013; Zhou 2017). It has been reported that administration of probiotic in diabetic and cardiac disease patients led to significant improvement in health and insulin response (Akbari and Hendijani 2016; Hendijani and Akbari 2018). These accumulative evidences highlighted the impact of different factors on the gut microbiota (Fig. 3.1).

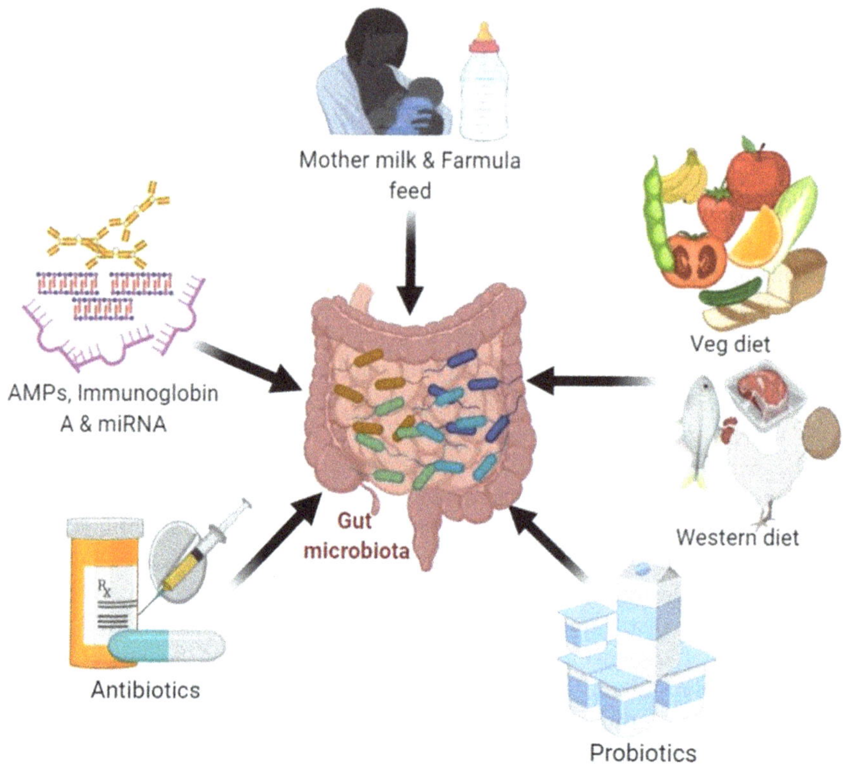

Fig. 3.1 Schematic representation of factors (Antibiotics, diet, and probiotics) that influence the Gut microbiota

3.4 The Gut Microbes and Brain Development

Adolescence and early adulthood in humans are the main phases for the development of the brain (Konrad et al. 2013). Both intrinsic and extrinsic factors are responsible for the complex process of neurodevelopment (Calof 1995). Several studies suggested that the gut and the brain cross talk in different neurological conditions including depression, anxiety, cognition, and autism spectrum disorder (ASD) (Du et al. 2020; Pulikkan et al. 2019). The key pre- and postnatal events, including molecular signals from the gut microbiome, are important in the development of a healthy and functional brain (Clarke et al. 2014). Several studies have strongly suggested that gut microbiome is crucial for development of central nervous system (Principi and Esposito 2016; Sharon et al. 2016). Microbiome supports the development of blood–brain barrier, neurogenesis, myelination, and microglia maturation (Abdel-Haq et al. 2019; Duncan and Watters 2019; Michel and Prat 2016). Animal behaviors are also influenced by gut microbiota (Ezenwa et al. 2012). The molecular universe of the gut is highly affected by various dietary components and is associated with brain development and functional alterations in the mature brain (Chang et al. 2009; Zeisel 2004). Accumulating evidences suggest the involvement of gut microbiome in brain developmental processes with long-term health implications (Forssberg 2019). Generally, gut microbiota absence, as demonstrated in germ-free animals, is associated with several CNS developmental problems (Uzbay 2019; Luczynski et al. 2016). Mice devoid of gut microbiota showed increased risk-taking behaviors, hyperactivity, and displaying deficits in learning and memory (Sharon et al. 2016; Clarke et al. 2013; Heijtz et al. 2011; Neufeld et al. 2011). In GF mice, changes in expression of the 5-hydroxytryptamine receptor (5-HT1A), neurotrophic factors (BDNF), and NMDA receptor subunits in the hippocampus have been observed (Heijtz et al. 2011; Bercik et al. 2011a; Bercik et al. 2011b; Sudo et al. 2004). Together with this, high myelination in the prefrontal cortex and damaged blood–brain barrier was also observed (Braniste et al. 2014; Hoban et al. 2016). Lately, association of the gut microbiota with neuropsychiatric conditions including depression and anxiety (Hsiao et al. 2013), ASD, schizophrenia, PD, and AD has been established (Krajmalnik-Brown et al. 2015; Severance et al. 2016; Keshavarzian et al. 2015).

In humans, after completion of gastrulation, the epiblast differentiates into neural stem cells. This first step in brain development occurs during the third week after conception and eventually would result in the brain. During cortical development, immense neuronal expansion and migration happen, and through the final weeks of maturation (38–41), ~50% of neurons undergo apoptosis. Survived cells that integrate into networks are only those supported by neurotrophic signals (Ceni et al. 2014). Brain-derived neurotrophic factor (BDNF, a neurotrophin) supports neuron survival, promotes maintenance and differentiation of numerous cell populations (Bercik et al. 2011b; Ichim et al. 2012), as well as aids in neuronal circuitry establishment (Ichim et al. 2012). One study observed increased neurogenesis in the dorsal hippocampus of adult GF mice as compared to wild-type mice (Ogbonnaya et al. 2015), which indicates that microbe's presence may influence neurogenesis.

The gut microbiome and its metabolites have a significant role in the formation of the selective barrier found between the brain and circulation, also called the blood–brain barrier (BBB), which is formed during gestation (Braniste et al. 2014). Increased BBB permeability was found in animals devoid of gut flora as compared to wild-type animals. Additionally, a defect in BBB can be rescued upon colonization of GF animals or by the administration of the short-chain fatty acids (SCFA) like butyrate, which is a metabolite produced during bacterial fermentation of dietary fibers in the gut (Braniste et al. 2014). Dynamic change in gut microbiota is observed during pregnancy (DiGiulio et al. 2015; Koren et al. 2012; MacIntyre et al. 2015; Nuriel-Ohayon et al. 2016), which affects neurodevelopment and offspring behavior (Jašarević et al. 2015). It has been found that non-absorbable antibiotics administration to pregnant rodent changes maternal and offspring's gut microbiota and induces hypoactivity and anxiety-like behavior (Degroote et al. 2016; Tochitani et al. 2016). In addition, both the microbial population and the behavior of offspring are associated with maternal diet (Buffington et al. 2016). Social deficit and repetitive behavior have been found in the offspring of mice fed a high-fat diet, compared to the control group fed normal chow. The replenishment of *Lactobacillus reuteri* (this bacterium was deficient in high-fat diet offspring in their gut microbiota) causes a reversal of social deficit in these mice (Buffington et al. 2016). Thus, these studies showed evidences that maternal gut microbiota can affect offspring's behavior. A fetus may not owe a microbiome but necessarily experience microbial products like fermentation products, secondary metabolites, lipopolysaccharides (LPS), and peptidoglycan. In mother GF mice, metabolites made by an auxotrophic strain of *Escherichia coli*, have accessibility to developing fetus which can influence certain developmental programs (Sharon et al. 2016). Peptidoglycan, which can access the fetal brain by crossing the placenta, induces the neuron multiplication in the frontal cortex. This is governed by high expression of the crucial neurogenesis regulator *FOXG1* (Humann et al. 2016). Therefore, the possible role of the gut microbiome in shaping offspring neurophysiology, behavior, and neuropathology has been demonstrated in mice (Sharon et al. 2016). Synaptic development and plasticity are associated with the development of the postnatal brain (Paolicelli et al. 2011; Zuchero and Barres 2015). Postnatal neurogenesis is highly restricted to the lateral ventricle subventricular zone and hippocampal dentate gyrus sub-granular zone. Proliferation, migration, and differentiation of glial cells occur continuously with postnatal development (Menn et al. 2006). Diminished hippocampal neurogenesis was observed in adult mice after long-term antibiotic treatment, and consequently, these mice were then found to be deficient in the novel object recognition task. Interestingly, probiotic treatment and voluntary exercise are enough to rescue these phenotypes (Möhle et al. 2016). Thus, it is conclusive that neurogenesis may also be influenced and governed by gut microbiota. Further research will give a clearer picture of the involvement of gut microbes in brain development.

Accurate conductance in neuronal axons is indispensable to communication with each other (axons), which is associated with myelination of neurons and is critical for the development of a healthy brain (Davison and Dobbing 1966). Several studies indicated that the gut microbiome modulates myelination of nerve cells (Hoban et al. 2016; Gacias et al. 2016). Similarly, to observations made with GF mice, antibiotic treatment of nonobese diabetic (NOD) mice resulted in increased expression of myelin-related genes in the prefrontal cortex (Hoban et al. 2016). The elevations in myelin-related transcripts were also observed in C57Bl/6 mice by transferring gut flora from antibiotic-treated NOD mice (Gacias et al. 2016). A correlation was found between beneficial bacteria and overall good mental health of animals which includes improvement in cases of depression, stress, anxiety, cognitive function, and decreased repetitive behavior (Bravo et al. 2011; Bercik et al. 2011a; Sudo et al. 2004; Ichim et al. 2012; Ait-Belgnaoui et al. 2012; Hsiao et al. 2013; Sun et al. 2016). It was also found that healthy human volunteers, who consume different probiotic bacteria containing fermented milk products, showed different brain activity during an emotional face's attention task in brain regions that control the processing of sensation and emotion, as tested by fMRI (Tillisch et al. 2013). Thus, these studies strongly suggest that there are close connections between gut microbiome, CNS development, and behavior.

3.5 The Gut Microbiota in Neurodegenerative Disorders

Human enteric nervous system (ENS) possesses approximately 200 to 600 million neurons (Furness et al. 2014), categorized into different classes on the basis of their morphology, electrophysiological chemical coding, and function-specific properties (Abraira and Ginty 2013). The ENS's interfaces close to luminal surface (the single cell layer of epithelium lining the large and small intestines) and its complexity can be imagined considering the size of the lumen and the exposure to the billions of microbes in its surrounding. It is also well connected with enteroendocrine cells, which secrete at least twenty hormones, and to the gut-allied immune system, which represent ~66% of host immune cells. Afferent neurons, immune cells, and enteroendocrine cells are encrypted to give sensory information in the gut. Hormones secreted through enteroendocrine cells and intestinal mechanical stimuli form the chief level association with CNS. Further, the ENS circuits optimize digestive functions and endocrine and paracrine signalling to vagal afferent controls CNS functions (Belkaid and Hand 2014; Mayer 2011). Thus, a balanced relationship between lumen and brain plays an important role in balancing the body's homeostasis.

The global incidences of neurological disorders are rapidly increasing (Bengmark 2013). A robust connection between GBA and dysbiosis has been established. Increased intestinal and BBB permeability is associated with a changed microbiome (dysbiosis) induced by many factors like pro-inflammatory cytokines, LPS, and immune cells. Dysbiosis promotes neuroinflammation, neural injury, and degeneration via misfolded proteins deposition, axonal damage, and neuronal demyelination. These processes further ease neurodegenerative disorders and related pathogenesis ranging from PD and AD to multiple sclerosis and amyotrophic lateral sclerosis (Sarkar and Banerjee 2019). A dysbiosis microbiome is thought to elicit progressive deposition of α-synuclein in AD as well as in the characteristic neuropathological features of PD due to the accumulation of misfolded amyloid-β in the neuronal cell body. This has been reported for periodontal, oral, and nasal microbiomes in AD and PD cases (Quigley 2017) and the presence of *Helicobacter pylori* in the gastric mucus layer has been linked to PD (Shen et al. 2017).

There have been some evidences of the higher abundance of rising serum LPS by *Enterobacteriaceae* (Forsythe et al. 2016) and increase in the inflammatory cytokines such as toll-like receptor (TLR) 4, tumor necrosis factor (TNF-α), interleukin (IL)-1β, IL-6, and nuclear factor-kappa B (NFκB) pathway, which further induce systemic inflammation (Guo et al. 2013; Block et al. 2007; Sui et al. 2014). These cytokines, in turn, critically affect the CNS (Hanke and Kielian 2011; Mousa and Bakhiet 2013). Certain other factors may interact and increase the chances of neurodegeneration, including aging and its attendant changes in the microbiota, immunity, gut barrier, and BBB functions (Quigley 2017).

Diet is another critical factor; consumption of industrially refined or manipulated diets is responsible for gut microbiome changes and could possibly influence mental health and disease conations (Bengmark 1998; Elinav et al. 2011). There is a critical connection between poor diet and decreased lumen microbial diversity with inflammation recognized in the elderly (Claesson et al. 2012). Link of nutritional deficiencies, dysphagia, and other gastrointestinal issues has been found with neurodegenerative disorders. Socio-personal factors like smoking were also found to influence neurological disease conditions (Scheperjans et al. 2015).

3.6 A Clinical Connection Between the Gut Microbes and Neurodegenerative Disorders

Neurodegenerative disorders have been linked to various types of infections. *H. pylori*, cause of gut infection and peptic ulcer, was found to be associated with PD through manipulating levodopa uptake (a medication for PD). The bacterium is also involved in severity and development of the disease (Wang et al. 2015; Pierantozzi et al. 2001; Dobbs et al. 2010; Tan et al. 2015). Although the literature is currently limited, antibiotics targeting *H. pylori* have shown clinical improvement in disease pathology (Hashim et al. 2014). Furthermore, increased growth of microbes in the small intestine was also found to be connected with PD conditions, which is indicated by a high prevalence of gut dysmotility (Cassani et al. 2015). The high prevalence of gastrointestinal dysfunction in PD (Quigley 2017) highlights gut functionality as an important factor in the onset and/or development of PD. Therefore, the microbiota of PD subjects has got considerable attention (Felice et al. 2016; Perez-Pardo et al. 2017). It has been also found that bacterial infections and derived factors are coupled with AD and PD conditions (Andreadou et al. 2017; Goldman et al. 2014). The technical advancement of molecular techniques like high-throughput sequencing, and omics (metagenomics, metatranscriptomics, metabolomics) has reformed the gut microbes and its derived metabolic products exploration.

Severe change in gut microbes has been already established in neurodegenerative diseased (AD and PD) individuals as compared to healthy volunteers. Elevation in pro-inflammatory bacterial genera like *Proteobacteria*, *Enterococcaceae*, *and Enterobacteriaceae (*Keshavarzian et al. 2015*;* Scheperjans et al. 2015*;* Hooper et al. 2012*)* and decreased anti-inflammatory genera like *Blautia*, *Coprococcus*, *Roseburia*, and *Faecalibacterium (*Quigley 2017*)* were found to be associated with postural instability and gait difficulty. In another study, multiple system atrophy was allied with an increase in pro-inflammatory microbes and damaged gut barrier function and inflammation (Engen et al. 2017). Additionally, a rise in pro-inflammatory cytokines and amyloid deposition in the brain of AD patients has been observed and correlated with a decrease in anti-inflammatory taxa (*Eubacterium rectale*) and rise in pro-inflammatory genera (*Escherichia* and *Shigella*) (Cattaneo et al. 2017). These evidences support the role of gut–brain axis in PD and AD but have certain confines, such as small-size experimental subjects and dependence solely on fecal sampling (Quigley 2017; Wu et al. 2016; Devkota 2016). It is difficult to assess the relative contribution of depression and disease conditions to the altered microbial patterns (Hill-Burns et al. 2017; Sherwin et al. 2016). Thus, the change in gut microbiota and disease conditions are bidirectional and there is possibility that disease conditions may change gut flora or vice versa (Quigley 2017; Elinav et al. 2011; Levy et al. 2015).

3.7 Routes of Communication

The precise molecular mechanistic details of the interactions between the gut microbiota and the brain are still unclear. Although there is anatomical separation, emerging evidence has indicated the existence of bidirectional communication between gut microbiota, through the nervous system (gut brain's neuroanatomical pathway) and by the nutritional, endocrine, immune, and metabolic systems. This bidirectional cross talk may have a vital impact during basic neuronal processes and in neurodegenerative disorders (Bravo et al. 2011; Quigley 2017; Fung et al. 2017). Several possibilities have been proposed regarding the gut microbes' interaction with CNS that influence neurochemistry brain and behavior. Information between gut and brain may be exchanged directly through ANS and vagus nerve (VN) in the spinal cord and/or via ENS in the gut and ANS and VN within the spinal cord. Another route of communication is our central stress response system, i.e., neuroendocrine–hypothalamic–pituitary–adrenal (HPA) axis, which has been reported to closely interact with the immune system, the intestinal barrier, and BBB, as well as the sensory ANS and a variety of microbial metabolites and gut hormones (Dinan and Cryan 2017; Farzi et al. 2018; Gao et al. 2020).

Many essential and vital neurotransmitters for the human host are produced by gut microbiota (Dinan and Cryan 2017). Multiple direct (e.g., VN) and indirect (e.g., SCFA, cytokines, and key amino acids, such as tryptophan derivatives) pathways exist for the gut–brain axis communication. These mechanisms involve interaction of microbial metabolites and immune mediators and signalling to the CNS directly via VN. The vagus nerve facilitates bidirectional pathway from the abdominal cavity to the brain, and gut microbiota activates this pathway to mediate behavioral and physiological effects on the brain. The VN is implicated in PD/AD, consequently, the communication through the gut–brain axis also via microbial metabolites, neurotransmitters, and neuromodulators, may be impaired (Bravo et al. 2011; Forsythe et al. 2016; Dinan and Cryan 2017). Furthermore, SCFAs, a wide category of microbially synthesized compounds, play a key role in psychological functioning, including affective and cognitive processes and their neural basis as well as in neuro-immunoendocrine regulation. Many metabolites synthesized by the host and essential amino acids such as tryptophan play a central role in the crosstalk between the gut microbiota and the brain (Farzi et al. 2018; Gao et al. 2020). Thus, there are numerous links between the gut and the brain and still a lot is needed to be identified.

3.8 Conclusions and Future Perspectives: A New Hope?

A healthy gut microbiome is critical for the normal physiological and metabolic functions of the host. Gut microbes provide a lot of benefits, such as extraction of energy from undigested food, synthesis of important metabolites, and prevention from pathogens. Moreover, the gut microbes' integrity is swayed by mental health (stress) and by extraneous factors including diet, prebiotics, probiotics, and drugs (antibiotics). Studies established a strong connection between gut microbiome and

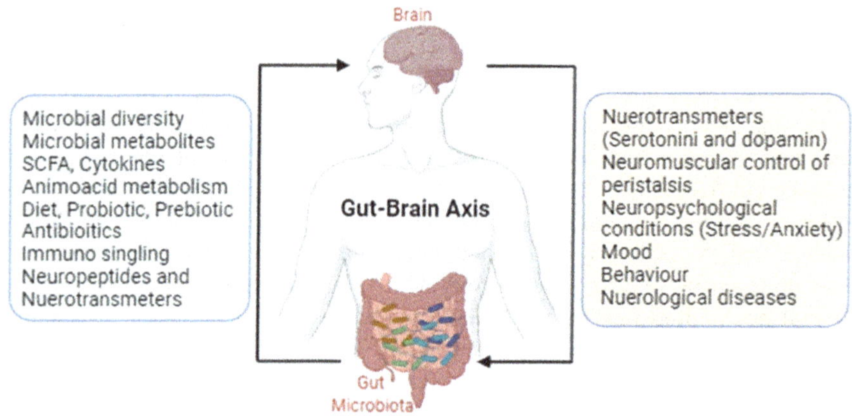

Fig. 3.2 Gut–brain axis: bidirectional talk between the gut flora and the brain which significantly impact each other

neurological diseases/disorders. The bidirectional interaction between gut microbiota and the brain regulates neurophysiological function, which further regulates neurotransmission, cognition, and behavior through the immune, neuroendocrine system, and bacterial metabolites pathways. Gut dysbiosis and associated host response were found to correlate with neurodegenerative disease conditions like AD and PD. Lumen microbe composition crucially depends on diet. Preclinical investigations of pre- and probiotics with other nutritional supplements have shown promising benefits for different psychiatric conditions including depression and anxiety (Pirbaglou et al. 2016). The microbiota–gut–brain axis may provide novel targets for the prevention and treatment of neuropsychiatric disorders.

There is an association between functional/behavioral changes in animal hosts and alteration in the normal gut microbiota. We have discussed above the interventions on specific gut bacterial taxa that can restore health in the sick individual. Thus, the identification of microbes, responsible or play a role, in diseases, can be further exploited as a tool for therapeutic interventions. Similarly, gut microbe changes, responsible for neurological disease and stress-related disorders, may be rescued by replenishing healthy gut flora (selected taxa), with certain other therapeutic agents. Since probiotics have already been recognized for benefits for anxiety/depression conditions and metabolic disorders, thus both probiotics and prebiotics could be further explored to address the problem of neurological and metabolic disorders. Furthermore, because of the added benefits of a plant-based diet on human health, it seems that a more rigorous effort is necessary to explore the welfare of the plant-based diet therapeutically (Fig. 3.2).

Acknowledgments Critical reading and suggestions to improve the chapter by Prof. Joseph Shlomai, Ms. Avital Cher, and Ms. Chiara Mazzoni are highly appreciated.

References

Aagaard K, Ma J, Antony KM, Ganu R, Petrosino J, Versalovic J (2014) The placenta harbors a unique microbiome. Sci Transl Med 6(237):237ra65

Abdel-Haq R, Schlachetzki JC, Glass CK, Mazmanian SK (2019) Microbiome–microglia connections via the gut–brain axis. J Exp Med 216(1):41–59

Abraira VE, Ginty DD (2013) The sensory neurons of touch. Neuron 79(4):618–639

Ahmed FE, Jeffries CD, Vos PW, Flake G, Nuovo GJ, Sinar DR, Naziri W, Marcuard SP (2009) Diagnostic microRNA markers for screening sporadic human colon cancer and active ulcerative colitis in stool and tissue. Cancer Genomics-Proteomics 6(5):281–295

Ait-Belgnaoui AE, Durand H, Cartier C, Chaumaz G, Eutamene H, Ferrier L, Houdeau E, Fioramonti J, Bueno L, Theodorou V (2012) Prevention of gut leakiness by a probiotic treatment leads to attenuated HPA response to an acute psychological stress in rats. Psychoneuroendocrinology 37(11):1885–1895

Akbari V, Hendijani F (2016) Effects of probiotic supplementation in patients with type 2 diabetes: systematic review and meta-analysis. Nutr Rev 74(12):774–784

Andreadou E, Pantazaki AA, Daniilidou M, Tsolaki M (2017) Rhamnolipids, microbial virulence factors, in Alzheimer's disease. J Alzheimers Dis 59(1):209–222

Artis D, Wang ML, Keilbaugh SA, He W, Brenes M, Swain GP, Knight PA, Donaldson DD, Lazar MA, Miller HR, Schad GA (2004) RELMβ/FIZZ2 is a goblet cell-specific immune-effector molecule in the gastrointestinal tract. Proc Natl Acad Sci 101(37):13596–13600

Bäckhed F, Ley RE, Sonnenburg JL, Peterson DA, Gordon JI (2005) Host-bacterial mutualism in the human intestine. Science 307(5717):1915–1920

Bäumler AJ, Sperandio V (2016) Interactions between the microbiota and pathogenic bacteria in the gut. Nature 535(7610):85–93

Beilharz JE, Kaakoush NO, Maniam J, Morris MJ (2016) The effect of short-term exposure to energy-matched diets enriched in fat or sugar on memory, gut microbiota and markers of brain inflammation and plasticity. Brain Behav Immun 57:304–313

Belkaid Y, Hand TW (2014) Role of the microbiota in immunity and inflammation. Cell 157(1): 121–141

Bengmark S (1998) Ecological control of the gastrointestinal tract. The role of probiotic flora. Gut 42(1):2–7

Bengmark S (2013) Gut microbiota, immune development and function. Pharmacol Res 69(1): 87–113

Bercik P, Denou E, Collins J, Jackson W, Lu J, Jury J, Deng Y, Blennerhassett P, Macri J, McCoy KD, Verdu EF (2011a) The intestinal microbiota affect central levels of brain-derived neurotropic factor and behavior in mice. Gastroenterology 141(2):599–609

Bercik P, Park AJ, Sinclair D, Khoshdel A, Lu J, Huang X, Deng Y, Blennerhassett PA, Fahnestock M, Moine D, Berger B (2011b) The anxiolytic effect of Bifidobacterium longum NCC3001 involves vagal pathways for gut–brain communication. Neurogastroenterol Motility 23(12):1132–1139

Block ML, Zecca L, Hong JS (2007) Microglia-mediated neurotoxicity: uncovering the molecular mechanisms. Nat Rev Neurosci 8(1):57–69

Bowden K, Ross WJ (1965) A comparison of anthelmintic and antibacterial activity of some phloroglucinol derivatives. J Pharm Pharmacol 17(4):239–242

Braniste V, Al-Asmakh M, Kowal C, Anuar F, Abbaspour A, Tóth M, Korecka A, Bakocevic N, Ng LG, Kundu P, Gulyás B (2014) The gut microbiota influences blood-brain barrier permeability in mice. Sci Transl Med 6(263) 263ra158

Bravo JA, Forsythe P, Chew MV, Escaravage E, Savignac HM, Dinan TG, Bienenstock J, Cryan JF (2011) Ingestion of Lactobacillus strain regulates emotional behavior and central GABA receptor expression in a mouse via the vagus nerve. Proc Natl Acad Sci 108(38):16050–16055

Buffington SA, Di Prisco GV, Auchtung TA, Ajami NJ, Petrosino JF, Costa-Mattioli M (2016) Microbial reconstitution reverses maternal diet-induced social and synaptic deficits in offspring. Cell 165(7):1762–1775

Calof AL (1995) Intrinsic and extrinsic factors regulating vertebrate neurogenesis. Curr Opin Neurobiol 5(1):19–27

Cardona F, Andrés-Lacueva C, Tulipani S, Tinahones FJ, Queipo-Ortuño MI (2013) Benefits of polyphenols on gut microbiota and implications in human health. J Nutr Biochem 24(8): 1415–1422

Cassani E, Barichella M, Cancello R, Cavanna F, Iorio L, Cereda E, Bolliri C, Maria PZ, Bianchi F, Cestaro B, Pezzoli G (2015) Increased urinary indoxyl sulfate (indican): new insights into gut dysbiosis in Parkinson's disease. Parkinsonism Relat Disord 21(4):389–393

Cattaneo A, Cattane N, Galluzzi S, Provasi S, Lopizzo N, Festari C, Ferrari C, Guerra UP, Paghera B, Muscio C, Bianchetti A (2017 Jan) Association of brain amyloidosis with pro-inflammatory gut bacterial taxa and peripheral inflammation markers in cognitively impaired elderly. Neurobiol Aging 49:60–68

Ceni C, Unsain N, Zeinieh MP, Barker PA (2014) Neurotrophins in the regulation of cellular survival and death. In: Neurotrophic factors. Springer, Berlin, Heidelberg, pp 193–221

Chang CY, Ke DS, Chen JY (2009) Essential fatty acids and human brain. Acta Neurol Taiwanica 18(4):231–241

Chen T, You Y, Xie G, Zheng X, Zhao A, Liu J, Zhao Q, Wang S, Huang F, Rajani C, Wang C (2018) Strategy for an association study of the intestinal microbiome and brain metabolome across the lifespan of rats. Anal Chem 90(4):2475–2483

Claesson MJ, Jeffery IB, Conde S, Power SE, O'connor EM, Cusack S, Harris HM, Coakley M, Lakshminarayanan B, O'Sullivan O, Fitzgerald GF (2012) Gut microbiota composition correlates with diet and health in the elderly. Nature 488(7410):178–184

Clarke G, Grenham S, Scully P, Fitzgerald P, Moloney RD, Shanahan F, Dinan TG, Cryan JF (2013) The microbiome-gut-brain axis during early life regulates the hippocampal serotonergic system in a sex-dependent manner. Mol Psychiatry 18(6):666–673

Clarke G, O'Mahony SM, Dinan TG, Cryan JF (2014) Priming for health: gut microbiota acquired in early life regulates physiology, brain and behaviour. Acta Paediatr 103(8):812–819

Clemente JC, Ursell LK, Parfrey LW, Knight R (2012) The impact of the gut microbiota on human health: an integrative view. Cell 148(6):1258–1270

Cleusix V, Lacroix C, Vollenweider S, Duboux M, Le Blay G (2007) Inhibitory activity spectrum of reuterin produced by Lactobacillus reuteri against intestinal bacteria. BMC Microbiol 7(1): 101

David LA, Maurice CF, Carmody RN, Gootenberg DB, Button JE, Wolfe BE, Ling AV, Devlin AS, Varma Y, Fischbach MA, Biddinger SB (2014) Diet rapidly and reproducibly alters the human gut microbiome. Nature 505(7484):559–563

Davison AN, Dobbing J (1966) Myelination as a vulnerable period in brain development. Br Med Bull 22(1):40–44

Degroote S, Hunting DJ, Baccarelli AA, Takser L (2016) Maternal gut and fetal brain connection: increased anxiety and reduced social interactions in Wistar rat offspring following periconceptional antibiotic exposure. Prog Neuro-Psychopharmacol Biol Psychiatry 71:76–82

den Besten G, van Eunen K, Groen AK, Venema K, Reijngoud DJ, Bakker BM (2013) The role of short-chain fatty acids in the interplay between diet, gut microbiota, and host energy metabolism. J Lipid Res 54(9):2325–2340

Desbonnet L, Clarke G, Traplin A, O'Sullivan O, Crispie F, Moloney RD, Cotter PD, Dinan TG, Cryan JF (2015) Gut microbiota depletion from early adolescence in mice: implications for brain and behaviour. Brain Behav Immun 1(48):165–173

Dethlefsen L, Relman DA (2011) Incomplete recovery and individualized responses of the human distal gut microbiota to repeated antibiotic perturbation. Proc Natl Acad Sci 108(Suppl. 1): 4554–4561

Devkota S (2016) Prescription drugs obscure microbiome analyses. Science 351(6272):452–453

DiGiulio DB, Callahan BJ, McMurdie PJ, Costello EK, Lyell DJ, Robaczewska A, Sun CL, Goltsman DS, Wong RJ, Shaw G, Stevenson DK (2015) Temporal and spatial variation of the human microbiota during pregnancy. Proc Natl Acad Sci 112(35):11060–11065

Dinan TG, Cryan JF (2017) Gut instincts: microbiota as a key regulator of brain development, ageing and neurodegeneration. J Physiol 595(2):489–503

Dinleyici EC, Eren M, Ozen M, Yargic ZA, Vandenplas Y (2012) Effectiveness and safety of *Saccharomyces boulardii* for acute infectious diarrhea. Expert Opin Biol Ther 12(4):395–410

Dobbs SM, Dobbs RJ, Weller C, Charlett A, Bjarnason IT, Lawson AJ, Letley D, Harbin L, Price AB, Ibrahim MA, Oxlade NL (2010) Differential effect of *Helicobacter pylori* eradication on time-trends in brady/hypokinesia and rigidity in idiopathic parkinsonism: report on completion of a randomized, double-blind, placebo-controlled efficacy study. Helicobacter 15(4):279–294

Du Y, Gao XR, Peng L, Ge JF (2020) Crosstalk between the microbiota-gut-brain axis and depression. Heliyon 6(6):e04097

Duncan ID, Watters JJ (2019) Remyelination and the gut–brain axis. Proc Natl Acad Sci 116(50): 24922–24924

Eckburg PB, Bik EM, Bernstein CN, Purdom E, Dethlefsen L, Sargent M, Gill SR, Nelson KE, Relman DA (2005) Diversity of the human intestinal microbial flora. Science 308(5728): 1635–1638

Elinav E, Strowig T, Kau AL, Henao-Mejia J, Thaiss CA, Booth CJ, Peaper DR, Bertin J, Eisenbarth SC, Gordon JI, Flavell RA (2011) NLRP6 inflammasome regulates colonic microbial ecology and risk for colitis. Cell 145(5):745–757

Engen PA, Dodiya HB, Naqib A, Forsyth CB, Green SJ, Voigt RM, Kordower JH, Mutlu EA, Shannon KM, Keshavarzian A (2017) The potential role of gut-derived inflammation in multiple system atrophy. J Parkinsons Dis 7(2):331–346

Everard A, Belzer C, Geurts L, Ouwerkerk JP, Druart C, Bindels LB, Guiot Y, Derrien M, Muccioli GG, Delzenne NM, De Vos WM (2013) Cross-talk between *Akkermansia muciniphila* and intestinal epithelium controls diet-induced obesity. Proc Natl Acad Sci 110(22):9066–9071

Ezenwa VO, Gerardo NM, Inouye DW, Medina M, Xavier JB (2012) Animal behavior and the microbiome. Science 338(6104):198–199

Farzi A, Fröhlich EE, Holzer P (2018) Gut microbiota and the neuroendocrine system. Neurotherapeutics 15(1):5–22

Felice VD, Quigley EM, Sullivan AM, O'Keeffe GW, O'Mahony SM (2016) Microbiota-gut-brain signalling in Parkinson's disease: implications for non-motor symptoms. Parkinsonism Relat Disord 27:1–8

Ferreiro A, Crook N, Gasparrini AJ, Dantas G (2018) Multiscale evolutionary dynamics of host-associated microbiomes. Cell 172(6):1216–1227

Forssberg H (2019) Microbiome programming of brain development: implications for neurodevelopmental disorders. Dev Med Child Neurol 61(7):744–749

Forsythe P, Kunze W, Bienenstock J (2016) Moody microbes or fecal phrenology: what do we know about the microbiota-gut-brain axis? BMC Med 14(1):58

Fröhlich EE, Farzi A, Mayerhofer R, Reichmann F, Jačan A, Wagner B, Zinser E, Bordag N, Magnes C, Fröhlich E, Kashofer K (2016) Cognitive impairment by antibiotic-induced gut dysbiosis: analysis of gut microbiota-brain communication. Brain Behav Immun 56:140–155

Fung TC, Olson CA, Hsiao EY (2017) Interactions between the microbiota, immune and nervous systems in health and disease. Nat Neurosci 20(2):145

Furness JB, Callaghan BP, Rivera LR, Cho HJ (2014) The enteric nervous system and gastrointestinal innervation: integrated local and central control. In: Microbial endocrinology: the microbiota-gut-brain axis in health and disease. Springer, New York, NY, pp 39–71

Gacias M, Gaspari S, Santos PM, Tamburini S, Andrade M, Zhang F, Shen N, Tolstikov V, Kiebish MA, Dupree JL, Zachariou V (2016) Microbiota-driven transcriptional changes in prefrontal cortex override genetic differences in social behavior. elife 5:e13442

Gao K, Mu CL, Farzi A, Zhu WY (2020) Tryptophan Metabolism: a link between the gut microbiota and brain. Adv Nutr 11(3):709–723

Ge X, Ding C, Zhao W, Xu L, Tian H, Gong J, Zhu M, Li J, Li N (2017) Antibiotics-induced depletion of mice microbiota induces changes in host serotonin biosynthesis and intestinal motility. J Transl Med 15(1):13

Gensollen T, Iyer SS, Kasper DL, Blumberg RS (2016) How colonization by microbiota in early life shapes the immune system. Science 352(6285):539–544

Gill SR, Pop M, DeBoy RT, Eckburg PB, Turnbaugh PJ, Samuel BS, Gordon JI, Relman DA, Fraser-Liggett CM, Nelson KE (2006) Metagenomic analysis of the human distal gut microbiome. Science 312(5778):1355–1359

Goldman SM, Kamel F, Ross GW, Jewell SA, Marras C, Hoppin JA, Umbach DM, Bhudhikanok GS, Meng C, Korell M, Comyns K (2014) Peptidoglycan recognition protein genes and risk of Parkinson's disease. Mov Disord 29(9):1171–1180

Gómez-Pinilla F (2008) Brain foods: the effects of nutrients on brain function. Nat Rev Neurosci 9(7):568–578

Groer MW, Luciano AA, Dishaw LJ, Ashmeade TL, Miller E, Gilbert JA (2014) Development of the preterm infant gut microbiome: a research priority. Microbiome 2:38

Grönlund MM, Arvilommi H, Kero P, Lehtonen OP, Isolauri E (2000) Importance of intestinal colonisation in the maturation of humoral immunity in early infancy: a prospective follow up study of healthy infants aged 0–6 months. Arch Dis Child Fetal Neonatal Ed 83(3):F186–F192

Guaraldi F, Salvatori G (2012) Effect of breast and formula feeding on gut microbiota shaping in newborns. Front Cell Infect Microbiol 2:94

Guo S, Al-Sadi R, Said HM, Ma TY (2013) Lipopolysaccharide causes an increase in intestinal tight junction permeability in vitro and in vivo by inducing enterocyte membrane expression and localization of TLR-4 and CD14. Am J Pathol 182(2):375–387

Hanke ML, Kielian T (2011) Toll-like receptors in health and disease in the brain: mechanisms and therapeutic potential. Clin Sci 121(9):367–387

Hashim H, Azmin S, Razlan H, Yahya NW, Tan HJ, Manaf MR, Ibrahim NM (2014) Eradication of *Helicobacter pylori* infection improves levodopa action, clinical symptoms and quality of life in patients with Parkinson's disease. PLoS One 9(11):e112330. https://doi.org/10.1371/journal.pone.0112330

Heijtz RD, Wang S, Anuar F, Qian Y, Björkholm B, Samuelsson A, Hibberd ML, Forssberg H, Pettersson S (2011) Normal gut microbiota modulates brain development and behavior. Proc Natl Acad Sci 108(7):3047–3052

Hendijani F, Akbari V (2018) Probiotic supplementation for management of cardiovascular risk factors in adults with type II diabetes: a systematic review and meta-analysis. Clin Nutr 37(2): 532–541

Hill-Burns EM, Debelius JW, Morton JT, Wissemann WT, Lewis MR, Wallen ZD et al (2017) Parkinson's disease and Parkinson's disease medications have distinct signatures of the gut microbiome. Mov Disord 32:739–749

Hoban AE, Stilling RM, Ryan FJ, Shanahan F, Dinan TG, Claesson MJ, Clarke G, Cryan JF (2016) Regulation of prefrontal cortex myelination by the microbiota. Transl Psychiatry 6(4) e774-

Hooper LV, Littman DR, Macpherson AJ, Aj M (2012) Interactions between the Microbiota and the immune system. Science 336:1268–1273

Hoyles L, Snelling T, Umlai UK, Nicholson JK, Carding SR, Glen RC, McArthur S (2018) Microbiome–host systems interactions: protective effects of propionate upon the blood–brain barrier. Microbiome 6(1):1–3

Hsiao EY, McBride SW, Hsien S, Sharon G, Hyde ER, McCue T, Codelli JA, Chow J, Reisman SE, Petrosino JF, Patterson PH (2013) Microbiota modulate behavioral and physiological abnormalities associated with neurodevelopmental disorders. Cell 155(7):1451–1463

Humann J, Mann B, Gao G, Moresco P, Ramahi J, Loh LN, Farr A, Hu Y, Durick-Eder K, Fillon SA, Smeyne RJ (2016) Bacterial peptidoglycan traverses the placenta to induce fetal neuroproliferation and aberrant postnatal behavior. Cell Host Microbe 19(3):388–399

Ichim G, Tauszig-Delamasure S, Mehlen P (2012) Neurotrophins and cell death. Exp Cell Res 318(11):1221–1228

Jakobsson HE, Jernberg C, Andersson AF, Sjölund-Karlsson M, Jansson JK, Engstrand L (2010) Short-term antibiotic treatment has differing long-term impacts on the human throat and gut microbiome. PLoS One 5(3):e9836. https://doi.org/10.1371/journal.pone.0009836

Jašarević E, Rodgers AB, Bale TL (2015) A novel role for maternal stress and microbial transmission in early life programming and neurodevelopment. Neurobiol Stress 1:81–88

Jernberg C, Löfmark S, Edlund C, Jansson JK (2007) Long-term ecological impacts of antibiotic administration on the human intestinal microbiota. ISME J 1(1):56–66

Keshavarzian A, Green SJ, Engen PA, Voigt RM, Naqib A, Forsyth CB, Mutlu E, Shannon KM (2015) Colonic bacterial composition in Parkinson's disease. Mov Disord 30(10):1351–1360

Klingensmith NJ, Coopersmith CM (2016) The gut as the motor of multiple organ dysfunction in critical illness. Crit Care Clin 32(2):203–212

Koenig JE, Spor A, Scalfone N, Fricker AD, Stombaugh J, Knight R, Angenent LT, Ley RE (2011) Succession of microbial consortia in the developing infant gut microbiome. Proc Natl Acad Sci 108(Suppl. 1):4578–4585

Konrad K, Firk C, Uhlhaas PJ (2013) Brain development during adolescence: neuroscientific insights into this developmental period. Dtsch Arztebl Int 110(25):425

Koren O, Goodrich JK, Cullender TC, Spor A, Laitinen K, Bäckhed HK, Gonzalez A, Werner JJ, Angenent LT, Knight R, Bäckhed F (2012) Host remodeling of the gut microbiome and metabolic changes during pregnancy. Cell 150(3):470–480

Krajmalnik-Brown R, Lozupone C, Kang DW, Adams JB (2015) Gut bacteria in children with autism spectrum disorders: challenges and promise of studying how a complex community influences a complex disease. Microb Ecol Health Dis 26(1):26914

Kumbhare SV, Patangia DV, Patil RH, Shouche YS, Patil NP (2019) Factors influencing the gut microbiome in children: from infancy to childhood. J Biosci 44(2):49

Kundu P, Blacher E, Elinav E, Pettersson S (2017) Our gut microbiome: the evolving inner self. Cell 171(7):1481–1493

Levy M, Thaiss CA, Zeevi D, Dohnalová L, Zilberman-Schapira G, Mahdi JA et al (2015) Microbiota-modulated metabolites shape the intestinal microenvironment by regulating NLRP6 inflammasome signaling. Cell 163:1428–1443

Link A, Becker V, Goel A, Wex T, Malfertheiner P (2012) Feasibility of fecal microRNAs as novel biomarkers for pancreatic cancer. PLoS One 7(8):e42933. https://doi.org/10.1371/journal.pone.0042933

Liu S, da Cunha AP, Rezende RM, Cialic R, Wei Z, Bry L, Comstock LE, Gandhi R, Weiner HL (2016) The host shapes the gut microbiota via fecal microRNA. Cell Host Microbe 19(1):32–43

Luczynski P, McVey Neufeld KA, Oriach CS, Clarke G, Dinan TG, Cryan JF (2016) Growing up in a bubble: using germ-free animals to assess the influence of the gut microbiota on brain and behavior. Int J Neuropsychopharmacol 19(8):1–17. https://doi.org/10.1093/ijnp/pyw020

MacIntyre DA, Chandiramani M, Lee YS, Kindinger L, Smith A, Angelopoulos N, Lehne B, Arulkumaran S, Brown R, Teoh TG, Holmes E (2015) The vaginal microbiome during pregnancy and the postpartum period in a European population. Sci Rep 5:8988

Martin R, Makino H, Yavuz AC, Ben-Amor K, Roelofs M, Ishikawa E, Kubota H, Swinkels S, Sakai T, Oishi K, Kushiro A (2016) Early-life events, including mode of delivery and type of feeding, siblings and gender, shape the developing gut microbiota. PLoS One 11(6):e0158498. https://doi.org/10.1371/journal.pone.0158498

Mayer EA (2011) Gut feelings: the emerging biology of gut–brain communication. Nat Rev Neurosci 12(8):453–466

Maynard CL, Elson CO, Hatton RD, Weaver CT (2012) Reciprocal interactions of the intestinal microbiota and immune system. Nature 489(7415):231

Menn B, Garcia-Verdugo JM, Yaschine C, Gonzalez-Perez O, Rowitch D, Alvarez-Buylla A (2006) Origin of oligodendrocytes in the subventricular zone of the adult brain. J Neurosci 26(30):7907–7918

Michel L, Prat A (2016) One more role for the gut: microbiota and blood brain barrier. Ann Transl Med 4 (1)15. https://doi.org/10.3978/j.issn.2305-5839.2015.10.16

Milani C, Duranti S, Bottacini F, Casey E, Turroni F, Mahony J, Belzer C, Palacio SD, Montes SA, Mancabelli L, Lugli GA (2017) The first microbial colonizers of the human gut: composition, activities, and health implications of the infant gut microbiota. Microbiol Mol Biol Rev 81(4): e00036–e00017

Möhle L, Mattei D, Heimesaat MM, Bereswill S, Fischer A, Alutis M, French T, Hambardzumyan D, Matzinger P, Dunay IR, Wolf SA (2016) Ly6Chi monocytes provide a link between antibiotic-induced changes in gut microbiota and adult hippocampal neurogenesis. Cell Rep 15(9):1945–1956

Mousa A, Bakhiet M (2013) Role of cytokine signaling during nervous system development. Int J Mol Sci 14(7):13931–13957

Natividad JM, Verdu EF (2013) Modulation of intestinal barrier by intestinal microbiota: pathological and therapeutic implications. Pharmacol Res 69(1):42–51

Neish AS (2009) Microbes in gastrointestinal health and disease. Gastroenterology 136(1):65–80

Neufeld KM, Kang N, Bienenstock J, Foster JA (2011) Reduced anxiety-like behavior and central neurochemical change in germ-free mice. Neurogastroenterol Motil 23(3):255–264. e119-

Nuriel-Ohayon M, Neuman H, Koren O (2016) Microbial changes during pregnancy, birth, and infancy. Front Microbiol 14(7):1031

O'Grady J, O'Connor EM, Shanahan F (2019) Dietary fibre in the era of microbiome science. Aliment Pharmacol Ther 49(5):506–515

Odamaki T, Kato K, Sugahara H, Hashikura N, Takahashi S, Xiao JZ, Abe F, Osawa R (2016) Age-related changes in gut microbiota composition from newborn to centenarian: a cross-sectional study. BMC Microbiol 16(1):90

Ogawa T, Hirose Y, Honda-Ogawa M, Sugimoto M, Sasaki S, Kibi M, Kawabata S, Ikebe K, Maeda Y (2018) Composition of salivary microbiota in elderly subjects. Sci Rep 8(1):1–9

Ogbonnaya ES, Clarke G, Shanahan F, Dinan TG, Cryan JF, O'Leary OF (2015) Adult hippocampal neurogenesis is regulated by the microbiome. Biol Psychiatry 78(4):e7–e9

Paolicelli RC, Bolasco G, Pagani F, Maggi L, Scianni M, Panzanelli P, Giustetto M, Ferreira TA, Guiducci E, Dumas L, Ragozzino D (2011) Synaptic pruning by microglia is necessary for normal brain development. Science 333(6048):1456–1458

Perez-Pardo P, Kliest T, Dodiya HB, Broersen LM, Garssen J, Keshavarzian A, Kraneveld AD (2017) The gut-brain axis in Parkinson's disease: possibilities for food-based therapies. Eur J Pharmacol 817:86–95

Pierantozzi M, Pietroiusti A, Sancesario G, Lunardi G, Fedele E, Giacomini P, Frasca S, Galante A, Marciani MG, Stanzione P (2001) Reduced L-dopa absorption and increased clinical fluctuations in *Helicobacter pylori*-infected Parkinson's disease patients. Neurol Sci 22(1):89–91

Pirbaglou M, Katz J, de Souza RJ, Stearns JC, Motamed M, Ritvo P (2016) Probiotic supplementation can positively affect anxiety and depressive symptoms: a systematic review of randomized controlled trials. Nutr Res 36(9):889–898

Podolsky DK, Lynch-Devaney K, Stow JL, Oates P, Murgue B, DeBeaumont M, Sands BE, Mahida YR (1993) Identification of human intestinal trefoil factor. Goblet cell-specific expression of a peptide targeted for apical secretion. J Biol Chem 268(9):6694–6702

Principi N, Esposito S (2016) Gut microbiota and central nervous system development. J Infect 73(6):536–546

Pulikkan J, Mazumder A, Grace T (2019) Role of the Gut Microbiome in Autism Spectrum Disorders. Adv Exp Med Biol 1118:253–269. https://doi.org/10.1007/978-3-030-05542-4_13

Quigley EM (2017) Microbiota-brain-gut axis and neurodegenerative diseases. Curr Neurol Neurosci Rep 17(12):94

Reza MM, Finlay BB, Pettersson S (2019) Gut microbes, ageing & organ function: a chameleon in modern biology? EMBO Mol Med 11(9):e9872

Rodríguez JM, Murphy K, Stanton C, Ross RP, Kober OI, Juge N, Avershina E, Rudi K, Narbad A, Jenmalm MC, Marchesi JR (2015) The composition of the gut microbiota throughout life, with an emphasis on early life. Microb Ecol Health Dis 26(1):26050

Rutayisire E, Huang K, Liu Y, Tao F (2016) The mode of delivery affects the diversity and colonization pattern of the gut microbiota during the first year of infants' life: a systematic review. BMC Gastroenterol 16(1):86

Sánchez B, Delgado S, Blanco-Míguez A, Lourenço A, Gueimonde M, Margolles A (2017) Probiotics, gut microbiota, and their influence on host health and disease. Mol Nutr Food Res 61(1):1600240

Sarkar SR, Banerjee S (2019) Gut microbiota in neurodegenerative disorders. J Neuroimmunol 328: 98–104

Scheperjans F, Aho V, Pereira PA, Koskinen K, Paulin L, Pekkonen E, Haapaniemi E, Kaakkola S, Eerola-Rautio J, Pohja M, Kinnunen E (2015) Gut microbiota are related to Parkinson's disease and clinical phenotype. Mov Disord 30(3):350–358

Sender R, Fuchs S, Milo R (2016) Revised estimates for the number of human and bacteria cells in the body. PLoS Biol 14(8):e1002533

Senghor B, Sokhna C, Ruimy R, Lagier JC (2018) Gut microbiota diversity according to dietary habits and geographical provenance. Hum Microbiome J 7:1–9

Servin AL (2004) Antagonistic activities of lactobacilli and bifidobacteria against microbial pathogens. FEMS Microbiol Rev 28(4):405–440

Severance EG, Yolken RH, Eaton WW (2016) Autoimmune diseases, gastrointestinal disorders and the microbiome in schizophrenia: more than a gut feeling. Schizophr Res 176(1):23–35

Sharon G, Sampson TR, Geschwind DH, Mazmanian SK (2016) The central nervous system and the gut microbiome. Cell 167(4):915–932

Shen X, Yang H, Wu Y, Zhang D, Jiang H (2017) Meta-analysis: association of *Helicobacter pylori* infection with Parkinson's diseases. Helicobacter 22(5):e12398

Sherwin E, Sandhu KV, Dinan TG, Cryan JF (2016) May the force be with you: the light and dark sides of the microbiota–gut–brain axis in neuropsychiatry. CNS Drugs 30(11):1019–1041

Spencer SJ, Korosi A, Layé S, Shukitt-Hale B, Barrientos RM (2017) Food for thought: how nutrition impacts cognition and emotion. NPJ Sci Food 1(1):1–8

Stark PL, Lee AD, Parsonage BD (1982) Colonization of the large bowel by Clostridium difficile in healthy infants: quantitative study. Infect Immun 35(3):895–899

Sudo N, Chida Y, Aiba Y, Sonoda J, Oyama N, Yu XN, Kubo C, Koga Y (2004) Postnatal microbial colonization programs the hypothalamic–pituitary–adrenal system for stress response in mice. J Physiol 558(1):263–275

Sui YT, Bullock KM, Erickson MA, Zhang J, Banks WA (2014) Alpha synuclein is transported into and out of the brain by the blood–brain barrier. Peptides 62:197–202

Sun J, Wang F, Ling Z, Yu X, Chen W, Li H, Jin J, Pang M, Zhang H, Yu J, Liu J (2016) Clostridium butyricum attenuates cerebral ischemia/reperfusion injury in diabetic mice via modulation of gut microbiota. Brain Res 1642:180–188

Tan AH, Mahadeva S, Marras C, Thalha AM, Kiew CK, Yeat CM, Ng SW, Ang SP, Chow SK, Loke MF, Vadivelu JS (2015) *Helicobacter pylori* infection is associated with worse severity of Parkinson's disease. Parkinsonism Relat Disord 21(3):221–225

Tang WW, Wang Z, Levison BS, Koeth RA, Britt EB, Fu X, Wu Y, Hazen SL (2013) Intestinal microbial metabolism of phosphatidylcholine and cardiovascular risk. N Engl J Med 368(17): 1575–1584

Tillisch K, Labus J, Kilpatrick L, Jiang Z, Stains J, Ebrat B, Guyonnet D, Legrain–Raspaud S, Trotin B, Naliboff B, Mayer EA (2013) Consumption of fermented milk product with probiotic modulates brain activity. Gastroenterology 144(7):1394–1401

Tochitani S, Ikeno T, Ito T, Sakurai A, Yamauchi T, Matsuzaki H (2016) Administration of non-absorbable antibiotics to pregnant mice to perturb the maternal gut microbiota is associated with alterations in offspring behavior. PLoS One 11(1):e0138293. https://doi.org/10.1371/journal.pone.0138293

Tomova A, Bukovsky I, Rembert E, Yonas W, Alwarith J, Barnard ND, Kahleova H (2019) The effects of vegetarian and vegan diets on gut microbiota. Front Nutr 6:47

Uzbay T (2019) Germ-free animal experiments in the gut microbiota studies. Curr Opin Pharmacol 49:6–10

Vissiennon C, Nieber K, Kelber O, Butterweck V (2012) Route of administration determines the anxiolytic activity of the flavonols kaempferol, quercetin and myricetin—are they prodrugs? J Nutr Biochem 23(7):733–740

Walker AW, Ince J, Duncan SH, Webster LM, Holtrop G, Ze X, Brown D, Stares MD, Scott P, Bergerat A, Louis P (2011) Dominant and diet-responsive groups of bacteria within the human colonic microbiota. ISME J 5(2):220–230

Wang D, Ho L, Faith J, Ono K, Janle EM, Lachcik PJ, Cooper BR, Jannasch AH, D'Arcy BR, Williams BA, Ferruzzi MG (2015) Role of intestinal microbiota in the generation of polyphenol-derived phenolic acid mediated attenuation of Alzheimer's disease β-amyloid oligomerization. Mol Nutr Food Res 59(6):1025–1040

Wu GD, Compher C, Chen EZ, Smith SA, Shah RD, Bittinger K, Chehoud C, Albenberg LG, Nessel L, Gilroy E, Star J (2016) Comparative metabolomics in vegans and omnivores reveal constraints on diet-dependent gut microbiota metabolite production. Gut 65(1):63–72

Wu Y, Zhang Q, Ren Y, Ruan Z (2017) Effect of probiotic Lactobacillus on lipid profile: a systematic review and meta-analysis of randomized, controlled trials. PLoS One 12(6): e0178868. https://doi.org/10.1371/journal.pone.0178868

Yoshioka H, Iseki KI, Fujita K (1983) Development and differences of intestinal flora in the neonatal period in breast-fed and bottle-fed infants. Pediatrics 72(3):317–321

Zeisel SH (2004) Nutritional importance of choline for brain development. J Am Coll Nutr 23 (sup6):621S–626S

Zhou K (2017) Strategies to promote abundance of *Akkermansia muciniphila*, an emerging probiotics in the gut, evidence from dietary intervention studies. J Funct Foods 33:194–201

Zuchero JB, Barres BA (2015) Glia in mammalian development and disease. Development 142(22):3805–3809

Chapter 4
Gut Microbiota Regulation of Cerebral Stroke

Kaushlendra Kumar, Hema Kumari, and Amit Kumar Tripathi

Abstract Microbiome involvement in human health is a recently emerging field in biomedical sciences. Biomedical research corroborations emphasize the significance of the gastrointestinal tract (GIT) microbiome in cancer, diabetes, skin disease, cardiovascular disease (CVD), neurodegenerative disorders (NDD), and stroke pathophysiology. Importantly, transient focal cerebral ischemia-reperfusion injury (TFCIRI) induced impairments of the gut microbiome regulated by the pleiotropic microbial metabolites such as nitrites, flavanol, short-chain fatty acids, sulfinic acid, and Trimethylamine N-oxide. These understandings speculate that targeting and manipulating the gut microbiome can be better therapeutic opportunities for various diseases, including cerebral stroke. In the present chapter, we will provide the basics of the microbiome, psychobiotics, neuroprotective potential of monobacteriotherapy, toll-like receptors (TLRs) such as TLR-2, TLR-4, and TLR-5, Nucleotide-binding oligomerization domain-containing protein (NOD)-like receptors (NLRs) such as NOD-1, NOD-2, and NOD-6, engineered microbiota for therapeutics, Western diet a related risk factor for stroke, Stroke-induced gut inflammatory immune response and microglial polarization (mainly M1 $>$ M2) under germ-free condition, the modulatory function of the gut microbiome in BBB breakdown, intervention strategies (diet, probiotics, prebiotics, antibiotics, fecal transplant, intragastric treatment) microbial metabolite, and seasonal changes (summer, winter, and raining) and environmental factors.

K. Kumar · H. Kumari
Clinical Research Division, School of Basic and Applied Science, Galgotias University, Greater Noida, India
e-mail: kaushalendra.kumar@galgotiasuniversity.edu.in;
hema.kumari@galgotiasuniversity.edu.in

A. K. Tripathi (✉)
Clinical Research Division, School of Basic and Applied Science, Galgotias University, Greater Noida, India

Molecular Biology Unit, Institute of Medical Science, Banaras Hindu University, Varanasi, India

A. K. Tripathi, M. Kotak (eds.), *Gut Microbiome in Neurological Health and Disorders*, Nutritional Neurosciences,
https://doi.org/10.1007/978-981-19-4530-4_4

Keywords Stroke · Microbiome · Neuroinflammation · Immunity · Short-chain fatty acid · Microbial diversity

Abbreviations

AIS	Acute ischemic stroke
BACCO	Bilateral Common Carotid Artery Occlusion
BBB	Blood–brain barrier
CIRI	Cerebral ischemia–reperfusion injury
CVD	Cardiovascular disease
GF	Germ Free
GFM	Germ-free mice
GIT	Gastrointestinal tract
MGBA	Microbiota gut–brain axis
NDD	Neurodegenerative disease
SCFA	Short-chain fatty acid
SDI	Stroke dysbiosis index
TJ	Tight junction
TMAO	Trimethylamine N-oxide
TMAO	Trimethyl amine oxide
TREM	Triggering receptor expressed on myeloid cells
WD	Western diet

4.1 Introduction

The human gut microbiota contains trillions of bacteria with more than three million genes and is considered the second genome of the body (Zhu et al. 2010). Microbiome characterization and analysis inspired the advancement of genomic technologies and provided useful information as the basis of genetic diversity, alteration in the pathophysiology of the disease, and a component of immunity (Grice and Segre 2012). The first report on gut bacteria was carried out in the nineteenth century (Winek et al. 2016a). This achievement was done when examining stool samples from healthy persons; Friedrich Escherich characterized Bacterium coli commune, today popularly known as *Escherichia coli*, as "a workhorse of molecular biology" (Shulman et al. 2007). Over the next few decades, thousands of microorganisms were separated, isolated, and characterized using new culture technologies such as sophisticated isolation culture methods. Human GIT contains a broad-spectrum, dynamic microbial community that contributes significantly to different biological processes in the host organism and reacts to variations in host pathophysiology (Shulman et al. 2007; Mathewson et al. 2016; Kaur et al. 2021). The microbial communities interact with the host organism via releasing different metabolites. Sometimes, these microbial metabolites are synthesized by carbohydrate fermentation inside the gut leading increase SCFAs concentration. Butyrate, an essential SCFA microbial metabolite, may influence the host body system by regulating histone-acetylation (Mathewson et al. 2016). However, TMAO, another metabolite, plays a vital role in NDD, CVD, and stroke (Aron-wisnewsky and Clement 2017).

Current theories unravel these microbiota–host interactions by releasing several proinflammatory cytokines, neuroendocrine messengers, and bacterial waste product help regulate many biological processes (Boziki et al. 2018). Biological molecules and neuromodulators can enter the blood vascular systems and finally alter neural connections carried by vagal and spinal afferent neurons to constantly communicate with the brain, possibly regulating mood and behavior. The above line of evidence suggests that microbiota influences mood behavior and alters the cognitive abilities of individuals by releasing growth factors, cytokine, and activation of lymphocytic cells.

The Human GIT microbiome consists of dominant bacterial species: Firmicutes, Bacteroidetes, Actinobacteria, and Proteobacteria (Monira et al. 2011; Dubourg et al. 2013; Hanfrey et al. 2011). However, mainly two bacterial phyla Bacteroidetes and Firmicutes consisting of the gut microbiota and Proteobacteria, Actinomyces, Fusobacterium, and Verrucomicrobia are relatively small. The implication of the gut microbiome for biomedical application, biomarker, and prognosis has been widely applicable for future therapeutic interventions. The microbiome has been a potential therapeutic target in neurological disorders such as Parkinson's disease (PD), Alzheimer's disease (AD), amyotrophic lateral sclerosis (ALS), Traumatic brain injury (TBI), autism, stroke, depression, and drug addiction (Wang et al. 2011; Mulak 2018; Nguyen et al. 2018; Houlden et al. 2016; Rice et al. 2019; Chen et al. 2019a; Ahmed et al. 2019; Yarandi et al. 2016). From the above line of evidence, it has been suggested that ischemic stroke alters gut microbiota composition and plays a vital role in pathomechanisms and modified neurological outcomes. However, Patients and experimental studies have reported that the gut microbiome is connected with possible factors for stroke, hypertension, diabetes, and obesity (Suzuki et al. 2017; Nam 2019). Worldwide scientific communities involved in investigating novel microbiome biomarkers associated with the incidence of ischemic stroke are beneficial and advantageous and improve diagnosis before patients arrive at the hospital. The emerging technology of metabolomics combining both clinical and biochemical biomarkers is helpful for therapeutic intervention in the growing field (Suzuki et al. 2017; Wang and Wang 2016; Sidorov et al. 2019).

The purposes of this chapter were as follows: first, providing the basic information associated with microbiome studies; second, collecting data related to the gut–brain axis in the context of cerebral stroke; third, to emphasize stroke-induced microbiota regulatory T-cells alteration, interleukin and neutrophils infiltration in brain, stroke dysbiosis index (SDI), and its relationship with BBB breakage; and fourth, compile the information related tools involved in microbiota studies.

4.2 Microbiota Gut–Brain Axis and Its Environmental Axis

The communicating messages between gut and brain through two neuroanatomical pathways explained gut–brain axis. First known communication connects the gut and brain via autonomic nervous system (ANS) and vagus nerves (VN) through spinal cord. Bidirectional communication connects gut–brain through communication between enteric nervous system (ENS) in the gut and ANS and VN within the spinal cord (Wang and Wang 2016). However, the underlying mechanism of this bidirectional communication in MGBA remains, as yet, unresolved. Inspite of the above facts, other communication such as neural mechanisms, immune response, neurotransmitter and neuropeptide release, and microbial metabolite (Bercik et al. 2012).

Environmental factors such as smoking, anxiety, and inactive lifestyle can significantly influence gut microbial diversity (Bressa et al. 2017). Smoking has an excessive impact on gut microbial diversity by increasing bacterial counts of *Bacteroides-Prevotella* (Guo et al. 2014). Stress and anxiety are other factors that impact motor activity by gut–brain axis connecting neuroendocrine pathways. It is associated with an altered microbiota profile suggesting potential decreasing counts of useful beneficial bacteria *Lactobacillus* (Lutgendorff et al. 2008; Grenham et al. 2011). Other factors such as human diet containing protein, carbohydrates, and fat are widely associated with microbiota diversity. Gut microbiota has a promising role in degradation of dietary components and produces amino acids, amine, ammonia, and SCFAs. Cysteine and threonine amino acid molecules can cause an incredible increase in beneficial microbial communities such as *lactobacilli* or *bifidobacteria* and a decrease in *Clostridiaceae* (Magee et al. 2000). A notable alteration in these bacterial counts can be associated with different types of bowel disease.

4.3 Stroke-Induced Gut Dysfunction and Translocation of Gut Microbiota

Emerging literature conferred gut microbiota has a significant role in pathological conditions affecting the CNS and cognitive disability. According to Global Burden of Disease (GBD) study, stroke remains to be an important disability with a high impact predicted to increase due to increased life expectancy and population aging (Murray and Lopez 2013). The major gut anomalies related to both ischemic and hemorrhagic stroke patients exhibit dysphagia, dysbiosis, dysmotility, hemorrhage, constipation, and gastrointestinal bleeding (Fig. 4.1) (Camara-Lemarroy et al. 2014; Schaller et al. 2006; Harari et al. 2004).

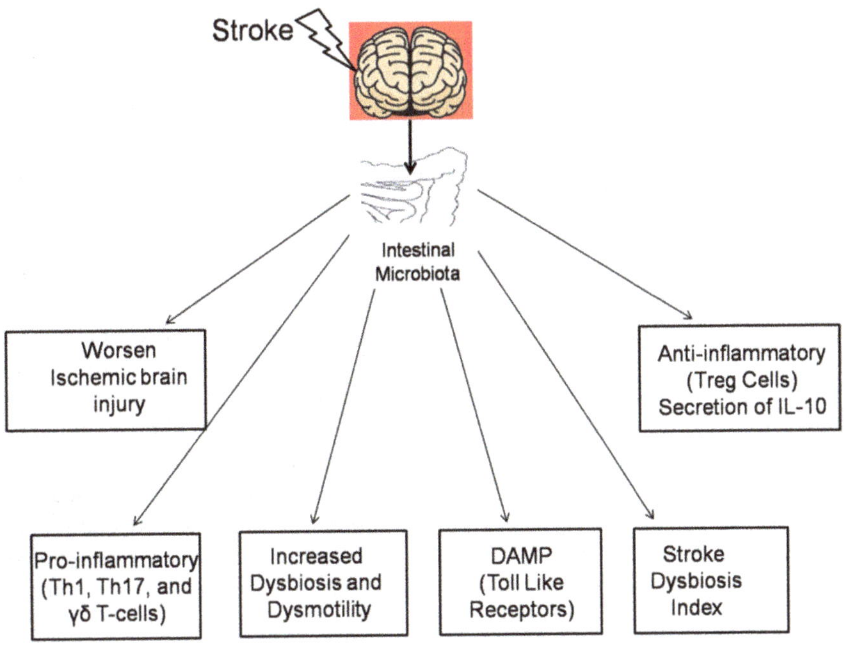

Fig. 4.1 Stroke-induced intestinal microbiome complication and its brain–gut–immune signaling pathway. Gut microbiome might contribute to the pathophysiology of vascular disorders and potential therapeutic targets for preventing the stroke risk factors, proinflammatory (Th1, Th17, and γδ T-Cells) and anti-inflammatory (Treg cells) and release of IL-10 responses and DAMP (Toll-like receptors)

4.4 Western Dietary Pattern and Related Risk Factor for Stroke-Induced Gut Microbiota Alteration

Dietary pattern, geographical differences, seasonal variation, and environmental conditions have an insightful role in regulating the gut microbiome in real-time manner. However, several studies reported that the abundance of gut microbiota is independent of age, diabetes, cardiovascular diseases, hypertension, and environmental factors (Pérez-López et al. 2009; Lindsberg and Grau 2003). The other common cause of gut dysbiosis are antibiotics treatment, psychological stress, aging, infection, alcohol intake, diet, reduced gut motility, and uses of NSAIDs may also play important role in stroke pathophysiology. The triangle association between microbiota, diet, and stroke injury suggests specific bacterial population and release of metabolite has attention for future research. Psychobiotics help in reestablishing the gastrointestinal microbiome, for example, *bifidobacteria* and *lactobacilli* (Gibson and Roberfroid 1995). Western Diet (WD) is characterized by a highly refined diet and energy dense and composed of animal protein, high amount of saturated fats, and simple sugars but low amount of fiber-rich plant-based food such as green vegetables. WD comprising a high-fat diet (HFD) associated with obesity and related metabolic diseases tends to induce considerable alterations in the composition of microbiota by increasing *Rikenellaceae* and decreasing *Ruminococcaceae* (Daniel et al. 2014). The WD has a detrimental role on microbial diversity by releasing several ROS and chemicals. High in *Bacteroides* and low in *Prevotella*, count is found in individuals on a long-term habit of WD (Wu et al. 2011). Reverse association has been found with a plant-based diet rich in fiber suggested that typical dietary routine and its changes the gut microbial diversity at the genus and species level. However, studies conducted to explore the mice fed with HFD having more choline as a substrate to increase TMAO formation, that increase the levels of proinflammatory Ly6C^{high} monocytes were higher (Haghikia et al. 2018). WD consumption is associated with specific increases in bacterial counts in the gut that are characterized by *Bacteroides* (Clarke et al. 2012). Investigation of fecal samples of 98 individuals characterized that protein and animal fat (high *Bacteroides*) versus sugars (*Prevotella*) (Purushe et al. 2010). The microbiome diversity changed noticeably within 24 h of initiating a high-fat/low-fiber or low-fat/high-fiber diet. The microbiota composition diversity in plant-based diet compared with individual habitual to WD differs in beneficial and detrimental bacteria (Frank et al. 2007). The Mediterranean diet includes fruit, vegetable, mono and polyunsaturated fats, and grains, is considered as a standard eating routine for a healthy lifestyle. The Mediterranean diet has decreased counts of *Bacillaceae* and *Proteobacteria* but higher *Clostridium* and *Bacteroidetes* populations (De Filippis et al. 2016). Furthermore, vegetarian diets could decrease *Clostridium cluster*XIVa, but undoubtedly increase the number of *Faecalibacterium prausnitzii*, *Clostridium clostridioforme*, and *Bacteroides Prevotella* (Ferrocino et al. 2015).

4.5 Stroke-Induced Gut Inflammatory Immune Response and Brain Infiltration

Gut inflammation and immune response play an important task in the pathophysiology of cerebral stroke and may become a future therapeutic option for this problem. The stroke-induced damaged brain tissue produces damaged-associated molecular patterns (DAMP) to activate innate and adaptive immune responses nearby and systemically through TLR. Following a cerebrovascular accident, white blood cells related to innate immunity such as neutrophils, microglia, mast cells, monocytes, eosinophils, basophiles, and NK cells within hours. The most abundant intestinal commensal, symbiotic environment appeared as a potent regulator of the lymphocyte population, including T regulatory (Treg) and Gamma delta T-cells (γδ T-Cells) and involved in the invasion of the ischemic brain. γδ T-cells, a significant lymphocyte subpopulation with innate immunity properties, are present on intestinal digestive system epithelial surfaces (Prinz et al. 2013). These cells have the property of infiltration in ischemic brain by releasing IL-17 and producing a chemotactic signal for peripheral myeloid neutrophils and monocyte cells

(Gelderblom et al. 2012; Shichita et al. 2009). According to Iadecola and another, IL-23 and IL-17 have essential roles in expanding cerebral infarction and associated neurological deficits (Shichita et al. 2009; Iadecola and Anrather 2011). However, according to Benakis et al., IL-10 and IL-17 are required for the neuroprotection afforded by intestinal dysbiosis (Benakis et al. 2016). Altered gut microbiota suppresses the translocation of effector T-cells from the gut to the meninges after stroke. Dysbiosis modulates the immune response in the intestine's epithelial surface, leading to an increased immunomodulator subpopulation of lymphocyte Treg cells and a decrease in IL-17–positive $\gamma\delta$ T-cells through altered dendritic cell (DC) activity (Benakis et al. 2016; Winer et al. 2016). T-Cells play a role in regulating stroke injury. Liesz et al. recommend that Treg cell invasion in the brain contributes to inhibiting postischemic inflammation (Liesz et al. 2015). However, fecal microbiota transplantation (FMT) with normal gut microbiota induces a proinflammatory T-cell polarization before filtering in the ischemic brain in mice during ischemic stroke (Singh et al. 2016; Sadler et al. 2017). It has been reported that 24 h after stroke injury, T- and B cell counts and IFNγ secreting lymphocytes are reduced in the Peyer's patches (Winek et al. 2016b; Schulte-Herbrüggen et al. 2009). Studies also suggested that the signaling mechanism related to MGBA might be connected to the highly innervated nature of the intestinal nerves by both extrinsic and intrinsic nerve fibers (Mawe and Hoffman 2013).

4.6 SCFAs Contributes to Protection Against Cerebral Ischemic Stroke

Gut microbes produce different microbial metabolites such as trimethylamine oxide (TMAO), SCFAs, acetic acid, propionic acid, butyric acid, and *n*-valeric acid in a fecal sample of stroke patients and a rodent model of stroke. Various techniques were used to characterize and identify SCFA from a fecal sample of stroke patients and a stroke animal model. The alteration in the functioning of the cardiovascular system results in a change of SCFAs contributing to vascular dysfunction via unknown mechanisms. SCFA concentration exacerbates neurological severity and stroke outcomes (Tang et al. 2017). The variations in SCFAs concentration in the gut resulted from the altered microbiota after stroke (Schroeder and Bäckhed 2016). Inulin, fructooligosaccharides, and galactooligosaccharides are well-established prebiotic constituents that microbial enzymes can degrade into SCFA, a vital energy resource for microbiota (Marchesi et al. 2016). The altered microbiota diversity after stroke, evidenced by increased counts of *Lactobacillus ruminis*, positively correlated with molecular microbial markers of stroke patients (Yamashiro et al. 2017). Reports suggested that molecular signaling related to the action of SCFAs on Tregs was mediated through GPCR43, a receptor for SCFAs, which is expressed on colonic Tregs (Smith et al. 2013). However, it is also possible that stroke patients taking dietary supplements such as prebiotics and probiotics following hospital admission may have contributed to the observed differences in microbiota and metabolite.

Another suggested that atherosclerotic stroke patients show both asymptomatic and without carotid atherosclerotic deposition (Yin et al. 2015). It has been observed that gut microbiome diversity is similar with and without atherosclerotic plaque. The stroke and transient ischemic attack (TIA) altered the gut microbiota and exhibited an increased abundance of various opportunistic pathogens and decreased abundance of beneficial genera representing the unfavorable effects of stroke (Durgan et al. 2019). Studies suggest that rodent animal models of cerebral stroke have provided additional confirmation of stroke-induced gut impairment and provided support for identifying molecular mechanisms through which stroke affects the gut mucosal epithelium and residing microbiome (Wekerle 2017). Producing large cerebral infarction by occluding blood flow at the base of the middle cerebral artery (MCA) for 60 min had gut dysbiosis, intestinal paralysis, increased gut permeability, a loss of cholinergic innervation in the ileum, and increased sympathetic activity (Durgan et al. 2019; Stanley et al. 2018). Following proximal middle cerebral artery occlusion (MCAO), gut permeability and bacteria translocation increased in young and aged mice. Investigators reported that bacterial family members typically present in the small intestine had been shown to shift in the lungs after the post-stroke. The negative outcome is related to stress before the experimental stroke MCAO model (Caso et al. 2009). Hayakawa et al. analyzed an unexpected reduction in commensal organisms and an increase in harmful bacteria after the brain injury (Hayakawa et al. 2011). Karlson et al. reported that atherosclerotic patients have altered GIT microbiome and the richness of the *Collinsella* genus and decreased plasma levels of β-carotene (Karlsson et al. 2012). Studies reported that MCAO, a filament rodent stroke model, show more than 60% translocation of microbiota to the lung (Wen and Wong 2017; Stanley et al. 2016a; Stanley et al. 2016b). However, all studies have not observed augmented bacterial translocation following stroke incidence from the gut to others (Singh et al. 2016).

Cerebrovascular incidence triggers microbiome alterations by enhancing pathogens and opportunistic bacteria such as *Bacteroides*, *Escherichia*, *Shigella*, *Haemophilus*, and others (Chen et al. 2019a). In general, cerebral ischemic stroke-induced gastrointestinal alteration seems to activate a signaling pathway that results in bacterial infections, prolonged hospitalizations, and death. Various studies reported considerable age-related changes in the gut bacterial population responsible for developing vascular changes that activate inflammatory responses in various NDD (Zapata and Quagliarello 2015; Langille et al. 2014). The gut microbiome in aged people is rich in endotoxin-releasing bacteria such as Bacteroides and Gammaproteobacteria, including Escherichia and Enterobacter (Claesson et al. 2012). Post-stroke GIT problems have altered patient improvement times and increased mortality rates by worsening neurobehaviour outcomes (Crumeyrolle-Arias et al. 2014). These gut microbiome-related issues are associated with an increased hospital stay, advancement of complications, dependency on others, poor neurological effects, and even increased mortality.

Impaired GIT microbiome abnormalities display modified neurological problems (Winek et al. 2016a; Wang et al. 2011; Yarandi et al. 2016; Wang and Wang 2016; Wu et al. 2011; Winer et al. 2016; Winek et al. 2016b; Yamashiro et al. 2017; Yin et al. 2015; Wekerle 2017; Wen and Wong 2017; Zapata and Quagliarello 2015), which is confirmed in experimental rodent stroke models (Winek et al. 2016a; Singh et al. 2016; Winek et al. 2016b). Winek et al. reported that the survival rate decreased in the MCAO microbiota-depleted mice when the antibiotic cocktail was stopped 3 days before surgery (Winek et al. 2016b). Moreover, a depleted animal's microbiota in which antibiotic treatment was terminated develops severe acute colitis (Schroeder and Bäckhed 2016). The above study raised essential questions about whether microbial establishment or specific pathogen-free microbiota is critical for stroke intervention. Stroke patients often display significant changes in microbial diversity and bacterial counts in fecal samples independent of Comorbidities such as age, hypertension, and type 2 diabetes. Emerging indication suggest that gut inflammation and immune response plays a vital role in the pathophysiology of stroke and may become a critical therapeutic target for its treatment. The studies to date determining gut microbiota can impact stroke outcomes have been performed only in rodent models. All these studies have been conducted to explore that gut microbiota can be manipulated to either therapeutically recover or exacerbate stroke outcomes. In general, a gut flora stabilizes the gut wall. It is significant in relation to inflammation and so favorable influence on the gut microbiota ecosystem can be a strategy to mitigate the inflammation.

4.7 Stroke Dysbiosis Index Scale for Diagnosis and Prognosis of Stroke Incidence

The stroke severity and modified Rankin Scale were done using the National Institutes of Health Stroke Scale (NIHSS) to assess the neurological outcome of stroke patients. The associations between clinical outcomes and microbiota were evaluated by an innovative tool popularly known as the stroke dysbiosis index (SDI) (Xia et al. 2019). However, the microbiota is clinically associated with AIS, diagnosis, and prognosis. With extensive data and complex changes in one index, stroke dysbiosis would greatly simplify the clinical application of gut microbiome results in basic research and development of new therapeutic interventions and targets (Tan et al. 2019). Xia et al. designed the mathematical formula as an SDI tool for stroke patients based on their gut microbiota dysbiosis configuration. They confirmed whether the index was associated with injury and early neurological outcomes (Xia et al. 2019). Patients were divided according to the stroke severity into two groups: mild stroke (NIHSS score < 8) and severe stroke (NIHSS score ≥ 8). Several studies confirmed that SDI was designed based on the analysis of gut bacterial population from 104 human subjects with 94 AIS and compared with 90 healthy controls (HC) (Table 4.2). On the other hand, studies were conducted

Table 4.1 Basic terminologies used for stroke-induced intestinal microbiota complication

Microbiota	The collection of microbes (bacteria, archaea, or lower eukaryotes, etc.) in gastrointestinal tract. GIT microbiome is a multifaceted diversified microorganism that helps to sustain active metabolic ecological equilibrium
Microbiota diversity	Genetic and species diversity of microbes present in intestine
Microbiome	The totality of microbes (bacteria, algae, amoebas, viruses, fungi, and protozoa) and their complete set of genes are called as microbiome. The term *microbiome* was coined by **Joshua Lederberg**
Metagenomics	Genetic material and its encoded functional attributes of the gut microbiome in humans or environmental samples is known as metagenomics
Metabolomics	Secreted metabolite composition of the microbiota community
Eubiosis	Balanced gut microbial ecosystem
Dysbiosis	Impaired gut microbial disproportion
Fecal microbiota transplantation	Transplantation of fecal bacteria from healthy individuals into recipient patients
Stroke Dysbiosis Index	Stroke Dysbiosis Index in gut microbiome is associated with ischemic brain injury and prognosis of stroke
Germ-Free mice	Germ-Free mice are produced by hysterectomy rederivation and must be maintained in isolators under very stringent management to keep them germ free
Short-chain fatty acids	SCFA, a microbial metabolite and fermentation product of carbohydrate and noncarbohydrate released by intestinal microbiota
Operational taxonomic units (OTUs)	Most metagenomic examinations are done utilizing 16S rRNA sequencing to assign operational taxonomic units (OTUs) in the form of species, genera, or phyla
Alpha diversity	Alpha diversity determines microbial diversity richness can be performed based on the number of OTUs observed and the distance between taxa.
Prebiotics, probiotics, and postbiotics	**Prebiotics**—Molecules present in food that help in the growth of beneficial microbes **Probiotics**—Live microbes good for your health and digestive system **Postbiotics**—Consists of metabolites and cell wall components released by probiotics
Microbiota–gut–brain axis (MGBA)	Bidirectional communication between GIT microbiome and the brain, which may provide a therapeutic intervention to protect against CNS disorder in the coming future

to analyze fecal bacteria counts, metabolites, and different biomarker samples of 40 stroke patients and 40 HC (Table 4.1) by using QRT-PCR and HPLC techniques showing a positive correlation between gut dysbiosis and stroke incidence (Yamashiro et al. 2017). Apart from these tools and other technologies of 16S rRNA sequencing of the human gut microbiome, we can determine the incidence of stroke and prognosis (Fig. 4.2).

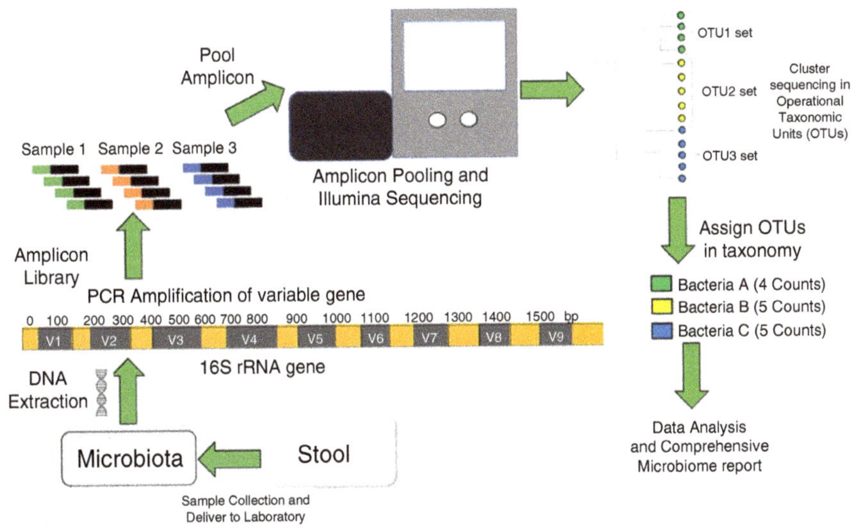

Fig. 4.2 Quantitative metagenomics employed the sequencing of DNA sample isolated from stool. The total DNA extracted from a stool sample, PCR amplification of variable gene led to amplicon library followed by amplicon pooling and Illumina sequencing to generate millions of read. Amplicon sequence analysis leads to operational taxonomic units (OTUs) assign in taxonomic

4.8 Regulatory Role of Gut Microbiome in Blood–Brain Barrier Breakage After Stroke

Blood vessels are a critical component for oxygen and nutrient transfer, and their vascularization around CNS possesses a unique property called Blood–Brain Barrier (BBB). BBB performs as a watchman for foreign and unknown molecules to infiltrate blood circulation (Stamatovic et al. 2008). The BBB architecture is made up of endothelial cells, astrocytes, and pericytes, but these cells' relative contributions are largely unknown (Daneman and Prat 2015). TJ proteins are present between endothelial cells and prevent paracellular transport of molecules from blood to the brain and its extracellular fluid (Zapata and Quagliarello 2015). Gut bacteria and factors translocated into the circulation and microbiome can modify the peripheral immune system to encourage communications between BBB and the neurovascular unit (Logsdon et al. 2018). However, the critical impact of ischemic stroke is the increased GIT permeability, altered gut barrier, and triggered microbial translocation in the bloodstream (Krezalek and Alverdy 2016). Tight junction (TJ) proteins such as claudin-5, occludin, and ZO-1 present in BBB show high-level protein expression and negatively correlate to ischemic stroke injury (Wolburg and Lippoldt 2002). However, an earlier report showed that germ-free mice showed increased BBB permeability compared with pathogen-free with normal gut microbiome (Braniste et al. 2014). The above evidence confirmed that GF mice with removed microbiota have an essential role in regulating BBB integrity. Another

Table 4.2 Various Bacterial, families, genera, and species (S) related to protective and disease progression in ischemic stroke and other cardiovascular diseases

Microbiota	Animal/subjects (n, M/F), sample type (country)	Preventive/disease progression	Microbiome analysis methods	Novel research findings	References
Fermicutes (P) to *Bacteroidetes (P) ratio*	Mice	Preventive	16S ribosomal RNA analysis	F:B ratio increases in young and decrease in aged	Spychala et al. (2018)
SCFAs producer, *Odoribacter*, *Akkermansia*	Patients ($n = 30$, 21 M/ 9F) HC(18 M/12F) Fecal.	Preventive	16S rRNA V1-V2 NGS	Correlation of dysbiosis with neurological outcomes	Li et al. (2019)
Reduced bacterial diversity Overgrowth of *Bacteroidetes*	Mice	Pathogenic Progression	16S rRNA NGS	Stroke altered the physiology of Gut	Singh et al. (2016)
Atopobium (G)	Patients ($n = 41$)	Increases and involved stroke pathology	16S and 23S rRNA-targeted qRT-PCR and HPLC analyses	Stroke patients had decreased numbers	Yamashiro et al. (2017)
Lactobacillus (G) ruminis (S)	Patients ($n = 41$)	No correlation. Directly. Correlated with IL-6 downregulation	16S and 23S rRNA-targeted qRT-PCR and HPLC analyses	No association of microbiome with stroke	
Lactobacillus (G) sakei (S)	Patients ($n = 41$)	Age, hypertension, and type 2 diabetes independent	16S and 23S rRNA-targeted qRT-PCR and HPLC analyses	Protective effect by inhibiting 2 bacteria	
Helicobacter (G)	Patients CVD ($n = 240$, 128 HP-seropositive, 62 M/66F), 77HP-seronegative	Involved in CVD	Carotid Ultrasonography	Metabolic disorder, hypertension, and high total cholesterol in CVD patients	Longo-Mbenza et al. (2012)
Clostridiaceae (G) S24–7 Bacteroidetes	Mice	Involved in protection	Fecal microbiota transplantation	Bacterial families involved in neuroprotection	Benakis et al. (2016)
Akkermansia (G)	Mice	microbiota	V3-V4 region of 16S rRNA sequencing	Stroke-induced alteration of microbiome	Stanley et al. (2018)

Prevotellaceae (*G*) *Peptococcaceae* (*G*)	Mice	Altered microbiota diversity	16S amplicon pyrosequencing	Noradrenaline released from ANS and mucoprotein production	(Houlden et al. 2016)
Clostridia (*G*) *butyricum*	Mice	Protective	16S rRNA sequencing	Microbial monotherapy against vascular dementia. (Increase BDNF/BcL2/p-A	Liu et al. (2015)
Escherichia (*G*) (for young) *Enterobacter* (*G*) (*for aged*)	Mice	These bacteria are not the signature of aged and young	16S rRNA Sequencing	Aged mice developed septic response, weight loss, and immune dysfunction	Tripathi et al. (2022)
Bacteroides (*G*) *Prevotella* (*G*) *Enterococcuc* (*G*)	Mice	Reduced the mortality	Shotgun sequencing techniques and 16S rRNA sequencing	Specific microbiota crucial for stroke outcome.	Winek et al. (2016b)
Enterobacter, *Megasphaera*, *Oscillibacter*, *Desulfovibrio*	Patients	Dysbiosis is correlated with severity of disease	16S rRNA V4 Sequencing	Ischemic stroke patients showed significant dysbiosis	(Yin et al. 2015)
Bacteroides vulgates *Bacteroides dorei*	Patients and Mice	Attenuate the athero-sclerotic lesions formation	Human fecal 16S rRNA sequencing	Decreased gut microbial LPS production and inhibiting proinflammatory response	Gevers et al. (2012)
Bacteroides, *Prevotella*, and *Faecalibacterium*.	Patients ($n = 78$, 69%M/31%F)	Maybe preventive	16S rRNA V4 Sequencing	Dysbiosis correlated with the severity of the disease	Gibson and Roberfroid (1995)

NGS next generation sequencing, *HC* healthy control, *rRNA* ribosomal RNA study suggested that SCFA releasing by intestinal microbiota and absorbed by the gut, and transported to the blood before crossing to BBB and modulates the 5-HT (Sherwin et al. 2016). SCFAs have reversing property of BBB breakage and increase the TJ protein expression. SCFAs have reversing properties of BBB breakage and increase the TJ protein expression. The increased permeability was sustained in germ-free mice and associated with the downregulated expression of BBB-related TJ proteins (Ballabh et al. 2004). The combinatorial treatment of Puerariae Lobatae Radix (PLR) and Chuanxiong Rhizoma (CXR) altered gut microecology and BBB leakage by modulating gut microbiota whose anti-inflammatory effects may decrease brain injury (Chen et al. 2019b). These altered endogenous bacteria are Akkermansia, Oscillospira, Ruminococcaceae, Alloprevotella, and Megasphaera, which protect the gut–brain barrier by upregulating the protein expression claudin-5 and ZO-1 levels and decreasing gut microbiome translocation by decreasing diamine oxidase, lipopolysaccharide, and d-lactate. Microbiota transplantation in GF mice has been watched the reduce BBB penetrability due to reduced expression of tight junction proteins (TJP) (Braniste et al. 2014).

Past examination recommended that intestine microbiota perform an essential part in physicochemical keenness of BBB through discharge SCFA such as acetic acid derivation, propionate, and butyrate (Luan et al. 2019). SCFA release by gut microbiota plays an essential role in BBB integrity via increasing expression of TJ protein (Delzenne et al. 2011). The authors suggested that the mucosal barrier reacted with SCFA signaling molecules and altered functioning. Oral administration of *Clostridium butyricum*, which is a butyrate producer bacterium, enhances the integrity of BBB. Another *Bacteroides thetaiotaomicron*, releasing the acetate and propionate SCFA, enhanced BBB integrity in the mice model (Braniste et al. 2014). Moreover, it was well known that *C. butyricum* administrations led to alteration of gut microbiota and given before stroke give rise to better neurological outcomes and decreased cell death resulting from an animal model of stroke (Sun et al. 2016). The beneficial effects of *C. butyricum* were correlated to increased butyrate levels and reduced oxidative stress in the brain (Sun et al. 2016). The above study was confirmed by TFCIRI, sodium butyrate a reversal of ischemic injury by decreasing BBB permeability and MMP-9 activity (Grice and Segre 2012; Wang et al. 2011; Li et al. 2016). Increased MMP-9 activity is associated with the degradation of several TJ proteins present in the BBB structure. The over data proposed that a prevalently set axis, the PGMBB axis, empowers neuroprotective metabolic impacts against stroke by diminishing the irritation and oxidative push and modifying BBB. However, the exact role of many SCFA molecules in stroke signaling is still undefined. These SCFA molecules transformed the mucosal gut barrier and improved BBB breakage by unknown molecular mechanisms.

4.9 Engineered Microbiota Used for Therapeutic Treatment of Ischemic Stroke

Recent advances in therapeutic intervention have revealed the control and prevention of diseases such as cancer, neurodegeneration, and stroke by using technologies of FMT a method of microbiota manipulation (Kootte et al. 2012; Dinan and Cryan 2017; Roy and Trinchieri 2017). Microbial diversity in different pathological conditions suggests a potential role in manipulating gut microbiota for future medicine. The administration of prebiotic, probiotic, and dietary supplements has been shown effective intervention of these disease conditions by regulating microbiota diversity. The mechanism involved for preventing disease condition by regulating nutrients absorption and production of microbial metabolite (Shimizu et al. 2018). These microbial metabolites have the potential for regulating and interacting with the immune system and translocation of these metabolites via an altered mucosal barrier communicate with body organ by pleiotropic signaling mechanism. Furthermore, the investigator conducted a long-term microbiome examination after stroke induction and suggested that alteration of some bacterial counts may be necessary for FMT technology, and bacterial monotherapy in NDD therapy, leading to a future scope for neurological therapeutic intervention. There is a prominent decrease of Bifidobacteriaceae bacteria found in probiotics up to 4-week post-stroke incidence. Also, the bacteria Helicobacteraceae was increased post-stroke suggesting a specific role of bacteria in neurological therapy (Longo-Mbenza et al. 2012). The Bifidobacteriaceae can be used for the therapeutic role against stroke by attenuation in infarction and neurological recovery.

Post-stroke patients suffer large changes in the Firmicutes to Bacteroidetes ratio. It was reported that an altered ratio was seen in a variety of metabolic disorders including diabetes and obesity (Longo-Mbenza et al. 2012). It was reported that bacterial population was more worsened after 2 weeks of stroke incidence. Another bacterium *Clostridium butyricum* has an important role to attenuate cognitive disability and histopathological alteration and decreases neuronal cell death. Furthermore, *C. butyricum* normalize the gut microbiome and re-established butyrate level in the fecal sample of patients (Liu et al. 2015). It has been noticed and strongly influences the specific role of *C. butyricum in* neuroprotection via unknown specific mechanisms. Another group of bacteria such as *Peptococcaceae* and *Prevotellaceae* were changes in gut during experimental stroke. *Enterococcus* spp., *Escherichia coli*, and *Morganella morganii* are most common in stroke patients who developed bacterial infections (Wekerle 2017; Stanley et al. 2016b). Studies reported that an increase in the proportion of *Peptococcaceae* in the gut microbiome was associated with stroke injury. The communication between gut microbiome and exacerbation of stroke was correlated with the release of noradrenaline from ANS (Zhao et al. 2018). Another study suggested that transient ischemic attack (TIA) patients had plenty of *Enterobacter*, *Megasphaera*, *Oscillibacter*, and *Desulfovibrio*, and a smaller number of beneficial genera including *Bacteroides*, *Prevotella*, and *Faecalibacterium* (Yin et al. 2015). Interestingly it has been reported that microbial α-diversity is similar in both ischemic stroke patients and normal healthy individuals suggesting basic role of microbiota is undisturbed. And stroke patients' gut microbiota has more SCFA producing bacteria such as *Odoribacter*, *Akkermansia*, *Ruminococcaceae*_UCG_005, and *Victivallis* (Li et al. 2019). *Several studies reported that impact of microbiota has impact on acute stroke brain injury. The studies reported that antibiotic-induced modification in the intestinal microflora attenuated the ischemic brain injury* (Benakis et al. 2016). Yoshida et al. demonstrated that 16S rRNA sequencing of human fecal revealed that lower abundance of *Bacteroides vulgatus* and *Bacteroides dorei* in patients with CAD (Yoshida et al. 2018). In describing this study oral gavage with live bacteria attenuate the atherosclerotic lesions in mice. Also suggested the fecal liposaccharide levels in CVD patients were significantly higher.

Studies suggest that the oral administration of antibiotics decreased neurological impairment and decreased serum triglyceride levels by altering the gut microbial diversity (Chen et al. 2019a). The ischemic stroke condition decreases intestinal SCFA. FMT is the method of administration of SCFA metabolites as a potent treatment for these situations. Butyric acid showed the maximum negative connection with cerebrovascular accidents (Chen et al. 2019a). *However, recently reported that transplantation of young microbiome using technique of fecal microbiota gavage into aged mice decreased the incidence of ischemic brain infarction* (Spychala et al. 2018). Inversely, an aged microbiota, when introduced in young mice, produces increased infarction and decreased neurological outcomes after stroke. Studies explore rational connection of microbiota and its immunological response in different mice with same genetic background C57BL6J, implying microbiota alteration between commercial breeders influences post-stroke immunological characteristics and neurological outcomes (Sadler et al. 2017). Furthermore, the recent application of microbiome in biomedical research recommends that targeting the specific microbiome through various interventions may be a potent approach for treating neurodevelopment disorders, neurodegenerative diseases, and stroke.

Fascinatingly, the microbiome can function as both a preventive and a harmful role in various diseases. The percentage infarction after the MCAO in animal model was more significant in germ free mice model than normal and single specific pathogen-free (Singh et al. 2016). Spychala et al. swapped gut microbiota from 3-month young mice and transplanted them with 20 month aged mice suggesting a specific role of immature microbiota in neuroprotection. However, aging is only a single component of gut microbiome alteration (Spychala et al. 2018). A young microbiome in gut mice shows increased survival, increased locomotor activity and anxiety, and increased motor strength during recovery MCAO after the ischemic insult. It is well established that MCAO-induced infarction in aged mice is smaller than in young naïve mice. From the above studies, it is confirmed that microbial monotherapy would be a better therapeutic opportunity for neuroprotective strategy confirmed by another study, using Prevotella histicola for Multiple sclerosis patients (Mangalam and Murray 2019). However, the cerebral infarction in aged mice was similar regardless diversified microbiome and remained smaller than that in the young mice after the MCAO surgery. Lee et al. established the supportive effect of a young microbiome on the healing of stroke occurs even when the FMTs were initiated 72 h after the MCAO surgery (Lee et al. 2020).

4.10 Tools for Regulating Microbiome Gene Expression

The microbiome study was principally focused on 16S rDNA sequencing and shotgun metagenomics, allowing for the explanation of gut microbiota (Gevers et al. 2012). A new method of using ^{1}H NMR in microbiome profiling of fecal samples as a tool to assess microbiome (Jacobs et al. 2008). On the other hand, various studies are running to explore the regulation of the microbiome by exploiting promoter and inducer interaction efficiency. Using promoter efficiency and translational tuning strategy attained an extraordinary level of gene expression that enables the imaging of fluorescent tagged protein more accurate and reliable for microbiome studies (Whitaker et al. 2017). The real-time gene expression studies will be regulated by the inducible promoter which is a key tool for gene regulation (Whitaker et al. 2017). Lim et al. developed the tunable expression platform for the prominent genus *Bacteroides* in which gene expression is regulated by a synthetic inducer (Lim et al. 2017). The bacterial mRNA expression inside the gut can be regulated by an inducer which is provided in drinking water or diet for mice. Sialidases an enzyme released by commensal microbes are used by their enzymatic action on sialic acid released from gut mucosa [(Tailford et al. 2015). Sialic acid is a nine-carbon atom sugar molecule present at the terminal residue of glycoprotein on the cell surface. The major role of sialic acid is to accelerate the development and repair of nerve cells, epithelial cells, and immune cells. Goodman elucidates that *Bacteroides* spp. colonization in mice whose sialidase gene expression was controlled by the artificial inducer. It can able to measure the relationship between sialidase activity and changes in the host environment. Particularly, this method enabled the researcher to quantify that sialic acid concentrations in the gut are limited by substrate availability, and not by commensal sialidase activity (Quigley 2017).

4.11 Psychobiotics

The psychobiotics are live beneficial bacteria when taken in an appropriate quantity to minimize neurological impairment by altering the host gut microbiota (Sarkar et al. 2016). Psychobiotics exert anxiolytics and antidepressant effects shown by changes in emotional, neurocognitive, and neural features. Psychobiotics exert the neuroprotective effect on ischemic stroke via ENS. Several bacterial strains have been used as a psychobiotic supplement. Psychobiotics strains will be used to support the gut microorganism. Some examples of these bacterial strains are *Lactobacillus helveticus*, *Bifidobacterium longum*, *Lactobacillus casei*, *Lactobacillus*

Plantarum, *Lactobacillus acidophilus*, *Lactobacillus delbrueckii* subsp. Bulgaricus, *Bifidobacterium breve*, and *Bifidobacterium infantis*.

4.12 Neuroprotective Potential of Monobacteriotherapy

Probiotics are a potential prospect for the interventions of different pathophysiological impairments (Kazemi et al. 2020; Zhang et al. 2020; Bonfili et al. 2018; Bonfili et al. 2017). A recent study confirmed that AD treatment by using *Lactobacillus lactis* strain carrying a plasmid that contains eukaryotic p62 gene expression which is similar to humans (Cecarini et al. 2020). Oral administration of *L. lactis* showed an increased expression of endogenous p62 in the brain. A personalized diet and orally treated beneficial bacteria can latest options for various treatments. However, there are several other dietary supplements such as vitamins and phytochemicals that are used to be a neuroprotective agents against stroke (Tripathi et al. 2022). It is strongly postulated that stroke-induced gut impairment reflects the misbalance of microbial diversity. This is a potential area of intervention in stroke and cardiovascular diseases.

4.13 Fecal MicroRNA Regulation of Gut Microbiota

MicroRNA are normal constituents present in human feces and play a significant role in between species gene regulation of host regulation of gut microbiome (Liu et al. 2016). Major sources of miRNA present in human fecal samples are epithelial cells present in the intestine and Hopx+ cells which are present in extracellular vesicles. IEC-miRNA-deficient ($Dicer1^{\Delta IEC}$) mice show impaired gut microbiota and exacerbated colitis. However, wild-type (WT) fecal miRNA transplantation restores the gut microbiota and decreases the colitis.

4.14 Conclusion and Future Directions

This chapter has attempted to highlight the recent advancement in the field of stroke-induced microbiome alteration in general including its microbiota manipulation for therapeutic intervention, the role of beneficial and detrimental bacteria, and FMT for treating ischemic stroke. Also, we are discussing the potential role of engineered microbes used in the therapeutic intervention of stroke pathophysiology and SCFA signaling in BBB integrity regulation. The brain-gut communication after stroke, it is important to consider potential opportunities and pitfalls during the experimental investigation. In conclusion, microbiota dysbiosis is induced in mouse models of ischemic stroke. The key questions that remain are (1) Comparative alteration in the GIT tract morphology of the human and rodent that would not helpful and identical for clinical and preclinical results, (2) Compositional differences in human and rodent microbiome and immune system, (3) Growing instabilities in GF animals because of other microbial populations, (4) Potential role of antibiotics in depletion microbiota on CNS and immunology, (5) Regulatory impact of gut virome on microbiota diversity. Potential role of housing condition and animal husbandry for GF mice development, (6) Impact of genetic diversity between humans on gut microbiota, and (7) Sample preparation and bioinformatics of gut metagenomics. In nutshell, gut microbiome investigation is a fascinating area with rational exciting discoveries and the potential role of gut bacteria in neurological impairment. Apart from these inquiries, others are related to new beneficial bacteria involved in NDD and stroke by releasing SCFA molecules.

References

Ahmed S, Busetti A, Fotiadou P et al (2019) In vitro characterization of gut microbiota-derived bacterial strains with neuroprotective properties. Front Cell Neurosci. https://doi.org/10.3389/fncel.2019.00402

Aron-wisnewsky J, Clement K (2017) Dietary choline derived TMAO: new role in thrombosis. AME Med J 2:112

Ballabh P, Braun A, Nedergaard M (2004) The blood–brain barrier: an overview. Neurobiol Dis 16: 1–13

Benakis C, Brea D, Caballero S et al (2016) Commensal microbiota affects ischemic stroke outcome by regulating intestinal γδ T cells. Nat Med 22:516–523

Bercik P, Collins SM, Verdu EF (2012) Microbes and the gut-brain axis. Neurogastroenterol Motil 24:405–413

Bonfili L, Cecarini V, Berardi S, Scarpona S, Suchodolski JS, Nasuti C, Fiorini D, Boarelli MC, Rossi G, Eleuteri AM (2017) Microbiota modulation counteracts Alzheimer's disease progression influencing neuronal proteolysis and gut hormones plasma levels. Sci Rep 7:2426

Bonfili L, Cecarini V, Cuccioloni M, Angeletti M, Berardi S, Scarpona S, Rossi G, Eleuteri AM (2018) SLAB51 probiotic formulation activates SIRT1 pathway promoting antioxidant and neuroprotective effects in an AD mouse model. Mol Neurobiol 55:7987–8000

Boziki M, Polyzos SA, Deretzi G et al (2018) A potential impact of *Helicobacter pylori*-related galectin-3 in neurodegeneration. Neurochem Int 113:137–151

Braniste V, Al-Asmakh M, Kowal C et al (2014) The gut microbiota influences blood-brain barrier permeability in mice. Sci Transl Med. 6:263ra158-263ra158

Bressa C, Bailén-Andrino M, Pérez-Santiago J, González-Soltero R, Pérez M, Montalvo-Lominchar MG, Maté-Muñoz JL, Domínguez R, Moreno D, Larrosa M (2017) Differences in gut microbiota profile between women with active lifestyle and sedentary women. PLoS One 12. e0171352

Camara-Lemarroy CR, Ibarra-Yruegas BE, Gongora-Rivera F (2014) Gastrointestinal complications after ischemic stroke. J Neurol Sci 346:20–25

Caso JR, Hurtado O, Pereira MP, García-Bueno B, Menchén L, Alou L, Gómez-Lus ML, Moro MA, Lizasoain I, Leza JC (2009) Colonic bacterial translocation as a possible factor in stress-worsening experimental stroke outcome. Am J Physiol Integr Comp Physiol 296:R979–R985

Cecarini V, Bonfili L, Gogoi O et al (2020) Neuroprotective effects of p62(SQSTM1)-engineered lactic acid bacteria in Alzheimer's disease: a pre-clinical study. Aging (Albany NY) 12:15995–16020
Chen R, Xu Y, Wu P et al (2019a) Transplantation of fecal microbiota rich in short chain fatty acids and butyric acid treat cerebral ischemic stroke by regulating gut microbiota. Pharmacol Res 148: 104403
Chen R, Wu P, Cai Z, Fang Y, Zhou H, Lasanajak Y, Tang L, Ye L, Hou C, Zhao J (2019b) Puerariae lobatae radix with chuanxiong rhizoma for treatment of cerebral ischemic stroke by remodeling gut microbiota to regulate the brain–gut barriers. J Nutr Biochem 65:101–114
Claesson MJ, Jeffery IB, Conde S et al (2012) Gut microbiota composition correlates with diet and health in the elderly. Nature 488:178–184
Clarke SF, Murphy EF, Nilaweera K, Ross PR, Shanahan F, O'Toole PW, Cotter PD (2012) The gut microbiota and its relationship to diet and obesity. Gut Microbes 3:186–202
Crumeyrolle-Arias M, Jaglin M, Bruneau A, Vancassel S, Cardona A, Daugé V, Naudon L, Rabot S (2014) Absence of the gut microbiota enhances anxiety-like behavior and neuroendocrine response to acute stress in rats. Psychoneuroendocrinology 42:207–217
Daneman R, Prat A (2015) The blood–brain barrier. Cold Spring Harb Perspect Biol 7. a020412
Daniel H, Gholami AM, Berry D et al (2014) High-fat diet alters gut microbiota physiology in mice. ISME J 8:295–308
De Filippis F, Pellegrini N, Vannini L et al (2016) High-level adherence to a Mediterranean diet beneficially impacts the gut microbiota and associated metabolome. Gut 65:1812–1821
Delzenne NM, Neyrinck AM, Bäckhed F, Cani PD (2011) Targeting gut microbiota in obesity: effects of prebiotics and probiotics. Nat Rev Endocrinol 7:639–646
Dinan TG, Cryan JF (2017) Gut instincts: microbiota as a key regulator of brain development, ageing and neurodegeneration. J Physiol 595:489–503
Dubourg G, Lagier J-C, Armougom F, Robert C, Audoly G, Papazian L, Raoult D (2013) High-level colonisation of the human gut by Verrucomicrobia following broad-spectrum antibiotic treatment. Int J Antimicrob Agents 41:149–155
Durgan DJ, Lee J, McCullough LD, Bryan RM (2019) Examining the role of the microbiota-gut-brain axis in stroke. Stroke 50:2270–2277
Ferrocino I, Di Cagno R, De Angelis M, Turroni S, Vannini L, Bancalari E, Rantsiou K, Cardinali G, Neviani E, Cocolin L (2015) Fecal microbiota in healthy subjects following omnivore, vegetarian and vegan diets: culturable populations and rRNA DGGE profiling. PLoS One 10:e0128669
Frank DN, St. Amand AL, Feldman RA, Boedeker EC, Harpaz N, Pace NR (2007) Molecular-phylogenetic characterization of microbial community imbalances in human inflammatory bowel diseases. Proc Natl Acad Sci 104:13780–13785
Gelderblom M, Weymar A, Bernreuther C et al (2012) Neutralization of the IL-17 axis diminishes neutrophil invasion and protects from ischemic stroke. Blood 120:3793–3802
Gevers D, Knight R, Petrosino JF, Huang K, McGuire AL, Birren BW, Nelson KE, White O, Methé BA, Huttenhower C (2012) The human microbiome project: a community resource for the healthy human microbiome. PLoS Biol 10:e1001377
Gibson GR, Roberfroid MB (1995) Dietary modulation of the human colonic microbiota: introducing the concept of prebiotics. J Nutr 125:1401–1412
Grenham S, Clarke G, Cryan JF, Dinan TG (2011) Brain? Gut? microbe communication in health and disease. Front Physiol. https://doi.org/10.3389/fphys.2011.00094
Grice EA, Segre JA (2012) The human microbiome: our second genome. Annu Rev Genomics Hum Genet 13:151–170
Guo AY, Stevens BW, Wilson RG et al (2014) Early life environment and natural history of inflammatory bowel diseases. BMC Gastroenterol 14:216
Haghikia A, Li XS, Liman TG et al (2018) Gut microbiota–dependent trimethylamine N-oxide predicts risk of cardiovascular events in patients with stroke and is related to proinflammatory monocytes. Arterioscler Thromb Vasc Biol 38:2225–2235

Hanfrey CC, Pearson BM, Hazeldine S, Lee J, Gaskin DJ, Woster PM, Phillips MA, Michael AJ (2011) Alternative spermidine biosynthetic route is critical for growth of campylobacter jejuni and is the dominant polyamine pathway in human gut microbiota. J Biol Chem 286:43301–43312

Harari D, Norton C, Lockwood L, Swift C (2004) Treatment of constipation and fecal incontinence in stroke patients. Stroke 35:2549–2555

Hayakawa M, Asahara T, Henzan N et al (2011) Dramatic changes of the gut flora immediately after severe and sudden insults. Dig Dis Sci 56:2361–2365

Houlden A, Goldrick M, Brough D, Vizi ES, Lénárt N, Martinecz B, Roberts IS, Denes A (2016) Brain injury induces specific changes in the caecal microbiota of mice via altered autonomic activity and mucoprotein production. Brain Behav Immun 57:10–20

Iadecola C, Anrather J (2011) The immunology of stroke: from mechanisms to translation. Nat Med 17:796–808

Jacobs DM, Deltimple N, van Velzen E, van Dorsten FA, Bingham M, Vaughan EE, van Duynhoven J (2008) 1 H NMR metabolite profiling of feces as a tool to assess the impact of nutrition on the human microbiome. NMR Biomed 21:615–626

Karlsson FH, Fåk F, Nookaew I, Tremaroli V, Fagerberg B, Petranovic D, Bäckhed F, Nielsen J (2012) Symptomatic atherosclerosis is associated with an altered gut metagenome. Nat Commun 3:1245

Kaur MJ, Shruti G, Hema K, Kaushalendra K, Gaurav K (2021) Association of gut microbiota with ischemic stroke: a review. Int J Curr Res Rev 13:77–83

Kazemi A, Soltani S, Ghorabi S, Keshtkar A, Daneshzad E, Nasri F, Mazloomi SM (2020) Effect of probiotic and synbiotic supplementation on inflammatory markers in health and disease status: a systematic review and meta-analysis of clinical trials. Clin Nutr 39:789–819

Kootte RS, Vrieze A, Holleman F, Dallinga-Thie GM, Zoetendal EG, de Vos WM, Groen AK, Hoekstra JBL, Stroes ES, Nieuwdorp M (2012) The therapeutic potential of manipulating gut microbiota in obesity and type 2 diabetes mellitus. Diabetes Obes Metab 14:112–120

Krezalek MA, Alverdy JC (2016) The role of the microbiota in surgical recovery. Curr Opin Clin Nutr Metab Care 19:347–352

Langille MG, Meehan CJ, Koenig JE, Dhanani AS, Rose RA, Howlett SE, Beiko RG (2014) Microbial shifts in the aging mouse gut. Microbiome 2:50

Lee J, D'Aigle J, Atadja L et al (2020) Gut microbiota–derived short-chain fatty acids promote poststroke recovery in aged mice. Circ Res 127:453–465

Li H, Sun J, Wang F, Ding G, Chen W, Fang R, Yao Y, Pang M, Lu Z-Q, Liu J (2016) Sodium butyrate exerts neuroprotective effects by restoring the blood-brain barrier in traumatic brain injury mice. Brain Res 1642:70–78

Li N, Wang X, Sun C et al (2019) Change of intestinal microbiota in cerebral ischemic stroke patients. BMC Microbiol 19:191

Liesz A, Hu X, Kleinschnitz C, Offner H (2015) Functional role of regulatory lymphocytes in stroke. Stroke 46:1422–1430

Lim B, Zimmermann M, Barry NA, Goodman AL (2017) Engineered regulatory systems modulate gene expression of human commensals in the gut. Cell 169:547–558. e15

Lindsberg PJ, Grau AJ (2003) Inflammation and infections as risk factors for ischemic stroke. Stroke 34:2518–2532

Liu J, Sun J, Wang F et al (2015) Neuroprotective effects of *Clostridium butyricum* against vascular dementia in mice via metabolic butyrate. Biomed Res Int 2015:1–12

Liu S, da Cunha AP, Rezende RM, Cialic R, Wei Z, Bry L, Comstock LE, Gandhi R, Weiner HL (2016) The host shapes the gut microbiota via fecal micro RNA. Cell Host Microbe 19:32–43

Logsdon AF, Erickson MA, Rhea EM, Salameh TS, Banks WA (2018) Gut reactions: how the blood–brain barrier connects the microbiome and the brain. Exp Biol Med 243:159–165

Longo-Mbenza B, Nkondi M, Gombet I, Ellenga-Mbolla V, Mbungu (2012) Helicobacter pylori infection is identified as a cardiovascular risk factor in central Africans. Vasc Health Risk Manag:455

Luan H, Wang X, Cai Z (2019) Mass spectrometry-based metabolomics: targeting the crosstalk between gut microbiota and brain in neurodegenerative disorders. Mass Spectrom Rev 38:22–33
Lutgendorff F, Akkermans L, Soderholm J (2008) The role of microbiota and probiotics in stress-induced gastrointestinal damage. Curr Mol Med 8:282–298
Magee EA, Richardson CJ, Hughes R, Cummings JH (2000) Contribution of dietary protein to sulfide production in the large intestine: an in vitro and a controlled feeding study in humans. Am J Clin Nutr 72:1488–1494
Mangalam AK, Murray J (2019) Microbial monotherapy with Prevotella histicola for patients with multiple sclerosis. Expert Rev Neurother 19:45–53
Marchesi JR, Adams DH, Fava F et al (2016) The gut microbiota and host health: a new clinical frontier. Gut 65:330–339
Mathewson ND, Jenq R, Mathew AV et al (2016) Gut microbiome–derived metabolites modulate intestinal epithelial cell damage and mitigate graft-versus-host disease. Nat Immunol 17:505–513
Mawe GM, Hoffman JM (2013) Serotonin signalling in the gut—functions, dysfunctions and therapeutic targets. Nat Rev Gastroenterol Hepatol 10:473–486
Monira S, Nakamura S, Gotoh K et al (2011) Gut microbiota of healthy and malnourished children in Bangladesh. Front Microbiol. https://doi.org/10.3389/fmicb.2011.00228
Mulak A (2018) A controversy on the role of short-chain fatty acids in the pathogenesis of Parkinson's disease. Mov Disord 33:398–401
Murray CJL, Lopez AD (2013) Measuring the global burden of disease. N Engl J Med 369:448–457
Nam HS (2019) Gut microbiota and ischemic stroke: the role of trimethylamine N-oxide. J Stroke 21:151–159
Nguyen TTT, Fujimura Y, Mimura I, Fujii Y, Nguyen NL, Arakawa K, Morita H (2018) Cultivable butyrate-producing bacteria of elderly Japanese diagnosed with Alzheimer's disease. J Microbiol 56:760–771
Pérez-López FR, Chedraui P, Haya J, Cuadros JL (2009) Effects of the mediterranean diet on longevity and age-related morbid conditions. Maturitas 64:67–79
Prinz I, Silva-Santos B, Pennington DJ (2013) Functional development of γδ T cells. Eur J Immunol 43:1988–1994
Purushe J, Fouts DE, Morrison M, White BA, Mackie RI, Coutinho PM, Henrissat B, Nelson KE (2010) Comparative genome analysis of *Prevotella ruminicola* and *Prevotella bryantii*: insights into their environmental niche. Microb Ecol 60:721–729
Quigley EMM (2017) Gut microbiome as a clinical tool in gastrointestinal disease management: are we there yet? Nat Rev Gastroenterol Hepatol 14:315–320
Rice MW, Pandya JD, Shear DA (2019) Gut microbiota as a therapeutic target to ameliorate the biochemical, neuroanatomical, and behavioral effects of traumatic brain injuries. Front Neurol. https://doi.org/10.3389/fneur.2019.00875
Roy S, Trinchieri G (2017) Microbiota: a key orchestrator of cancer therapy. Nat Rev Cancer 17: 271–285
Sadler R, Singh V, Benakis C, Garzetti D, Brea D, Stecher B, Anrather J, Liesz A (2017) Microbiota differences between commercial breeders impacts the post-stroke immune response. Brain Behav Immun 66:23–30
Sarkar A, Lehto SM, Harty S, Dinan TG, Cryan JF, Burnet PWJ (2016) Psychobiotics and the manipulation of bacteria–gut–brain signals. Trends Neurosci 39:763–781
Schaller BJ, Graf R, Jacobs AH (2006) Pathophysiological changes of the gastrointestinal tract in ischemic stroke. Am J Gastroenterol 101:1655–1665
Schroeder BO, Bäckhed F (2016) Signals from the gut microbiota to distant organs in physiology and disease. Nat Med 22:1079–1089
Schulte-Herbrüggen O, Quarcoo D, Meisel A, Meisel C (2009) Differential affection of intestinal immune cell populations after cerebral ischemia in mice. Neuroimmunomodulation 16:213–218
Sherwin E, Sandhu KV, Dinan TG, Cryan JF (2016) May the force be with you: the light and dark sides of the microbiota–gut–brain axis in neuropsychiatry. CNS Drugs 30:1019–1041

Shichita T, Sugiyama Y, Ooboshi H et al (2009) Pivotal role of cerebral interleukin-17–producing γδT cells in the delayed phase of ischemic brain injury. Nat Med 15:946–950

Shimizu K, Yamada T, Ogura H et al (2018) Synbiotics modulate gut microbiota and reduce enteritis and ventilator-associated pneumonia in patients with sepsis: a randomized controlled trial. Crit Care 22:239

Shulman ST, Friedmann HC, Sims RH (2007) Theodor escherich: the first pediatric infectious diseases physician? Clin Infect Dis 45:1025–1029

Sidorov E, Sanghera DK, Vanamala JKP (2019) Biomarker for ischemic stroke using metabolome: a clinician perspective. J Stroke 21:31–41

Singh V, Roth S, Llovera G, Sadler R, Garzetti D, Stecher B, Dichgans M, Liesz A (2016) Microbiota dysbiosis controls the neuroinflammatory response after stroke. J Neurosci 36: 7428–7440

Smith PM, Howitt MR, Panikov N, Michaud M, Gallini CA, Bohlooly-Y M, Glickman JN, Garrett WS (2013) The microbial metabolites, short-chain fatty acids, regulate colonic treg cell homeostasis. Science 80–341:569–573

Spychala MS, Venna VR, Jandzinski M et al (2018) Age-related changes in the gut microbiota influence systemic inflammation and stroke outcome. Ann Neurol 84:23–36

Stamatovic S, Keep R, Andjelkovic A (2008) Brain endothelial cell-cell junctions: how to “Open” the blood brain barrier. Curr Neuropharmacol 6:179–192

Stanley D, Mason LJ, Mackin KE et al (2016a) Translocation and dissemination of commensal bacteria in post-stroke infection. Nat Med 22:1277–1284

Stanley D, Hughes RJ, Geier MS, Moore RJ (2016b) Bacteria within the gastrointestinal tract microbiota correlated with improved growth and feed conversion: challenges presented for the identification of performance enhancing probiotic bacteria. Front Microbiol. https://doi.org/10.3389/fmicb.2016.00187

Stanley D, Moore RJ, Wong CHY (2018) An insight into intestinal mucosal microbiota disruption after stroke. Sci Rep 8:568

Sun J, Ling Z, Wang F, Chen W, Li H, Jin J, Zhang H, Pang M, Yu J, Liu J (2016) *Clostridium butyricum* pretreatment attenuates cerebral ischemia/reperfusion injury in mice via anti-oxidation and anti-apoptosis. Neurosci Lett 613:30–35

Suzuki T, Heaney LM, Jones DJL, Ng LL (2017) Trimethylamine N-oxide and risk stratification after acute myocardial infarction. Clin Chem 63:420–428

Tailford LE, Owen CD, Walshaw J, Crost EH, Hardy-Goddard J, Le Gall G, de Vos WM, Taylor GL, Juge N (2015) Discovery of intramolecular trans-sialidases in human gut microbiota suggests novel mechanisms of mucosal adaptation. Nat Commun 6:7624

Tan BQ, Paliwal P, Sharma V (2019) Gut microbiota and stroke. Ann Indian Acad Neurol. https://doi.org/10.4103/aian.AIAN_483_19

Tang WHW, Kitai T, Hazen SL (2017) Gut microbiota in cardiovascular health and disease. Circ Res 120:1183–1196

Tripathi AK, Ray AK, Mishra SK (2022) Molecular and pharmacological aspects of piperine as a potential molecule for disease prevention and management: evidence from clinical trials. Beni-Suef Univ J Basic Appl Sci 11:16

Wang H-X, Wang Y-P (2016) Gut microbiota-brain axis. Chin Med J 129:2373–2380

Wang L, Christophersen CT, Sorich MJ, Gerber JP, Angley MT, Conlon MA (2011) Low relative abundances of the mucolytic bacterium akkermansia muciniphila and *Bifidobacterium* spp. in feces of children with autism. Appl Environ Microbiol 77:6718–6721

Wekerle H (2017) Brain autoimmunity and intestinal microbiota: 100 trillion game changers. Trends Immunol 38:483–497

Wen SW, Wong CHY (2017) An unexplored brain-gut microbiota axis in stroke. Gut Microbes 8: 601–606

Whitaker WR, Shepherd ES, Sonnenburg JL (2017) Tunable expression tools enable single-cell strain distinction in the gut microbiome. Cell 169:538–546. e12

Winek K, Dirnagl U, Meisel A (2016a) The gut microbiome as therapeutic target in central nervous system diseases: implications for stroke. Neurotherapeutics 13:762–774

Winek K, Meisel A, Dirnagl U (2016b) Gut microbiota impact on stroke outcome: fad or fact? J Cereb Blood Flow Metab 36:891–898

Winer DA, Luck H, Tsai S, Winer S (2016) The intestinal immune system in obesity and insulin resistance. Cell Metab 23:413–426

Wolburg H, Lippoldt A (2002) Tight junctions of the blood–brain barrier. Vasc Pharmacol 38:323–337

Wu GD, Chen J, Hoffmann C et al (2011) Linking long-term dietary patterns with gut microbial enterotypes. Science 80-334:105–108

Xia G-H, You C, Gao X-X et al (2019) Stroke dysbiosis index (SDI) in gut microbiome are associated with brain injury and prognosis of stroke. Front Neurol. https://doi.org/10.3389/fneur.2019.00397

Yamashiro K, Tanaka R, Urabe T, Ueno Y, Yamashiro Y, Nomoto K, Takahashi T, Tsuji H, Asahara T, Hattori N (2017) Gut dysbiosis is associated with metabolism and systemic inflammation in patients with ischemic stroke. PLoS One 12. e0171521

Yarandi SS, Peterson DA, Treisman GJ, Moran TH, Pasricha PJ (2016) Modulatory effects of gut microbiota on the central nervous system: how gut could play a role in neuropsychiatric health and diseases. J Neurogastroenterol Motil 22:201–212

Yin J, Liao S, He Y et al (2015) Dysbiosis of gut microbiota with reduced trimethylamine-N-oxide level in patients with large-artery atherosclerotic stroke or transient ischemic attack. J Am Heart Assoc. https://doi.org/10.1161/JAHA.115.002699

Yoshida N, Emoto T, Yamashita T et al (2018) *Bacteroides vulgatus* and *Bacteroides dorei* reduce gut microbial lipopolysaccharide production and inhibit atherosclerosis. Circulation 138:2486–2498

Zapata HJ, Quagliarello VJ (2015) The microbiota and microbiome in aging: potential implications in health and age-related diseases. J Am Geriatr Soc 63:776–781

Zhang H, Chen Y, Chen Y, Li Y, Jia P, Ji S, Zhou Y, Wang T (2020) Dietary pterostilbene supplementation attenuates intestinal damage and immunological stress of broiler chickens challenged with lipopolysaccharide. J Anim Sci. https://doi.org/10.1093/jas/skz373

Zhao L, Xiong Q, Stary CM, Mahgoub OK, Ye Y, Gu L, Xiong X, Zhu S (2018) Bidirectional gut-brain-microbiota axis as a potential link between inflammatory bowel disease and ischemic stroke. J Neuroinflammation 15:339

Zhu B, Wang X, Li L (2010) Human gut microbiome: the second genome of human body. Protein Cell 1:718–725

Chapter 5
Aging: Impact of Gut Microbiota

Santosh Anand, Ranoji Rao Narasinga Rao Lakshmikanth, Kannasandra Ramaiah Manjula, Doddhakathanahalli Ramanath Jayashree, and Tekupalli Ravikiran

Abstract The worldwide population of elderlies is rapidly accelerating and the health status of this elderly population warrants an immediate attention. Age-associated gut microbiota modulations are linked to immunosenescence and inflammaging which perturbs the gut microbiotal ecosystem. Several studies have correlated the human gut microbiota with the aging process and found that various factors are associated with variation in the gut microbial composition during aging. Among them, the crucial role players are diet, geographical origin, and intervention of pre/probiotics. To conclude with, the gut microbiota can be a potential target contributing toward the health status of the elderly population. In this chapter, we have attempted to highlight the updates regarding the possible factors associated with changes in GM composition during the aging process.

Keywords Aging · Gut microbiota · Diet · Pre/probiotics · Health

Abbreviations

GIT Gastrointestinal tract
GM Gut microbiota

S. Anand (✉) · K. R. Manjula
Department of Biotechnology, School of Applied Sciences, REVA University, Bangalore, India
e-mail: santosh.anand@reva.edu.in; manjulakr@reva.edu.in

R. R. N. R. Lakshmikanth · D. R. Jayashree
Department of Biotechnology and Genetics, Ramaiah College of Arts, Science & Commerce, Bengaluru, India

T. Ravikiran
Department of Biotechnology, Bangalore University, Bengaluru, India
e-mail: ravikiran@bub.ernet.in

A. K. Tripathi, M. Kotak (eds.), *Gut Microbiome in Neurological Health and Disorders*, Nutritional Neurosciences,
https://doi.org/10.1007/978-981-19-4530-4_5

5.1 Introduction

The global population of older adults is rapidly mounting and the health status of this elderly population is therefore posing an alarming issue to be addressed (Biagi et al. 2013). Aging is a multifaceted process accompanied by several physiological changes affecting metabolic, genomic, and immunological functions (Nagpal et al. 2018; Varricchi et al. 2020). These physiological alterations lead to an augmented susceptibility to various infections and diseases thereby increasing mortality. Despite the fact that, the etiologies of age-associated ailments are relatively divergent, significant reports suggest chronic, low-grade inflammation (inflammaging) among the most reliable physiological alteration in age-linked diseases (Buford 2017; Clements and Carding 2018).

Healthy aging and microbes correlation was reported for the first time by Elie Metchnikoff, one of the founding fathers of modern-day immunology and microbiology (Bischoff 2016). The gut microbiota (GM) is modulated during the aging process in relation to composition and functionality (Tiihonen et al. 2010). Several studies support the importance of a favorable GM behind healthy aging and any physiological functions alteration upon aging has the ability to influence the composition and activities of microbial species inhabiting GM (Kumar et al. 2016). Maintaining a healthy living style and intake of nutritional diet have been considered beneficial for healthy aging (Choi and Lee 2016). The dietary constituents play a pivotal role in articulating the chemostatic environment in the gut. In recent days, drastic processes undertaken during food processing have resulted in microbiome alteration imparting harmful health consequences, particularly in elderly population (Cotillard et al. 2013; Keenan et al. 2015). Alterations in physiological parameters like appetite reduction, masticatory abnormalities, and constipation have been associated with advancing age (Candela et al. 2014). In addition, variations in dietary patterns associated with geographical region-specific differences may also influence the GM constitution during aging. Escobar et al. (2014)) have highlighted the significance of geographical origin as a chief factor governing the GM composition. Despite factors associated with dietetic, lifestyle, and geographical status, supplementation of pre- and probiotics have been suggested to curtail the variations in GM in the elderlies (Bischoff 2016). Thereby, based on the above facts, managing intestinal homeostasis in the aging process can help in prolonging life span (Clark and Walker 2018). In this chapter, an attempt has been made to discuss the updates regarding the factors associated with changes in GM composition during the aging process.

5.2 Aging Gut Microbiota: Composition

The gastrointestinal tract of the humans is inhabited by more than 100 trillion microbes, comprising greater than 1000 species of identified bacteria (Clements and Carding 2018; Gill et al. 2006). The microbial diversity in the gut mainly consists of members of Firmicutes, Actinobacteria, Bacteroidetes, and proteobacteria (Maukonen and Saarela 2015; Ple et al. 2015). Microbial diversification of the GM is observed during different stages of life, wherein an increase in the proteolytic and reduction in the saccharolytic members has been documented during the aging process (Buford 2017; Bischoff 2016; Biagi et al. 2012; Odamaki et al. 2016). The dysbiosis of the gastrointestinal tract (GIT) leads to progression of comorbidities in the aging population (Wu et al. 2020).

During the development of the fetus, the GM is usually sterile (Choi and Lee 2016; Bourlioux et al. 2003). Nearing the birth, it is inhabited by the microbes from the diet, environment, and maternal origin as substantiated by the presence of microbes in the placenta (Collado et al. 2016). In the early stages of life, diversity and complexity are less, and the main members are the actinobacteria, proteobacteria, and Firmicutes (Kumar et al. 2016; Clark and Walker 2018; Koenig et al. 2011; Turroni et al. 2012). By the age of three years, an increase in the diversity and complexity of GM has been observed (Koenig et al. 2011; Claesson et al. 2011; Voreades et al. 2014). During adulthood, the GM composition varies with dominant species being Firmicutes and Bacteroidetes (80–90%), and characterized by high variability among individuals (Thomas and Percival 2009; Human Microbiome Project Consortium 2012). Bifidobacteria, an actinobacterial member with many species are widely habituated in the human gut (Backhed et al. 2005; Duranti et al. 2016; Arboleya et al. 2016).

GM diversity has been observed to be reduced in the elderlies, with low bifidobacteria and augmented proteobacteria documented enhanced *E. coli* and Bacteroidetes levels in the older population (Hopkins et al. 2002; Wilson 2005; Mariat et al. 2009). Biagi et al. (Biagi et al. 2012) have reported the variation of GM among the elderly population residing in different geographical regions. The composition of GM at different stages in human life is illustrated in Table 5.1.

In the late stages of life, a reduction in the density of advantageous bacteria and increase in harmful bacteria have been reported (Biagi et al. 2013). The dysbiosis and change in the density and composition modify the immune capacity and causes inflammation of the organs associated with the gut (Belkaid and Naik 2013). The fortification of beneficial bacteria in the form of pre/probiotics enhances the health in the elderlies (Biagi et al. 2013). Tabouy et al. (Tabouy et al. 2018) have demonstrated that supplementation of *Lactobacillus reuteri* has the ability to curtail GM dysbiosis.

Table 5.1 Factors affecting gut microbial composition during the aging process in humans

Phase	Microbial species	References	Factors affecting GM
Infancy	Bifidobacteria ↑	Duranti et al. (2016), Arboleya et al. (2016)	Mode of delivery Gestation age at birth Type of lactation
Children	Bifidobacteria ↑ Enterobacter↑	Thomas and Percival, (2009), Koenig et al. (2011), Turroni et al. (2012), Biagi et al. (2012), Bischoff (2016)	Type of lactation
Adults	Bacteroides ↑ Actinobacteria↑, Firmicutes↑, Proteobacteria↑	Thomas and Percival (2009), Choi and Lee (2016), Backhed et al. (2005), Clements and Carding (2018)	Diet Environment Immunity Sanitation
Elderly	Veillonella ↓, Bifidobacteria↓ Clostridia↑, Enterobacteria ↑, Lactobacilli↑ Bacteroides↑ Bifidobacteria↓, Bacteroides↑, Proteobacteria↑ Gamma Proteobacteria ↑	Wilson (2005), Odamaki et al. (2016), Bischoff (2016), Biagi et al. (2012), Hopkins and MacFarlane (2005), Choi and Lee (2016), Claesson et al. (2011), Mariat et al. (2009), Clark and Walker (2018)	Diet Environment Immunity Sanitation

↑ increased levels; ↓ decreased levels

5.3 Aging Gut Microbiota: Diet

Healthy GM has a significant recognition in the field of nutritional diet, metabolic activity of human health, and a challenging influence on aging. The human GM composition, diversity, function, modulation by environmental factors, and lifestyle have been revealed in the established studies and with the modern Omics approach (Lagier et al. 2015; Suzuki and Worobey 2014). In the last decade, GM has been an important spotlight of many studies because it plays a vital role in the prevention and treatment of various ailments (Turnbaugh et al. 2009; Qin et al. 2010; Karlsson et al. 2013). The symbiotic relationship between the host and GM is established during birth and develops with age (Vemuri et al. 2018).

Aging is accompanied with variations in the GM that are commonly associated with GIT modifications, as well as, changing dietary habits, and concomitant cognitive and immune system impairment resulting in frailty. Therefore, supplementation of nutritional diet to restore the GM in the elderlies have to be considered from a global point of view taking microbiota and extraintestinal targets into account (Salazar et al. 2017). GM influences the host housekeeping function by metabolizing specific nutrients in the GIT. Healthier aging can be promoted by improving the dietary constituents that can modulate GM (Ghosh et al. 2020). An increased uptake of plant foods rich in proteins, vitamins, fibers, carbohydrates, and polyunsaturated fatty acids with a synergistic reduction in alcohol, fat, and sugar forms the basis of diet and health associated with advantageous alterations in the gut metabolism (Berendsen et al. 2018).

Variation in diet and microbiome status between different countries at baseline has been demonstrated by Ghosh et al. (Ghosh et al. 2020). The study reported that intake of the Mediterranean diet by elderly individuals (65–79 years) across five countries exhibited gender- as well as country-specific diversity with regard to adherence, cognitive ability, and metabolism, with reduction in bone loss in osteoporotic individuals and augmented immunity and blood pressure (Ghosh et al. 2020).

Metagenomic findings have revealed changes in the composition of intestinal microbiota with age and diet for the maintenance of a healthy microbial ecosystem (Salazar et al. 2014). Functional metagenomics-based studies also suggested a "functional core microbiome" having the ability to substitute another microbiome to restore the functioning of various physiological pathways (Lozupone et al. 2012; Lloyd-Price et al. 2016; Biagi et al. 2016). The core microbiome consists of enterotypes and is distinguished by the predominant genera belonging to the Bacteroidetes and Firmicutes phylum (Arumugam et al. 2011). An experimental study revealed a functional core microbiome common among four different age groups, young, elderly, centenarian, and semi-supercentenarian, from different geographical origins wherein similar core functions in their GM was evident, belonging to enterotypes. Therefore, the importance of GM for healthy aging could be attributed to a compositional alteration in the functional core microbes (Kim and Jazwinski 2018).

Nutrition is an important environmental aspect that interacts with the host genes linked to nutrient signaling pathways, that orchestrate the host's GM, connecting the gut microbiome with the host genome, for instance, folate and choline (methyl donors), can influence DNA methylation (Anderson et al. 2012). Health and life span both can be extended by appropriate diet or caloric restriction as evidenced in *C. elegans* and *D. melanogaster* models (Alic and Partridge 2011; Alcedo and Kenyon 2004; Libert et al. 2006). Further, studies in *C. elegans* have documented the relationship between the gut microbiome and the host IIS and TOR signaling cascades influencing directly at the functional level to modulate the host's health and longevity (Kim and Jazwinski 2018). Based on these facts and reports, nutrition remains the common factor interconnecting the gut microbiome and the host nutrient signaling pathways.

Higher intake of plant-based diet (Mediterranean diet) has been accompanied with low mortality in the elderly population preferably by elevating the abundance of beneficial species and thereby preventing gut-related diseases (Kumar et al. 2016; Bamia et al. 2007). Diets rich in fiber content supplement the Bacteroidetes loads and reduces the Firmicutes abundance (De Filippo et al. 2010; Davenport et al. 2014). Hence, consumption of fiber is very crucial for the elderlies as it substitutes prebiotics for maintaining a beneficial GM. Based on the effect of plant-based diets on

GM and reported evidences, a healthy diet is considered as one of the approaches employed to promote healthy aging and has invited researchers to design more prospective natural supplements in regular diets to promote the existence of beneficial bacterial species of GM (Roberfroid et al. 2010; Hill et al. 2014).

5.4 Aging Gut Microbiota: Pre/Probiotics

Humans contain 90% of bacteria, of which the microbiota is shown to cause beneficial effects in many conditions like cancer, hypertension, and oral health. The role of probiotics, prebiotics, and symbiotics influence the healthy microbiota. There is strong experimental evidence to support this hypothesis. Probiotics are composed of lactic acid bacteria and bifidobacteria and prebiotics are composed of nondigestible oligosaccharides. Aging shows a reduction in microbiota biodiversity, increased pathobionts, and increased interindividual viability. Figure 5.1 depicts the schematic representation of the ways by which host GM aids in prolonging life span and boosts healthy aging.

Studies in *C. elegans* revealed that *E. coli* breaks down mechanically prior to entering the gut and becomes a source of nutrition and also secretes metabolites and small RNAs which help healthy aging. In older animals, *E. coli* population suppresses the infection and helps to prolong lifespan (Zhang and Hou 2013). Some free radicals like NO affect posttranslational modifications of proteins and also act as signaling molecule and have been associated with functions such as immunity, cardiovascular, and neurotransmission. Some organisms like *C. elegans* cannot produce NO on their own since they lack NO synthase, hence they utilize NO produced by the microbiota in their microenvironment (Gusarov et al. 2013). Some experiments have shown that NO acts as a longevity signal. RNA sequencing analysis of the worms supplemented with NO showed regulation of DAF-16/FOXO

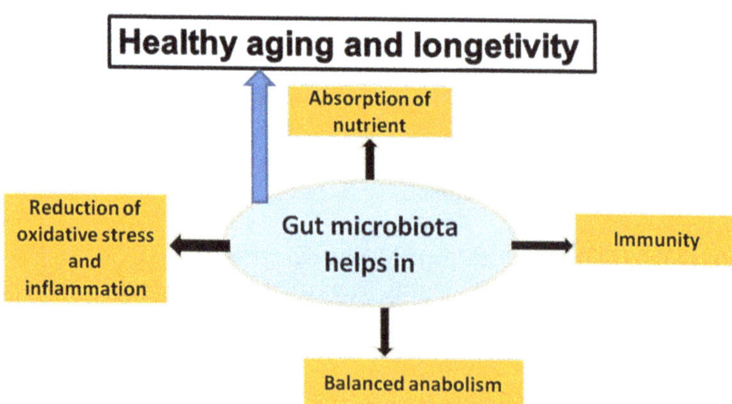

Fig. 5.1 Interrelationship between Gut microbiota and Aging

and HSF-1 transcription factors known for mediating the lifespan (Kenyon 2010). These studies influence holobiome impact as key to gene manipulations on longevity (Heintz and Mair 2014).

Studies in *D. melanogaster* fed with Lactobacilli show the inhibition of JAK/Stat signaling helps gut homeostasis which intern prolongs lifespan. This happens due to metaplasia of the gastric epithelium due to activation of JAK/Stat pathway. A low level of inflammation boosts age-related pathophysiological changes, which shows that gut compartmentalization reduces microbiota dysbiosis and helps in prolonging lifespan (Li et al. 2016). Earlier studies have shown that age-related immunosenescence has link with microbiota dysbiosis and loss of barrier function thereby triggering mortality (Clark et al. 2015).

Experiments on mammalian model organisms like mice and rat have given evidence that GM helps in healthy aging by improving muscle mass indirectly helping to reduce the symptoms of sarcopenia. These experiments revealed the association between GM-related parameters with measures like muscle mass, function in animal models and humans. Fecal bacterium and Bifidobacterium showed improved grip strength. Studies on Parkinson's patients have reported beneficial effects of probiotics. Aging and gut microbial composition when coorelated, a positive correlation has been noticed between the diversity of beneficial GM constitution during aging responsible not only for improving physical strength but also longevity (Ticinesi et al. 2019). There is a decline in diversity and homeostasis of microbiomes and gut flora with aging which can be related to health deficit accumulation. It is evident that gut dysbiosis leads to health deficits like sarcopenia and physical inactivity. In these situations, usage of probiotics and prebiotics comes handy, since it is cost-effective and widely available. Usage of synbiotics, which is a combination of probiotics and prebiotics, have proved to be more beneficial in older people. Synergistic effect is seen in different combinations in old age (Jayanama and Theou 2020).

There is increased inflammation associated with increased intestinal epithelial permeability (leaky gut) and abnormal (dysbiotic) GM in older adults (Rohr et al. 2020). Supplementation of a mixture of microbiota helps to reduce these complications in the older mice. It is observed that a combination of probiotics and prebiotics reduced inflammation by reducing salt hydrolase activity and increasing taurine abundance. Taurine is proven to be enhancing lifespan by reducing adiposity and leaky gut (Chander et al. 2018; Kerry et al. 2018; Sanders et al. 2019; Ahmadi et al. 2020).

Alteration in the brain lipid metabolism is associated with aging which leads to pathophysiological conditions. It is observed that GM has a positive effect by influencing host metabolism by increasing fatty acid profiles (MUFA and PUFA) in mice model (Varlamov et al. 2015; Tracey et al. 2018; Albouery et al. 2020). Considerable evidences are available to prove that consumption of probiotics improves brain aging. It helps in the reduction of cognitive and immune function loss (Sivamaruthi et al. 2018). Changes in the microbiota in older people, who are physically morbid suffer more negatively (Nagpal et al. 2018; O'Toole and Jeffery 2015). Benefits of probiotics in old age are prevention of diarrheal diseases, boosting immunity, and enhancing intestinal barrier function which indirectly helps in healthy aging (Landete et al. 2017).

5.5 Aging Gut Microbiota: Diseases

The GM in humans constitutes a large community of microorganisms. The prospective role of aging GM in various etiologies has become a focus for present-day researchers. Specifically, certain age-associated modulations in the phylogenetic constitution of the GM can pose a threat to the elderlies in developing various ailments via diverse mechanisms. A connection has been noticed between altered GM composition and clinical status of the elderly population in terms of physical infirmity, colorectal cancer, vulvovaginal complications, and atherosclerosis (Zapata and Quagliarello 2015). Further, it has been documented that, in the elderlies, there is a reduction in the gut microbial diversity with augmented prevalence of neurodegenerative ailments, namely Alzheimer's disease, Parkinson's disease, and dementia (Choi et al. 2018; Tripathi et al. 2022). Ding et al. (Ding et al. 2019) reported the involvement of the GM in cardiovascular disease. Therefore, modification of the GM in the aging process holds a promising strategy to decipher the age-linked comorbidities.

5.6 Conclusion and Future Prospects

The global population of the elderlies ($\geq$80 years) is expected to achieve the target of 370 million by 2050. The elderly population has been reported to be more vulnerable to several age-associated ailments. At present, an important objective of the scientific community is to investigate novel approaches for healthy aging, in order to prevent age-linked etiologies. Advancing age can influence the constitution and metabolic status of the gut microbiome leading to various gut-associated diseases. Interventions like diet and pre/probiotics have been suggested to sequester normal microbial species for healthy aging. Further studies are warranted on the relationship between aging and gut ecosystem for better understanding to enhance health status and longevity.

Acknowledgments We acknowledge Ramaiah College of Arts, Science and Commerce and REVA University for providing the necessary facilities.

References

Ahmadi S, Wang S, Nagpal R, Wang B, Jain S, Razazan A et al (2020) A human-origin probiotic cocktail ameliorates aging-related leaky gut and inflammation via modulating the microbiota/taurine/tight junction axis. JCI Insight 5(9) e132055

Albouery M, Buteau B, Gregoire S, Cherbuy C, Barros J, Martine L et al (2020) Age-related changes in the gut microbiota modify brain lipid composition. Front Cell Infect Microbiol 9:444

Alcedo J, Kenyon C (2004) Regulation of *C. elegans* longevity by specific gustatory and olfactory neurons. Neuron 41:45–55

Alic N, Partridge L (2011) Death and dessert: nutrient signalling pathways and ageing. Curr Opin Cell Biol 23:738–743

Anderson OS, Sant KE, Dolinoy DC (2012) Nutrition and epigenetics: an interplay of dietary methyl donors, one-carbon metabolism and DNA methylation. J Nutr Biochem 23:853–859

Arboleya S, Watkins C, Stanton C, Ross RP (2016) Gut bifidobacteria populations in human health and aging. Front Microbiol 7(1204):1–9

Arumugam M, Raes J, Pelletier E, Le Paslier D, Yamada T, Mende DR et al (2011) Enterotypes of the human gut microbiome. Nature 473:174–180

Backhed F, Levy RE, Sonnenburg JL, Peterson DA, Gordon JI (2005) Host-bacterial mutualism in the human intestine. Science 307:1915–1920

Bamia C, Trichopoulos D, Ferrari P, Overvad K, Bjerregaard L, Tjonnelang A et al (2007) Dietary patterns and survival of older Europeans: the EPIC-elderly study (European Prospective Investigation into Cancer and Nutrition). Public Health Nutr 10(6):590–598

Belkaid Y, Naik S (2013) Compartmentalized and systemic control of tissue immunity by commensals. Nat Immunol 14(7):646–653

Berendsen AAM, van de Rest O, Feskens EJM, Santoro A, Ostan R, Pietruszka B et al (2018) Changes in dietary intake and adherence to the NU-AGE diet following a one-year dietary intervention among European older adults: results of the NU-AGE randomized trial. Nutrients 10:E1905

Biagi E, Candela M, Fairweather-Tait S, Franceschi C, Brigidi P (2012) Ageing of the human metaorganism: the microbial counterpart. Age 34:247–267

Biagi E, Candela M, Turroni S, Garagnani P, Franceschi C, Brigidi P (2013) Ageing and gut microbes: perspectives for health maintenance and longevity. Pharmacol Res 69:11–20

Biagi E, Franceschi C, Rampelli S, Severgnini M, Ostan R, Turroni S et al (2016) Gut microbiota and extreme longevity. Curr Biol 26:1480–1485

Bischoff SC (2016) Microbiota and aging. Curr Opin Clin Nutr Metab Care 19(1):26–30

Bourlioux P, Koletzko B, Guarner F, Braesco V (2003) The intestine and its microflora are partners for the protection of the host: report on the Danone Symposium "The Intelligent Intestine", held in Paris, June 14, 2002. Am J Clin Nutr 78:675–683

Buford TW (2017) (Dis) Trust your gut: the gut microbiome in age-related inflammation, health, and disease. Microbiome 5(80):1–11

Candela M, Biagi E, Brigidi P, O'Toole PW, De Vos WM (2014) Maintenance of a healthy trajectory of the intestinal microbiome during aging: a dietary approach. Mech Ageing Dev 136–137:70–75

Chander AM, Yadav H, Jain S, Bhadada SK, Dhawan DK (2018) Cross-talk between gluten, intestinal microbiota and intestinal mucosa in celiac disease: recent advances and basis of autoimmunity. Front Microbiol 9:2597

Choi YJ, Lee DH (2016) Microbiota: a key for healthy aging. Ann Geriatr Med Res 20(4):168–176

Choi J, Hur TY, Hong Y (2018) Influence of altered gut microbiota composition on aging and aging-related diseases. J Lifestyle Med 8:1–7

Claesson MJ, Cusack S, O'Sullivan O, Greene-Diniz R, Weerd H, Flannery E et al (2011) Composition, variability, and temporal stability of the intestinal microbiota of the elderly. Proc Natl Acad Sci 108(1):4586–4591

Clark RI, Walker DW (2018) Role of gut microbiota in aging-related health decline: insights from invertebrate models. Cell Mol Life Sci 75(1):93–101
Clark RI, Salazar A, Yamada R, Fitz-Gibbon S, Morselli M, Alcaraz J et al (2015) Distinct shifts in microbiota composition during Drosophila aging impair intestinal function and drive mortality. Cell Rep 12:1656–1667
Clements SJ, Carding SR (2018) Diet, the intestinal microbiota, and immune health in aging. Crit Rev Food Sci Nutr 58(4):651–661
Collado MC, Rautava S, Aakko J, Isolauri E, Salminen S (2016) Human gut colonisation may be initiated in utero by distinct microbial communities in the placenta and amniotic fluid. Sci Rep 6: 23129
Cotillard A, Kennedy SP, Kong LC, Prifti E, Pons N, Le Chatelier E et al (2013) Dietary intervention impact on gut microbial gene richness. Nature 500:585–588
Davenport ER, Mizrahi-Man O, Michelini K, Barreiro LB, Ober C, Gilad Y (2014) Seasonal variation in human gut microbiome composition. PLoS One 9(3) e90731
De Filippo C, Cavalieri D, Di Paola M, Ramazzotti M, Poullet JB, Massart S et al (2010) Impact of diet in shaping gut microbiota revealed by a comparative study in children from Europe and rural Africa. Proc Natl Acad Sci U S A 107(33):14691–14696
Ding RX, Goh WR, Wu RN, Yue XQ, Luo X, Khine WWT et al (2019) Revisit gut microbiota and its impact on human health and disease. J Food Drug Anal 27(3):623–631
Duranti S, Milani C, Lugli GA, Mancabelli L, Turroni F, Ferrario C et al (2016) Evaluation of genetic diversity among strains of the human gut commensal Bifidobacterium adolescentis. Sci Rep 6:23971
Escobar JS, Klotz B, Valdes BE, Agudelo GM (2014) The gut microbiota of Colombians differs from that of Americans, Europeans and Asians. BMC Microbiol 14:311
Ghosh TS, Rampelli S, Jeffery IB, Santoro A, Neto M, Capri M et al (2020) Mediterranean diet intervention alters the gut microbiome in older people reducing frailty and improving health status: the NU-AGE 1-year dietary intervention across five European countries. Nutrition 69(7): 1218–1228
Gill SR, Pop M, Deboy RT, Eckburg PB, Turnbaugh PJ, Samuel BS et al (2006) Metagenomic analysis of the human distal gut microbiome. Science 312:1355–1359
Gusarov I, Gautier L, Smolentseva O, Shamovsky I, Eremina S, Mironov A et al (2013) Bacterial nitric oxide extends the lifespan of *C. elegans*. Cell 152:818–830
Heintz C, Mair W (2014) You are what you host: Microbiome modulation of the aging process. Cell 156(3):408–411
Hill C, Guarner F, Reid G, Gibson GR, Merenstein DJ, Pot B et al (2014) Expert consensus document: the international scientific association for probiotics and prebiotics consensus statement on the scope and appropriate use of the term probiotic. Nat Rev Gastroenterol Hepatol 11: 506–514
Hopkins MJ, Sharp R, Macfarlane GT (2002) Variation in human intestinal microbiota with age. Dig Liver Dis 34(2):S12–S18
Human Microbiome Project Consortium (2012) Structure, function and diversity of the healthy human microbiome. Nature 486(7402):207–214
Jayanama K, Theou O (2020) Effects of probiotics and prebiotics on frailty and ageing: a narrative review. Curr Clin Pharmacol 15:1–10
Karlsson FH, Tremaroli V, Nookaew I, Bergstrom G, Behre CJ, Fagerberg B et al (2013) Gut metagenome in European women with normal, impaired and diabetic glucose control. Nature 498:99–103
Keenan MJ, Marco ML, Ingram DK, Martin RJ (2015) Improving healthspan via changes in gut microbiota and fermentation. Age 37(5):98
Kenyon CJ (2010) The genetics of ageing. Nature 464:504–512
Kerry RG, Patra JK, Gouda S, Park Y, Shin HS, Das G (2018) Benefaction of probiotics for human health: a review. J Food Drug Anal 26(3):927–939

Kim S, Jazwinski SM (2018) The gut microbiota and healthy aging: a mini-review. Gerontology 64: 513–520

Koenig JE, Spor A, Scalfone N, Fricker AD, Stombaugh J, Knight R et al (2011) Succession of microbial consortia in the developing infant gut microbiome. Proc Natl Acad Sci U S A 108(1): 4578–4585

Kumar M, Babaei P, Ji B, Nielsen J (2016) Human gut microbiota and healthy aging: recent developments and future prospective. Nutr Healthy Aging 4(1):3–16

Lagier JC, Hugon P, Khelaifia S, Fourrnier PE, La Scola B, Raoult D et al (2015) The rebirth of culture in microbiology through the example of culturomics to study human gut microbiota. Clin Microbiol Rev 28:237–264

Landete JM, Gaya P, Rodriguez E, Langa S, Peiroten A, Medina M et al (2017) Probiotic bacteria for healthier aging: immunomodulation and metabolism of phytoestrogens. Biomed Res Int 2017:5939818

Li H, Qi Y, Jasper H (2016) Preventing age-related decline of gut compartmentalization limits microbiota dysbiosis and extends lifespan. Cell Host Microbe 19:240–253

Libert S, Chao Y, Chu X, Pletcher SD (2006) Tradeoffs between longevity and pathogen resistance in *Drosophila melanogaster* are mediated by NFkappaB signaling. Aging Cell 5:533–543

Lloyd-Price J, Abu-Ali G, Huttenhower C (2016) The healthy human microbiome. Genome Med 8: 51

Lozupone CA, Stombaugh JI, Gordon JI, Jansson JK, Knight R (2012) Diversity, stability and resilience of the human gut microbiota. Nature 489:220–230

Mariat D, Firmesse O, Levenez F, Guimaraes V, Sokol H, Dore J et al (2009) The Firmicutes/ Bacteroidetes ratio of the human microbiota changes with age. BMC Microbiol 9:123

Maukonen J, Saarela M (2015) Human gut microbiota: does diet matter? Proc Nutr Soc 74:23–36

Nagpal R, Mainali R, Ahmadi S, Wang S, Singh R, Kavanagh K et al (2018) Gut microbiome and aging: physiological and mechanistic insights. Nutr Healthy Aging 4(4):267–285

O'Toole PW, Jeffery IB (2015) Gut microbiota and aging. Science 350:1214–1215

Odamaki T, Kato K, Sugahara H, Hashikura N, Takahashi S, Xiao JZ et al (2016) Age-related changes in gut Microbiota composition from newborn to centenarian: a cross-sectional study. BMC Microbiol 16:90

Ple C, Breton J, Daniel C, Foligne B (2015) Maintaining gut ecosystems for health: are transitory food bugs stowaways or part of the crew? Int J Food Microbiol 213:139–143

Qin J, Li R, Raes J, Arumugam M, Burgdorf KS, Manichanh C et al (2010) A human gut microbial gene catalogue established by metagenomic sequencing. Nature 464:59–65

Roberfroid M, Gibson GR, Hoyles L, McCartney AL, Rastall R, Rowland I et al (2010) Prebiotic effects: metabolic and health benefits. Br J Nutr 104(2):S1–S63

Rohr MW, Narasimhulu CA, Rudeski-Rohr TA, Parthasarathy S (2020) Negative effects of a high-fat diet on intestinal permeability: a review. Adv Nutr 11(1):77–91

Salazar N, Arboleya S, Valdes L, Stanton C, Ross P, Ruiz L et al (2014) The human intestinal microbiome at extreme ages of life. Dietary intervention as a way to counteract alterations. Front Genet 5:406

Salazar N, Valdes-Varela L, Gonzalez S, Gueimonde M, Clara DLR-G (2017) Nutrition and the gut microbiome in the elderly. Gut Microbes 8(2):82–97

Sanders ME, Merenstein DJ, Reid G, Gibson GR, Rastall RA (2019) Probiotics and prebiotics in intestinal health and disease: from biology to the clinic. Nat Rev Gastroenterol Hepatol 16(10): 605–616

Sivamaruthi BS, Kesika P, Chaiyasut C (2018) A review on anti-aging properties of probiotics. Int J App Pharm 10(5):23–27

Suzuki TA, Worobey M (2014) Geographical variation of human gut microbial composition. Biol Lett 10(2):20131037

Tabouy L, Getselter D, Ziv O, Karpuj M, Tabouy T, Lukic I et al (2018) Dysbiosis of microbiome and probiotic treatment in a genetic model of autism spectrum disorders. Brain Behav Immun 73:310–319

Thomas JG, Percival SL (eds) (2009) Microbiology and aging. Indigenous microbiota and association with the host. Springer Science + Business Media, LLC, Cham
Ticinesi A, Nouvenne A, Cerundolo N, Catania P, Prati B, Tana C et al (2019) Gut microbiota, muscle mass and function in aging: a focus on physical fraity and sarcopenia. Nutrients 11:1633
Tiihonen K, Ouwehand AC, Rautonen N (2010) Human intestinal microbiota and healthy ageing. Ageing Res Rev 9(2):107–116
Tracey TJ, Steyn FJ, Wolvetang EJ, Ngo ST (2018) Neuronal lipid metabolism: multiple pathways driving functional outcomes in health and disease. Front Mol Neurosci 11:10
Tripathi AK, Ray AK, Mishra SK (2022) Molecular and pharmacological aspects of piperine as a potential molecule for disease prevention and management: evidence from clinical trials. Beni-Suef Univ J Basic Appl Sci 11(1):1–24
Turnbaugh PJ, Ridaura VK, Faith JJ, Rey FE, Knight R, Gordon JI (2009) The effect of diet on the human gut microbiome: a metagenomic analysis in humanized gnotobiotic mice. Sci Transl Med 1(6):6ra14
Turroni F, Peano C, Pass DA, Foroni E, Severgnini M, Claesson MJ et al (2012) Diversity of bifidobacteria within the infant gut microbiota. PLoS One 7. e36957
Varlamov O, Bethea CL, Roberts CT (2015) Sex-specific differences in lipid and glucose metabolism. Front Endocrinol 5:241
Varricchi G, Bencivenga L, Poto R, Pecoraro A, Shamji MH, Rengo G (2020) The emerging role of T follicular helper (TFH) cells in aging: influence on the immune frailty. Ageing Res Rev 61: 101071
Vemuri R, Gundamaraju R, Shastri MD, Shukla SD, Kalpurath K, Ball M et al (2018) Gut microbial changes, interactions, and their implications on human lifecycle: an ageing perspective. Biomed Res Int 2018:4178607
Voreades N, Kozil A, Weir TL (2014) Diet and the development of the human intestinal microbiome. Front Microbiol 5:494
Wilson M (2005) Microbial inhabitants of humans: their ecology and role in health and disease, 1st edn. Cambridge University Press, New York
Wu W, Kong Q, Tian P, Zhai Q, Wang G, Liu X et al (2020) Targeting gut microbiota dysbiosis: potential intervention strategies for neurological disorders. Engineering 6(4):415–423
Zapata HJ, Quagliarello VJ (2015) The microbiota and microbiomein aging: potential implications in health and age-related diseases. J Am Geriatr Soc 63(4):776–781
Zhang R, Hou A (2013) Host-Microbe interactions in *Caenorhabditis elegans*. ISRN Microbiol 2013:356451

Chapter 6
Gut Microbiome Regulation of Appetite and Role in Neurological Disorders

Ankita Singh, Om Prakash Verma, and Rajavashisth Tripathi

Abstract All human beings need food for energy and survival. Leptin and ghrelin are two of the most powerful peripheral hormones that manage your hunger and consumption of vitality. The level of circulating leptin acts as an energy reserve indicator and guides the nervous system to change food intake and consume energy. Ghrelin's functions include food intake, fat deposition, and growth hormone. To provide the nutrients required for their development, intestinal flora relies entirely on their moderator. The homeostatic pathway controls energy balance from energy stored in our body by increasing food consumption. This chapter discusses how bacterial growth is caused by nutrients, the dynamics of the gut bacteria population, and the mechanistic effect of bacteria in the gut. This chapter centralized the connections both the intestinal bacteria as well as the host are involved. This chapter explores the evidence supporting intestinal microbes' role in regulating human hunger and proposes an integrated homeostatic model of appetite suppression that takes both the host's and intestinal microbe's energy requirements into consideration. It also provides information on how the Western diet influences the gut microbiota. The Western diet encourages inflammation in the native intestinal bacteria that occurs through both functional and structural alterations. This chapter also attempts to present an outline of the bacteria of the gastrointestinal systems' possible role in the etiology of neurological diseases.

Keywords Appetite · Homeostatic Model · Leptin · Ghrelin · Arcuate nucleus · Growth hormone receptor · Melanocortin · Rostrocaudal

---

A. Singh
Jaypee University of Information Technology, Solan, India

O. P. Verma
Department of Molecular and Cellular Engineering, JIBB, SHUATS, Prayagraj, India
e-mail: om.verma@shiats.edu.in

R. Tripathi (✉)
Molecular Biology Unit, Institute of Medical Science, Banaras Hindu University, Varanasi, India

A. K. Tripathi, M. Kotak (eds.), *Gut Microbiome in Neurological Health and Disorders*, Nutritional Neurosciences,
https://doi.org/10.1007/978-981-19-4530-4_6

Abbreviations

5HT	5-Hydroxytryptamine receptors
ADI	Acceptable daily intake
AgRP	Agouti-related protein
ARC	Arcuate Nucleus
ASD	Autism spectrum disorder
CD1	Cluster of differentiation 1
CeA	Central nucleus of the amygdala
CLpB	Caseinolytic peptidase B homologous protein
CNS	Central nervous system
E. coli	*Escherichia coli*
EECs	Enteroendocrine cells
GF	Germ Free
GHSR1	Growth hormone secretagogue receptor 1
GLP1	Glucagon-like peptide 1
IL	Interleukin
LHA	Lateral hypothalamic region
MC4R	Melanocortin 4 receptor
NP	Nanoparticles
NPY	Neuropeptide tyrosine
NTS	Nucleus of Solitary tract
PBN	Parabrachial nucleus
POMC	Pro-opiomelanocortin
ppb	Part per billion
PVN	Paraventricular
PYY	Peptide tyrosine tyrosine
SCFAs	Short-chain fatty acids
TLR	Toll-like receptor
VMN	Ventromedial nuclei
VTA	Ventral Tegmental Area
WT	Wild type
α-MSH	α-Melanocyte-stimulating hormone

6.1 Introduction

In the contemporary era, there has been a rise in interest in the microbial community, particularly intestinal flora. The human microbiota is a collection of different types of microbes such as archaea, eukaryotes, protozoa, and viruses mainly bacteria that live symbiotically in our human gut or different parts of the human body (Zhao and Lukiw 2018; Rowan-Nash et al. 2019). The oral cavity, genital organs, respiratory tract, skin, and intestinal tract are sources of these communities (Lloyd-Price et al.

2016). The human microbiota approximates between 10^{13} and 10^{14} microbes, with a proportion of 1:1 microbes of human cells (Sender et al. 2016). These figures are taken from the total count of microbes in the colon, which is 3.8×10^{13} microbes, making it the organ with the most microbes (Sender et al. 2016; Sommer et al. 2017). The gastrointestinal microflora consists mainly of three main phyla of bacteria, such as Firmicutes, Bacteroidetes, and Actinobacteria (Tap et al. 2009). Due to technological drawbacks, particularly in trying to study non-cultivable bacteria and an absence of information at the population level that describes the mixtures and roles of the microbiota, the characteristics of the intestinal microbiota as well as the host–microbiota relationships were largely unknown until recent decades (Tap et al. 2009). The homeostatic and hedonic circuits regulate food intake in a complementary way. When energy resources are exhausted, the homeostatic pathway increases the interest in eating (Hartman et al. 2019; Lutter and Nestler 2009). This chapter explores the evidence supporting the intestinal microbe's role in regulating human hunger and offers an integrated hunger homeostatic model for suppression that takes both the host's and intestinal microbe's energy requirements into consideration. It also discusses the relationships between appetite and the gut microbiome. Finally, this chapter examines the symbiotic connections between intestinal microflora and the host, as well as how the Western diet affects the intestinal flora and the gut microbiome's role in neurological disorders.

6.2 Roles of Intestinal Bacteria

The interrelationship within both the intestinal flora and the host is controlled and stabilized via a huge system of metabolism, immunological, and neurological interactions. It is theoretically mediated by metabolites synthesized by microbes. It exhibits pleiotropic effects, including signalling molecules that regulate the neuroimmune–inflammatory axes of the host that could connect the gut with other organ systems physiologically (Zhao and Lukiw 2018; Roman et al. 2018).

6.3 Metabolism

For the study of human fecal samples by using 16S ribosomal RNA and metagenomic sequencing methods, it shows richness in the metabolism of gut microbial polysaccharides, amino acids, xenobiotics, and micronutrients, indicating that these microbes promote the production of host energy and metabolic efficiency (Gill et al. 2006; Tripathi et al. 2022). Germ-free (GF) animal experiments confirmed the findings, revealing that mice without germs had 40 percent less epididymal fat and needed an extra 10–30% of overall intake of food to keep their same body weight as rodents having normal microbiota (Tripathi et al. 2022; Bäckhed et al. 2004). Gut flora is required for the decomposition of insoluble dietary fiber and soluble starch. Mostly in the presence of SCFAs, fermented end-products exist. SCFA is an energy substrate for the host (together with butyrate, propionate, acetate, and pentanoate) (Salminen et al. 1998). Microbially generated SCFAs account for 70% of colonic ATP synthesis, and butyrate is the preferred fuel for colonocytes (Kong et al. 2021). They are quite important in the process co-metabolism of gastric acid with human intestinal bacteria. To help digestion, cholesterol, and lipid metabolisms, these cholesterol compounds are synthesized in the liver, then conjugated with taurine or glycine before being stored in the gall bladder and eventually secreted through the intestine (Krishnan et al. 2015; He et al. 2016).

6.4 Resistance to Colonization

Colonization resistance is another key component of the human microbiome, in which it defends against pathogenic colonization and inhibits pathogenic microbiota members from overgrowing (Zhao and Lukiw 2018).

6.5 Appetite Control in Homeostatic Model

Some of the strongest peripheral hormones are ghrelin and leptin. Leptin is produced by adipose tissue, and its level rises in ratio to body calories. Leptin rates are high and restrict consuming meal and activate metabolism mechanisms to burn off stored energy. Ghrelin, on the other hand, is an abdominal peptide whose levels increase in reaction to an adverse energy equilibrium, promoting nutrition and energy conservation (Zigman and Elmquist 2003).

6.6 Host Energy Homeostasis and Brain

The present perspective of appetite control is against gut microbes in host appetite control, which offers microorganisms that are interconnected with the host (Fetissov 2017). A homeostatic model for appetite control was developed based on morphological, molecular genetics, and histological research (Schwartz et al. 2000). The guideline of craving for all the particular genes put into code, basically peptide couriers and their receptors are communicated by cells through neuroendocrine and neuronal pathways (Hökfelt et al. 2003). The craving pathways in the cerebrum are arranged around the circumventricular organs like the hypothalamic arcuate nucleus (Langlet et al. 2013). The NTS, for example, gets vagal input (Lutter and Nestler 2009), and the afferent populations of the cranial nerves. Both ARC and NTS participate in the various hypothalamus–brain stem and frontal brain sites, forming a dynamic neural organization for autonomic desire regulation that is impacted by intellectual aspects (Gill et al. 2006; Tripathi et al. 2022; Waterson and Horvath 2015). The hypothalamic PVN and VMN, the LHA, the PBN, and the central amygdala nucleus (CeA) have all been assigned important placements throughout these regions (Waterson and Horvath 2015; Richard 2015; Berthoud 2002; McMinn et al. 2000).

GABA-ergic neurons release orexigenic NPY and AgRP (Loh et al. 2015; Ollmann et al. 1997; Hahn et al. 1998; Tatemoto et al. 1982). Another group is made up of an adjacent collection of pro-opiomelanocortin (POMC) glutamatergic neurons, an antecedent of alpha-melanocyte-animating hormone (alpha-melanocyte-animating hormone, alpha-melanocyte-animating hormone), and anorexigenic neuropeptides (Fan et al. 1997; Poggioli et al. 1986; Harris and Lerner 1957). POMC neuron activation is also tracked by αMSH discharge and attached to the MC4R, which promotes the passing of the pathway of melanocortin satiety (Harris and Lerner 1957; Cone 2005; Broberger et al. 1998). In the ARC and in their downstream projections, these communities are anatomically and functionally interconnected locally (Cone 2005; Broberger et al. 1998; Zhao et al. 2007), where hunger-related signals from humoral pathways are integrated (Fetissov 2017).

Cholinergic vagal neurons that discharge glutamate to energize NTS neurons are the basic concept of the neuronal flagging route in craving control, which can rotate gut satietogenic signs toward other regions of the cerebrum also, involving PVN and CeA (Zhao et al. 2007). Anorexigenic transmitters and peptides, as well as catecholamines, GLP1, and αMSH, are all considered by NTS neurons (Appleyard et al. 2005; Appleyard et al. 2007; Larsen et al. 1997). The connections between ARC, NTS, and POMC neurons are equal, suggesting that humoral and neurological paths in satiation are redundant (Wang et al. 2015; Mimee et al. 2014; Cone 1999).

Long-term hunger control involves the peptidergic hypothalamic network that begins with the satiety route in the same ARC neurons (Wang et al. 2015; Mimee et al. 2014; Cone 1999; Elmquist et al. 1998). The long-term discharge of peptide hormones from gastric mucosa and power stores, for example, is the basis for this movement. Leptin, a long-term anorexigenic sensor derived from body mass-equivalent adipose tissue, induces adverse energy equilibrium and promotes ARC and POMC (Hahn et al. 1998; Tatemoto et al. 1982; Friedman and Mantzoros 2015). The levels of anorexigenic and orexigenic neuropeptide gene expression, as well as their receptors, govern the action of this peptidergic network (Higuchi et al. 2008).

6.7 Hedonic Versus Homeostatic Regulation

The homeostatic model describes the enhanced hunger and food consumption generated by energy shortages as well as the sense of satisfaction that comes with feeding, which includes brain stimulation. It is present in all types of motivation, especially food consumption. Bulimia nervosa can cause homeostatic signals to be overwhelmed by an abnormal hedonic drive to eat (Higuchi et al. 2008; Simon et al. 2016; Avena et al. 2008). The mesolimbic dopamine framework which will begin in the VTA region includes the main pathways. The homeostatic craving host communicates with both the cerebrum and the dopamine, which is its source and in the cerebrum's target regions as well as its nerve center, the central frontal cortex, and it's also involved with both homeostatic and eating reactions (Friedman and Mantzoros 2015; Higuchi et al. 2008; Simon et al. 2016; Avena et al. 2008; Jerlhag et al. 2007). The blood–brain barrier secures VTA dopaminergic receptors from the possible impacts of intestinal microscopic species inferred by flowing atoms on dopamine should be handed over by other neurons like PBN (Meguid et al. 2000; Legrand et al. 2015; Barson et al. 2011; Norgren et al. 2006).

6.8 Bowel Transmission to the Brain

Homeostatic and hedonic mind frameworks commanding hunger are initiated by chemicals determined by organs and the tissues that give signal to different metabolic measures which can conclude energy and healthful status (Broberger 2005; Scott and of N and BSA 1992). In the gastrointestinal tract, mucosal endocrine cells are included, which can secrete several peptide chemicals into circulation (Scott and of N and BSA 1992; Wren and Bloom 2007; Janssen and Depoortere 2013). Cholecystokinin, it contributes to satiation, and PYY, it encourages postprandial satiety, are anorexigenic chemicals developed primarily in the small and inner organs (Feinle-Bisset 2016; Degen et al. 2001; Batterham et al. 2002). As a satietogenic chemical, GLP1, the result of the proglucagon quality, is reported in enteroendocrine neurons to decrease postprandial blood sugar levels by promoting insulin discharge (Drucker 2005).

The components responsible for craving control imply that the food status is dependent on the discharge of appetite and satiety chemicals from intestinal flora, including sensations related to cerebrum routes. Food taken is suggested to be set off by an increase in ghrelin plasma levels, which happens while feeding, triggering the promotion of NPY and AgRP cells that indicate GHSR1 (Drucker 2005; Nakazato et al. 2001; Wynne and Bloom 2006). After this, the increment in plasma levels of PYY initiated by a feast prompts the restraint of NPY and AgRP neurons by specific binding of PYY to the type 2 neuropeptide Y receptor (Y2RRP) (Wynne and Bloom 2006; Cowley et al. 2001). This restraint of NPY and AgRP cells eliminates their receptor GABA-ergic contribution to both the ARC and POMC cells, promoting the melanocortin satiety route (Cowley et al. 2001).

6.9 Gut Microbiome Regulates the Appetite

Gut bacteria create peptides that mimic and interfere with normal appetite regulation, causing food intake. Gut bacteria can also increase hunger by activating opioid and cannabinoid and altering taste buds. Probiotics help in reducing hunger by increasing the diversity of gut bacteria. The gut microbiota influences all host immunity and neural hunger, which, when combined, can affect human feeding behavior in metabolism and disordered eating like obesity and anemia (Batterham et al. 2002; Drucker 2005). Our hunger is related to the diversity of our microbiota. Satiety is linked with an increase in the microbial population. The anorexigenic pathways in our brain are active when we digest nutrients, and the intestinal satiety hormones secreted in the stomach stimulate the anorexigenic pathways (de Clercq et al. 2016). Microbes that affect our feelings of hunger and satiety have grown in number in recent years. When receptors are active, hormones like cholecystokinin, for example, regulate satiety and hunger by binding to bacterial products like lipopolysaccharides and flagellin (Sun et al. 2020).

6.10 Bacterial Growth Caused by Nutrients

Living microscopic creatures can determine their own pace of development and populace numbers. Microbial multiplication occurs during exponential growth, which is a rate of division dictated by the amount of time required for replication of DNA and cellular multiplication (Wang and Levin 2009) in case of *Escherichia coli* to copy after arrangement of supplement time taken is 20 min. The microbial growth enters a stationary growth process after microbial replication to a degree of $\sim 10^9$ cells per ml, which is characterized by supporting microbial count above a few hours. Therefore, bigger the initial community of bacteria, the less time it takes to arrive at the stationary. The stationary, or fixed, phase is preceded by a decline phase in which the count of microbes belonging to the natural cell lysis action is continuously reduced (Rice and Bayles 2008). There may be a connection between the number and synthesis of gut microorganisms as well as the host's craving for shorter and longer durations (Fetissov 2017).

The development of 10^9 microbes in 20 min requires an initial density of 10^9 cells, as only a single division of each microscopic organism is sensible in the period of 20 min, which depends on growth dynamics. As a minimum sum for acceptance of the fixed stage or stationary process, this cell number age is filled in as higher introductive amounts of *E. coli* also equivalent to $\sim 10^9$ per ml, which enhances microbial load after that in vitro supplement arrangement. Then it was all replicated in rodents in which the stationary process of intestinal microbes was initiated 20 min after the implantation of supplements into their enormous digestive systems (Breton et al. 2016). Previous studies (in 1983) indicated the doubling time of *E. coli* placed into the little or enormous digestive tract of mice was 2 h with a primary microbial convergence of 2.4×10^4 per ml cells (Freter et al. 1983). Supplement gives rise to bacterial development; hence, it shows special dynamics that are characterized by the strength of a microbial count (Kong et al. 2021).

6.11 Host Control

In vitro, microbial production is regulated entirely by supplements given, and bacterial population expansion is constantly impacted by diet (Flint et al. 2012). The balance between positive and negative influences is determined by the relative equilibrium of the microbial community in every region of the intestine (Fig. 6.1). The antibacterial ability of both chemical and physical host-derived factors may be reduced as the bacterial population gradient in the rostro-caudal gastrointestinal tract increases. Peristalsis is absent in the colon with the highest bacteria density. The time it takes for absorbed resources to reach the stationary stage after nutrient-induced bacterial development is quicker than the time it takes for bacteria to reach that stage (Table 6.1) (Fetissov 2017).

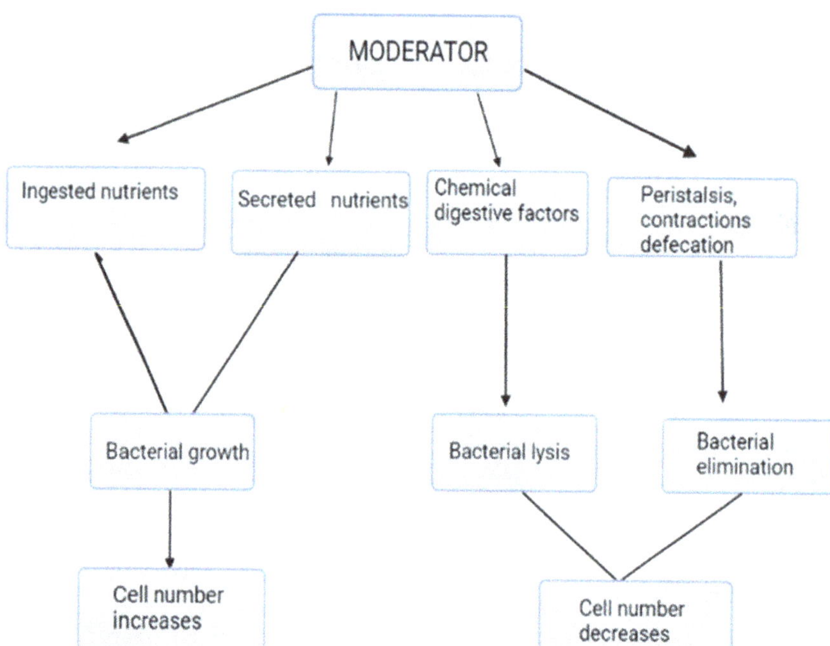

Fig. 6.1 A variety of bacterial cells in the gut are affected by the equilibrium between stimulation and inhibition caused by host-related substances. In contrast to in vitro circumstances, the extension of bacteria is controlled by the availability of supplements, the expansion of microbial pathogens in the intestine is constantly influenced by nutrition, resulting in bacterial regulation. The particular stability of each portion of a microbial species' gastrointestinal system is determined by a balance of positive and negative stimuli (Fetissov 2017)

Table 6.1. The presence of chemical and transit time along the digestive tract could explain the rising rostro-caudal gradient of bacterial content. The travel time in the upper bowel is less than the time required for the microbial community to reach the stationary growth phase (Stat), as calculated using the formula: t (min) = G (generation time, 20 min, assumed based on in vitro experiments and in vivo infusions) × 3.3 log (minimal microbial count in the Stat phase, that is 10^9)/microbial count before multiplication. Mucosal adhesion is one of the chemical factors that causes and prevents bacterial lysis in digestive juices containing gastric and bile acids, as well as numerous digestive enzymes (Fetissov 2017)

Host produces digestive chemicals	Bacterial content (cells per g)	Time to stationary phase	Transit time
Stomach HCl pH ~1.4 Pepsin Gastric lipase	Stomach 10^1	8.8 h	1–3 h
Liver Bile acids	Duodenum 10^3	6.6 h	30–60 min
Pancreas Trypsin Amylase Carboxypeptidase	Jejunum 10^4	5.5 h	1–5 h
Small intestine Brush border enzymes	Ileum 10^7	2.2 h	1–5 h
Colon No host digestive factors	Colon 10^{12}	20 min	10–60 h

6.12 Mechanistic Impact of Bacteria from the Gut

With each nutritionally microbial replication, a steady rise in microbial activity restricted by the stationary phase indicates that the intestinal sensory system can perceive a simple and repeatable pressure gradient of many pathogen compounds upstream to the channels of satiety. These microbial compounds that are made can be more or less precise antimicrobial substances for bigger population organisms, such as communities of gastrointestinal bugs. Usually, proteins' more complex signals are transmitted by their derived peptides, as well as lipids and sugars (Fetissov 2017; Wexler 2007).

Chemical sensory components located in the epithelium of the gut on enteroendocrine cells or in vagal afferents, specifically intestinal intestines, should be used to characterize bacteria molecules in this formation of short-term hunger routes (Broberger 2005). Enteroendocrine cells produce hormones both systemically and locally to cause vagal afferents. As a result of the intraluminal administration of lactobacillus, vagal afferent activity is increased (Perez-Burgos et al. 2013).

The intestinal barrier requires the synchronized activity of the central nervous system enabled by locally secreted transmitters like serotonin (Prins et al. 2007; Neunlist et al. 2013). Microorganisms create a number of modulator neuroactive chemicals, such as biological amines (Lyte 2014) that help the intestinal epithelium produce 5HT (Yano et al. 2015; Reigstad et al. 2015). Microbes use an indole as a chemical indicator to insulate themselves from stress during the release of the tryptophan receptor and 5HT during the formation of numerous bacteria, including *E. coli* (Lee et al. 2010). Indole has been demonstrated to affect GLP1 secretion based on its strength (Chimerel et al. 2014) and to be a ligand of the aryl hydrocarbon channel that can send microbial instructions to the host via this intestinal immune response (Hubbard et al. 2015). The existence of the route of the aryl hydrocarbon regulates the role of bacterial-derived indoles in core food intake is indicated in the hypothalamic NPY/AgRP and POMC neurons (Fetissov et al. 2004).

The most dominant bacterial compounds that could enable the TLRs (toll-like receptors) generated by the epithelium in the intestine, like enteroendocrine cells (Bogunovic et al. 2007; Palazzo et al. 2007). It was shown that lipopolysaccharide synchronizes gene expression in the gut epithelium (Mukherji et al. 2013). The taste response to sucrose was reduced and its ingestion (Zhu et al. 2014). Appearance in the vagal afferents of the TLR4 lipopolysaccharide receptors. The function of intestinal lipopolysaccharide is to indicate satiety, which is consistent with its anorexigenic nature (Hosoi et al. 2005; Topping and Clifton 2001).

Bacteria expose their cell components as a result of the enzymatic conversion of nondigestible and other resources from the host to metabolites. Butyrate, propionic acid, and acetate are three of the more important ones. SCFAs are formed from a variety of polymers, the majority of which are indigestible to gut microorganisms (Topping and Clifton 2001). In addition, propionic acid can directly activate PYY and GLP1 production from both in vitro L-cells and in situ in humans (Topping and Clifton 2001; Chambers et al. 2015a; Chambers et al. 2015b; Samuel et al. 2008).

In relation to their enzymatic function in nutritional catabolism, due to their molecular mimicry of cellular proteins, some microbial molecules can communicate directly with their host role as adhesion pathways. In *E. coli*, the first such compound was described as a conformational mimetic antigen of αMSH (Figure 6.2) (Tennoune et al. 2014).

6.13 The Western Diet Influences the Gut Microbiota

Obesity and related metabolic problems are tightly linked to the nutritional pattern that characterizes the Western diet, but the molecular mechanisms that underpin these links are unknown. Inflammation of the native microbiota is caused by both systemic and lifestyle diseases in the Western diet (Tripathi et al. 2022; Zinöcker and Lindseth 2018). It has implications for future research into public health recommendations and food production methods to better comprehend the function of the microbiome in the development of diet-related ailments. According to food trend research (Zinöcker and Lindseth 2018), whole foods appear to be a common diet denominator linked to a reduced level of nutrition sickness.

Gut microbial alterations in rodents in the context of dietary changes showed that the microbiota can change quickly and lead to negative health impacts (Zinöcker and Lindseth 2018; David et al. 2014). These modifications may involve the dysbiosis of

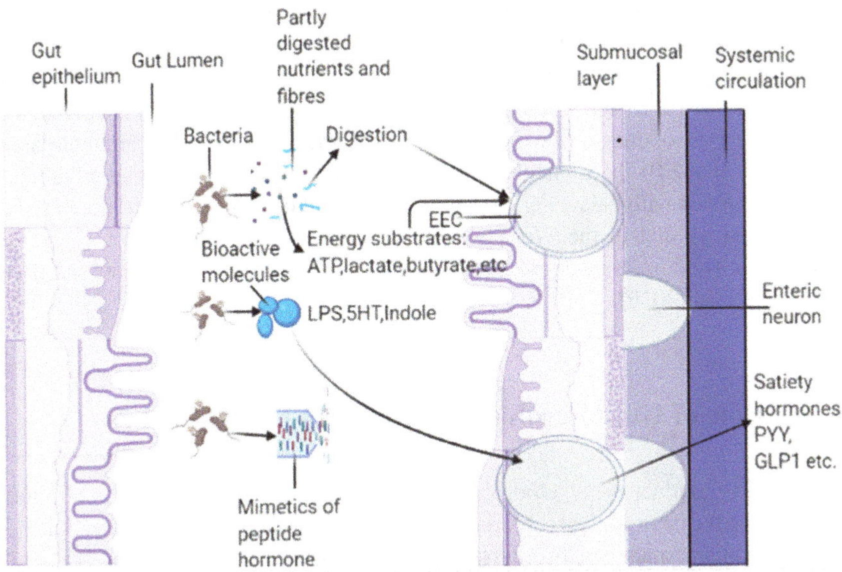

Fig. 6.2 Chemical signals derived from intestinal flora could cause intestinal routes of satiety. Bacteria metabolize nondigestible fibers throughout their existence in the intestine. Many energy substrates like ATP, lactate, and butyrate are provided by digestible nutrients. Bioactive substances like lipopolysaccharide (LPS), on bacterial lysis proteins that continue their enzyme reactions, are released by synthesizing bioactive metabolites such as 5HT, or by functioning directly as peptide hormones such as CLpB. Both chemical signals originating from bacteria and including nutrients, have direct interaction with the epithelium of the intestine that holds chemical agents. Bacterial signal activation of enteroendocrine cells (EECs) directly or indirectly (via enterocytes) activates systemic and local secretion of PYY and GLP1, then satiety signals occur. The paracrine behavior of pathogen molecules on 5HT-producing EEC and enterochromaffin cells may promote the enteric nervous system, controlling gastric microbiota and intestinal barrier permeability, including access to the gastrointestinal tract and gut barrier to vagal afferents, which occur in microbial indications (Fetissov 2017)

microbiota, contributing to inflammation and metabolic disorders (David et al. 2014; Suez et al. 2014; Chassaing et al. 2015). Dietary variables that change the metabolic activity of previously existing microorganisms can potentially have a substantial impact. Gut flora alters their metabolism in response to compounds produced by other microorganisms as well as the availability of nutrients, potentially affecting metabolic and inflammatory pathways (Turnbaugh et al. 2008; Stecher et al. 2013).

The degradation of microbial diversity due to dietary changes has been proven in recent animal and human research to be handed down to future generations with a drastic diversity reduction (Sonnenburg et al. 2016). Microbes may be permanently lost as a result of a Western diet. It is necessary for the microbiome's function and, through the epigenome, can cause heritable metabolic alterations. In conclusion, the environment generated in the stomach by ultra-processed nutrients may operate as an evolutionarily unique selection for pathogens, attracting various bacteria and disorders related to inflammation (Thaiss et al. 2016; Krautkramer et al. 2016).

A modern diet can boost endotoxin-producing microbes in the digestive systems of humans and animals, contributing to metabolic endotoxemia (Palazzo et al. 2007; Mukherji et al. 2013). Much of the research looking at the influence of heavy diet alterations on rodent microbiota has employed high-fat and unhealthy diets. Changes are also due to the fat content, like metabolic endotoxemia diseases (Chambers et al. 2015b; Sonnenburg et al. 2016). Processed diet-induced inflammation may interact with balancing resource systems and trigger feelings of hunger and hyperphagia. A diet in the Western style in mice triggered both anxiety and an appetite for sweet, energy-dense diets (Thaiss et al. 2016; Krautkramer et al. 2016; Ulrich-Lai et al. 2015).

6.14 Role of Gut Microbiome in Neurological Disorders

It's vital to conduct more clinical and preclinical research on gut microbiome therapeutics for neurological diseases such as Parkinson's disease, Anxiety, Schizophrenia, Autism spectrum disorders, Multiple sclerosis, Alzheimer's disease, Epilepsy, and Stroke.

6.15 Parkinson's Disease

Parkinson's disease is a CNS degenerative ailment that influences dopamine-producing cells in the brainstem nuclei. Parkinson's disease affects 7–ten million individuals globally, with men 1.5 times more likely to have it than women. PD is more common as people age, with roughly 1% of those aged 60 years and over suffering from the disease (Dutta et al. 2019a). Akinesia, muscle stiffness, tremors, bradykinesia, and difficulties walking and gait are some of the signs of the condition (DeMaagd and Philip 2015). These motor traits are accompanied by symptoms such as confusion, anxiety, and cognitive and autonomic dysfunction. Lewy bodies are eosinophilic cytoplasmic inclusions that form when α-synuclein insoluble polymers deposit inside the neural side as a result of synucleinopathy (Xu and Pu 2016). Neurodegeneration and neuronal death are caused by these Lewy bodies (Wolters and Braak 2006; Sulzer 2007). Certain abnormalities in the intestinal flora of people with Parkinson's disease have been discovered in certain studies, which could act as biomarkers and trigger the conformational changes of α-synuclein, which causes neurodegeneration in people with the disease (Dutta et al. 2019b). In PD patients, gastrointestinal dysmotility, which includes delayed emptying of the stomach and constipation, is prevalent (Goetze et al. 2005; Unger et al. 2011; Hardoff et al. 2001; Noyce et al. 2012; Cersosimo et al. 2013).

6.16 Anxiety

Anxiety is a mental illness marked by feelings of dread and terror, as well as physical symptoms like palpitations and perspiration. Muscle tension, restlessness, weariness, and attention difficulties are all common symptoms of anxiety. Anxiety symptoms can be long-term (or widespread) or short-term (or specific), and they can cause panic attacks. The quantity, strength, and frequency of symptoms can also vary from individual to individual (Rynn and Brawman-Mintzer 2004). The presence of toxic gut microbes was linked to anxiety, which causes oxidative stress once those germs outcompete healthy ones. The vagus nerve can be triggered when harmful gut bacteria gain control and cause inflammation, leading to cognitive problems (Breit et al. 2018). Some microorganisms can also create peptides that deliver stress signals to the brain, affecting gene expression and the CNS.

6.17 Schizophrenia

Hallucinations, delusions, apathy, recurring periods of psychosis, and profoundly disorganized thinking are all symptoms of schizophrenia, a complicated and devastating brain condition (Edition 2013; Owen et al. 2016). Mental illness affects over 21 million individuals globally and is linked to social and physical morbidity (Hjorth et al. 2017; Marwaha and Johnson 2004). Schizophrenia symptoms typically appear between the ages of 16 and 30 years, with males being more susceptible. In their study (Nemani et al. 2015) Nemani et al. discovered a relationship between altered intestinal flora and schizophrenia. Patients with first-episode psychosis had more abundant Lactobacillaceae, Halothiobacillaceae, Brucellaceae, and Micrococcineae, but Veillonellaceae were less abundant according to Schwarz et al. (Schwarz et al. 2018).

6.18 Autism Spectrum Disorder

ASD is a neurological ailment characterized by stereotyped behaviors, cognitive deficits, and language or interactional difficulties (Williams et al. 2012; El-Ansary et al. 2020). Patients with few symptoms are more able to manage freely, however, those with serious complications will involve significant assistance in their everyday tasks. It affects one out of every 68 children under the age of three and therefore is four times higher popular in boys as compared to girls (Tillmann et al. 2018; Rosenfeld 2015). Apart from its diversity, the disease is characterized by perplexing pathophysiology and etiological pathways connected to a defective intestinal flora, like glutamate excitotoxicity, oxidative stress, and neuroinflammation (El-Ansary and Bhat 2020). Some of the physiology that seems to underline the intestinal microbe's impact on autism are affective responses, meal responses, upper gastro-intestinal illness, unusual stool, autistic enterocolitis, leaky gut syndrome, excessive inflammation, aberrant glutathione levels, and irregular metal or mineral levels (Jepson and Johnson 2007). In children with autism, increased intestinal permeability can result in digestive problems, allowing neurotoxic substances to slip through an inflammatory gut membrane and cause neurological issues (Wakefield et al. 2002; Reddy and Saier 2015).

6.19 Multiple Sclerosis

Multiple sclerosis is a neurodegenerative state that causes persistent inflammatory conditions and demyelinating plaques in the brain and spinal cord (Peng et al. 2019; Wingerchuk et al. 2001; Calvo-Barreiro et al. 2018; Bhargava and Mowry 2014). Tiredness, coldness, lack of coordination, dizziness, visual loss, disorientation, discomfort, bladder and bowel problems, or even depression are all markers of this illness (Mielcarz and Kasper 2015). MS affects over 2.1 million individuals globally. MS also increases the number of CNS-targeting autoreactive immune cells (Zhang et al. 1994). It's fair to relate gut commensal flora to MS risk since the gut microbiota aids in body's defense education and has a critical role in a variety of autoimmune and metabolic illnesses. Vartanian et al. demonstrated person bowel colonization by *Clostridium perfringens* type B in a patient with her first relapse of MS, and this revealed that the epsilon toxin released by this pathogen caused microangiopathy; similar studies showed that this can further disrupt the BBB, causing neuronal and oligodendrocyte damage (Rumah et al. 2013; Mete et al. 2013; Dorca-Arévalo et al. 2008; Lonchamp et al. 2010; Finnie et al. 1999). Antibodies against epsilon toxin were shown to be more common in Multiple Sclerosis patients. In Jhangi et al., they found an increase in *Methanobrevibacter* levels in MS patients (Jhangi et al. n.d.). In MS patients, concentrations of *Butyricimonas, Lachnospiraceae,* and *Faecalibacterium* were shown to be lower than in controls, and *Faecalibacterium* was shown to be less abundant in the patients with inflammatory bowel illness (Mowry et al. n.d.; Machiels et al. 2014).

6.20 Alzheimer's Disease

Alzheimer's disease is a neurological disorder marked by memory loss and the accumulation of abnormal amyloid-beta (Aβ) protein, mostly in the intermembrane space of dopaminergic neurons (Hu et al. 2016; Querfurth and LaFerla 2010). AD affects around 44 million individuals worldwide, with adults over 65 being the most affected. Language issues, disorientation, mood swings, motivation, and self-care management are all signs of Alzheimer's disease that are known to increase over time. Modifications in intestinal flora can cause effects on the brain, which in turn can cause changes in host behavior (Sampson and Mazmanian 2015). AD has been the leading form of dementia, with 36 million cases reported in 2010, with forecasts of 66 million patients by 2030 and 115 million by 2050 (Prince et al. 2013). Alzheimer's disease is originated from a combination of inherited and environmental causes (Xu and Wang 2016). Because it promotes overactivity of the innate and adaptive immune cells, which induces inflammation and promotes intestinal absorption and microbial overgrowth, aging is a vital risk agent for AD (Vogt et al. 2017; Franceschi 2007; Ulluwishewa et al. 2011; Tran and Greenwood-Van Meerveld 2013).

6.21 Epilepsy

Epilepsy is a chronic issue that impacts more than 50 million people all over the world, accounting for 0.5% of all disease burdens (World Health Organization 2019). It's a brain disorder characterized by a long-term proclivity to produce seizures (Fisher et al. 2014).

Every year, over 2.4 million people suffer from epilepsy. Seizures can be caused by diseases that affect connections between neurons, ionic channel function, and neurotransmitter reception, among other things (Stafstrom and Carmant 2015). While the WHO estimates that 70% of people with epilepsy might become seizure-free with adequate medication, less than half of epileptic patients in developing countries have access to anti-epileptic drugs. In animal models, the microbiome takes part in an important role in brain growth and neurobehavioral function (Vuong et al. 2017).

6.22 Strokes

Variations in the makeup and activity of the intestinal flora have been connected to the development of both intestinal and extraintestinal illnesses, and all of these changes have an impact on cardiovascular risks like obesity, immune disorders, and atherosclerotic (Levy et al. 2017). The gut microbiota is important in bilateral intestine-brain connections. This connection is related to the microbiome–gut–brain axis (Cryan et al. 2019). New research reveals that the microbiome–gut–brain axis acts as a critical immunological controller following an acute ischemic stroke. Dysbiosis can have systemic negative consequences when the systemic inflammatory process develops after such an ischemic stroke (Battaglini et al. 2020).

6.23 Conclusion

The molecular processes of gut microbial life, satiety, and hunger in a novel method were the focus of this chapter. From this chapter, we can say that the gut microbiota's nutrient-induced growth dynamics may act in the host's control of appetite on a regular daily basis to preserve bacterial immune function. The energy required for intestinal bacterial multiplication could account for a significant portion of a healthy person's daily energy requirements. Chemical signals produced by bacteria that can tell the host about the gut microbiota's energy state provide biological pathways for appetite control. The microbiota's composition could influence how the host's energy metabolism is regulated. They provide a conceptual framework for future research into the impact of microbial messages on the maintenance of host energy balance. Among the neurological illnesses treated with the gut microbiota are Parkinson's disease, multiple sclerosis, anxiety, schizophrenia, Alzheimer's disease, autism spectrum disorders, epilepsy, and stroke.

References

Appleyard SM, Bailey TW, Doyle MW, Jin YH, Smart JL, Low MJ, Andresen MC (2005) Proopiomelanocortin neurons in nucleus tractus solitarius are activated by visceral afferents: regulation by cholecystokinin and opioids. J Neurosci 25(14):3578–3585

Appleyard SM, Marks D, Kobayashi K, Okano H, Low MJ, Andresen MC (2007) Visceral afferents directly activate catecholamine neurons in the solitary tract nucleus. J Neurosci 27(48): 13292–13302

Avena NM, Rada P, Hoebel BG (2008) Underweight rats have enhanced dopamine release and blunted acetylcholine response in the nucleus accumbens while bingeing on sucrose. Neuroscience 156:865–871

Bäckhed F, Ding H, Wang T, Hooper LV, Koh GY, Nagy A, Semenkovich CF, Gordon JI (2004) The gut microbiota as an environmental factor that regulates fat storage. Proc Natl Acad Sci 101(44):15718–15723

Barson JR, Morganstern I, Leibowitz SF (2011) Similarities in hypothalamic and mesocorticolimbic circuits regulating the overconsumption of food and alcohol. Physiol Behav 104:128–137

Battaglini D, Pimentel-Coelho PM, Robba C, Dos Santos CC, Cruz FF, Pelosi P, Rocco PR (2020) Gut microbiota in acute ischemic stroke: from pathophysiology to therapeutic implications. Front Neurol 598

Batterham RL, Cowley MA, Small CJ, Herzog H, Cohen MA, Dakin CL, Wren AM, Brynes AE, Low MJ, Ghatei MA, Cone RD (2002) Gut hormone PYY3-36 physiologically inhibits food intake. Nature 418(6898):650–654

Berthoud HR (2002) Multiple neural systems controlling food intake and body weight. Neurosci Biobehav Rev 26(4):393–428

Bhargava P, Mowry EM (2014) Gut microbiome and multiple sclerosis. Curr Neurol Neurosci Rep 14(10):1–8

Bogunovic M, Davé SH, Tilstra JS, Chang DTW, Harpaz N, Xiong H et al (2007) Enteroendocrine cells express functional Toll-like receptors. Am J Physiol Gastrointest Liver Physiol 292: G1770–G1783

Breit S, Kupferberg A, Rogler G, Hasler G (2018) Vagus nerve as modulator of the brain–gut axis in psychiatric and inflammatory disorders. Front Psychiatry 44

Breton J, Tennoune N, Lucas N, Francois M, Legrand R, Jacquemot J et al (2016) Gut commensal *E. coli* proteins activate host satiety pathways following nutrient-induced bacterial growth. Cell Metab 23:324–334

Broberger C (2005) Brain regulation of food intake and appetite: molecules and networks. J Intern Med 258:301–327

Broberger C, Johansen J, Johansson C, Schalling M, Hökfelt T (1998) The neuropeptide Y/agouti gene-related protein (AGRP) brain circuitry in normal, anorectic, and monosodium glutamate-treated mice. Proc Natl Acad Sci 95(25):15043–15048

Calvo-Barreiro L, Eixarch H, Montalban X, Espejo C (2018) Combined therapies to treat complex diseases: the role of the gut microbiota in multiple sclerosis. Autoimmun Rev 17(2):165–174

Cersosimo MG, Raina GB, Pecci C, Pellene A, Calandra CR, Gutiérrez C, Micheli FE, Benarroch EE (2013) Gastrointestinal manifestations in Parkinson's disease: prevalence and occurrence before motor symptoms. J Neurol 260(5):1332–1338

Chambers ES, Morrison DJ, Frost G (2015a) Control of appetite and energy intake by SCFA: what are the potential underlying mechanisms? Proceedings of the Nutrition Society. Cambridge University Press 74:328–336

Chambers ES, Viardot A, Psichas A, Morrison DJ, Murphy KG, Zac-Varghese SE, MacDougall K, Preston T, Tedford C, Finlayson GS, Blundell JE (2015b) Effects of targeted delivery of propionate to the human colon on appetite regulation, body weight maintenance and adiposity in overweight adults. Gut 64(11):1744–1754

Chassaing B, Koren O, Goodrich JK, Poole AC, Srinivasan S, Ley RE et al (2015) Dietary emulsifiers impact the mouse gut microbiota promoting colitis and metabolic syndrome. Nature 519:92–96

Chimerel C, Emery E, Summers DK, Keyser U, Gribble FM, Reimann F (2014) Bacterial metabolite indole modulates incretin secretion from intestinal enteroendocrine L cells. Cell Rep 9:1202–1208

Cone RD (1999) The central melanocortin system and energy homeostasis. Trends Endocrinol Metab 10:211–216

Cone RD (2005) Anatomy and regulation of the central melanocortin system. Nat Neurosci 8(5): 571–578

Cowley MA, Smart JL, Rubinstein M, Cerdán MG, Diano S, Horvath TL, Cone RD, Low MJ (2001) Leptin activates anorexigenic POMC neurons through a neural network in the arcuate nucleus. Nature 411(6836):480–484

Cryan JF, O'Riordan KJ, Cowan CS, Sandhu KV, Bastiaanssen TF, Boehme M, Codagnone MG, Cussotto S, Fulling C, Golubeva AV, Guzzetta KE (2019) The microbiota-gut-brain axis. Physiol Rev

David LA, Maurice CF, Carmody RN, Gootenberg DB, Button JE, Wolfe BE, Ling AV, Devlin AS, Varma Y, Fischbach MA, Biddinger SB (2014) Diet rapidly and reproducibly alters the human gut microbiome. Nature 505(7484):559–563

de Clercq NC, Groen AK, Romijn JA, Nieuwdorp M (2016) Gut microbiota in obesity and undernutrition. Adv Nutr 7(6):1080–1089

Degen L, Matzinger D, Drewe J, Beglinger C (2001) The effect of cholecystokinin in controlling appetite and food intake in humans. Peptides 22:1265–1269

DeMaagd G, Philip A (2015) Parkinson's disease and its management: part 1: disease entity, risk factors, pathophysiology, clinical presentation, and diagnosis. Pharmacol Ther 40(8):504

Dorca-Arévalo J, Soler-Jover A, Gibert M, Popoff MR, Martín-Satué M, Blasi J (2008) Binding of ε-toxin from *Clostridium perfringens* in the nervous system. Vet Microbiol 131(1–2):14–25

Drucker DJ (2005) Biologic actions and therapeutic potential of the proglucagon-derived peptides. Nat Clin Pract Endocrinol Metab 1(1):22–31

Dutta SK, Verma S, Jain V, Surapaneni BK, Vinayek R, Phillips L, Nair PP. Parkinson's disease: the emerging role of gut dysbiosis, antibiotics, probiotics, and fecal microbiota transplantation. Neurogastroenterol Motil 2019a;25(3):363

Dutta SK, Verma S, Jain V, Surapaneni BK, Vinayek R, Phillips L, Nair PP (2019b) Parkinson's disease: the emerging role of gut dysbiosis, antibiotics, probiotics, and fecal microbiota transplantation. J Neurogastroenterol Motil 25(3):363

Edition F (2013) Diagnostic and statistical manual of mental disorders. Am Psychiatric Assoc 21: 591–643

El-Ansary A, Bhat RS (2020) Targeting gut microbiota as a possible therapeutic intervention in autism. In: Autism 360. Academic Press, pp 301–327

El-Ansary A, Bhat RS, Zayed N (2020) Gut microbiome and sex bias in autism spectrum disorders. Curr Behav Neurosci Rep 7(1):22–31

Elmquist JK, Maratos-Flier E, Saper CB, Flier JS (1998) Unraveling the central nervous system pathways underlying responses to leptin. Nature 1:445–450

Fan W, Boston BA, Kesterson RA, Hruby VJ, Cone RD (1997) Role of melanocortinergic neurons in feeding and the agouti obesity syndrome. Nature 385(6612):165–168

Feinle-Bisset C (2016) Upper gastrointestinal sensitivity to meal-related signals in adult humans – relevance to appetite regulation and gut symptoms in health, obesity and functional dyspepsia. Physiol Behav 162:69–82

Fetissov SO (2017) Role of the gut microbiota in host appetite control: bacterial growth to animal feeding behaviour. Nat Rev Endocrinol 13(1):11–25

Fetissov SO, Huang P, Zhang Q, Mimura J, Fujii-Kuriyama Y, Rannug A et al (2004) Expression of hypothalamic neuropeptides after acute TCDD treatment and distribution of Ah receptor repressor. Regul Pept 119:113–124

Finnie JW, Blumbergs PC, Manavis J (1999) Neuronal damage produced in rat brains by Clostridium perfringens type D epsilon toxin. J Comp Pathol 120:415–420

Fisher RS, Acevedo C, Arzimanoglou A, Bogacz A, Cross JH, Elger CE, Engel J Jr, Forsgren L, French JA, Glynn M, Hesdorffer DC (2014) ILAE official report: a practical clinical definition of epilepsy. Epilepsia 55(4):475–482

Flint HJ, Scott KP, Louis P, Duncan SH (2012) The role of the gut microbiota in nutrition and health. Nat Rev Gastroenterol Hepatol 9(10):577–589

Franceschi C (2007) Inflammaging as a major characteristic of old people: can it be prevented or cured? Nutr Rev 65(suppl_3):S173–S176

Freter R, Brickner H, Fekete J, Vickerman MM, Carey KE (1983) Survival and implantation of Escherichia coli in the intestinal tract. Infect Immun 39(2):686–703

Friedman JM, Mantzoros CS (2015) 20 years of leptin: from the discovery of the leptin gene to leptin in our therapeutic armamentarium. Metab Clin Exp 64(1):1–4

Gill SR, Pop M, DeBoy RT, Eckburg PB, Turnbaugh PJ, Samuel BS, Gordon JI, Relman DA, Fraser-Liggett CM, Nelson KE (2006) Metagenomic analysis of the human distal gut microbiome. Science 312(5778):1355–1359

Goetze O, Wieczorek J, Mueller T, Przuntek H, Schmidt WE, Woitalla D (2005) Impaired gastric emptying of a solid test meal in patients with Parkinson's disease using 13C-sodium octanoate breath test. Neurosci Lett 375(3):170–173

Hahn TM, Breininger JF, Baskin DG, Schwartz MW (1998) Coexpression of Agrp and NPY in fasting-activated hypothalamic neurons. Nat Neurosci 1(4):271–272

Hardoff R, Sula M, Tamir A, Soil A, Front A, Badarna S, Honigman S, Giladi N (2001) Gastric emptying time and gastric motility in patients with Parkinson's disease. Mov Disord 16(6): 1041–1047

Harris JI, Lerner AB (1957) Amino-acid sequence of the α-melanocyte-stimulating hormone. Nature 179:1346–1347

Hartman ME, Ekkekakis P, Dicks ND, Pettitt RW (2019) Dynamics of pleasure–displeasure at the limit of exercise tolerance: conceptualizing the sense of exertional physical fatigue as an affective response. J Exp Biol 222(3):jeb186585

He X, Ji G, Jia W, Li H (2016) Gut microbiota and nonalcoholic fatty liver disease: insights on mechanism and application of metabolomics. Int J Mol Sci 17(3):300

Higuchi H, Niki T, Shiiya T (2008) Feeding behavior and gene expression of appetite-related neuropeptides in mice lacking for neuropeptide Y Y5 receptor subclass. World J Gastroenterol 14:6312–6317

Hjorth P, Medici CR, Juel A, Madsen NJ, Vandborg K, Munk-Jørgensen P (2017) Improving quality of life and physical health in patients with schizophrenia: a 30-month program carried out in a real-life setting. Int J Soc Psychiatry 63(4):287–296

Hökfelt T, Bartfai T, Bloom F (2003) Neuropeptides: opportunities for drug discovery. Lancet Neurol 2(8):463–472

Hosoi T, Okuma Y, Matsuda T, Nomura Y (2005) Novel pathway for LPS-induced afferent vagus nerve activation: possible role of nodose ganglion. Auton Neurosci 120:104–107

Hu X, Wang T, Jin F (2016) Alzheimer's disease and gut microbiota. Sci China Life Sci 59(10): 1006–1023

Hubbard TD, Murray IA, Bisson WH, Lahoti TS, Gowda K, Amin SG, Patterson AD, Perdew GH (2015) Adaptation of the human aryl hydrocarbon receptor to sense microbiota-derived indoles. Sci Rep 5(1):1–3

Janssen S, Depoortere I (2013) Nutrient sensing in the gut: new roads to therapeutics? Trends Endocrinol Metab 24:92–100

Jepson B, Johnson J (2007) Changing the course of autism: a scientific approach for parents and physicians. Sentient Publications

Jerlhag E, Egecioglu E, Dickson SL, Douhan A, Svensson L, Engel JA (2007) Preclinical study: ghrelin administration into tegmental areas stimulates locomotor activity and increases extracellular concentration of dopamine in the nucleus accumbens. Addict Biol 12(1):6–16

Jhangi S, Gandhi R, Glanz B, Cook S, Nejad P, Ward D, Li N, Gerber G, Bry L, Weiner H. Increased Archaea species and changes with therapy in gut microbiome of multiple sclerosis subjects (S24. 001)

Kong G, Ellul S, Narayana VK, Kanojia K, Ha HT, Li S, Renoir T, Le Cao KA, Hannan AJ (2021) An integrated metagenomics and metabolomics approach implicates the microbiota-gut-brain axis in the pathogenesis of Huntington's disease. Neurobiol Dis 148:105199

Krautkramer KA, Kreznar JH, Romano KA, Vivas EI, Barrett-Wilt GA, Rabaglia ME et al (2016) Diet-microbiota interactions mediate global epigenetic programming in multiple host tissues. Mol Cell 64:982–992

Krishnan S, Alden N, Lee K (2015) Pathways and functions of gut microbiota metabolism impacting host physiology. Curr Opin Biotechnol 36:137–145

Langlet F, Levin BE, Luquet S, Mazzone M, Messina A, Dunn-Meynell AA, Balland E, Lacombe A, Mazur D, Carmeliet P, Bouret SG (2013) Tanycytic VEGF-A boosts blood-hypothalamus barrier plasticity and access of metabolic signals to the arcuate nucleus in response to fasting. Cell Metab 17(4):607–617

Larsen PJ, Tang-Christensen M, Holst JJ, Ørskov C (1997) Distribution of glucagon-like peptide-1 and other preproglucagon-derived peptides in the rat hypothalamus and brainstem. Neuroscience 77:257–270

Lee HH, Molla MN, Cantor CR, Collins JJ (2010) Bacterial charity work leads to population-wide resistance. Nature 467(7311):82–85

Legrand R, Lucas N, Breton J, Déchelotte P, Fetissov SO (2015) Dopamine release in the lateral hypothalamus is stimulated by α-MSH in both the anticipatory and consummatory phases of feeding. Psychoneuroendocrinology 56:79–87

Levy M, Kolodziejczyk AA, Thaiss CA, Elinav E (2017) Dysbiosis and the immune system. Nat Rev Immunol 17(4):219–232

Lloyd-Price J, Abu-Ali G, Huttenhower C (2016) The healthy human microbiome. Genome Med 8(1):1

Loh K, Herzog H, Shi YC (2015) Regulation of energy homeostasis by the NPY system. Trends Endocrinol Metab 26(3):125–135

Lonchamp E, Dupont JL, Wioland L, Courjaret R, Mbebi-Liegeois C, Jover E, Doussau F, Popoff MR, Bossu JL, De Barry J, Poulain B (2010) *Clostridium perfringens* epsilon toxin targets granule cells in the mouse cerebellum and stimulates glutamate release. PLoS One 5(9) e13046

Lutter M, Nestler EJ (2009) Homeostatic and hedonic signals interact in the regulation of food intake. J Nutr 139(3):629–632

Lyte M (2014) Microbial endocrinology and the microbiota-gut-brain axis. Microbial Endocrinol:3–24

Machiels K, Joossens M, Sabino J, De Preter V, Arijs I, Eeckhaut V, Ballet V, Claes K, Van Immerseel F, Verbeke K, Ferrante M (2014) A decrease of the butyrate-producing species *Roseburia hominis* and *Faecalibacterium prausnitzii* defines dysbiosis in patients with ulcerative colitis. Gut 63(8):1275–1283

Marwaha S, Johnson S (2004) Schizophrenia and employment. Soc Psychiatry Psychiatr Epidemiol 39(5):337–349

McMinn JE, Wilkinson CW, Havel PJ, Woods SC, Schwartz MW (2000) Effect of intracerebroventricular α-MSH on food intake, adiposity, c-Fos induction, and neuropeptide expression. Am J Phys Regul Integr Comp Phys 279(2):R695–R703

Meguid MM, Fetissov SO, Varma M, Sato T, Zhang L, Laviano A et al (2000) Hypothalamic dopamine and serotonin in the regulation of food intake. Nutrition 16:843–857

Mete A, Garcia J, Ortega J, Lane M, Scholes S, Uzal FA (2013) Brain lesions associated with *Clostridium perfringens* type D epsilon toxin in a Holstein heifer calf. Vet Pathol 50(5):765–768

Mielcarz DW, Kasper LH (2015) The gut microbiome in multiple sclerosis. Curr Treat Options Neurol 17(4):1

Mimee A, Kuksis M, Ferguson AV (2014) α-MSH exerts direct postsynaptic excitatory effects on NTS neurons and enhances GABAergic signaling in the NTS. Neuroscience 262:70–82

Mowry E, Waubant E, Chehoud C, De Santis T, Kuczynski J, Warrington J. Gut bacterial populations in multiple sclerosis and in health (P05. 106)

Mukherji A, Kobiita A, Ye T, Chambon P (2013) Homeostasis in intestinal epithelium is orchestrated by the circadian clock and microbiota cues transduced by TLRs. Cell 153:812–827

Nakazato M, Murakami N, Kojima M, Matsuo H, Kangawa K, Matsukura S (2001) A role for ghrelin in the central regulation of feeding. Nature 409(6817):194–198

Nemani K, Ghomi RH, McCormick B, Fan X (2015) Schizophrenia and the gut–brain axis. Prog Neuro-Psychopharmacol Biol Psychiatry 56:155–160

Neunlist M, Van Landeghem L, Mahé MM, Derkinderen P, Des Varannes SB, Rolli-Derkinderen M (2013 Feb) The digestive neuronal–glial–epithelial unit: a new actor in gut health and disease. Nat Rev Gastroenterol Hepatol 10(2):90–100

Norgren R, Hajnal A, Mungarndee SS (2006) Gustatory reward and the nucleus accumbens. Physiol Behav 89:531–535

Noyce AJ, Bestwick JP, Silveira-Moriyama L, Hawkes CH, Giovannoni G, Lees AJ, Schrag A (2012) Meta-analysis of early nonmotor features and risk factors for Parkinson disease. Ann Neurol 72(6):893–901

Ollmann MM, Wilson BD, Yang YK, Kerns JA, Chen Y, Gantz I, Barsh GS (1997) Antagonism of central melanocortin receptors in vitro and in vivo by agouti-related protein. Science 278(5335): 135–138

Owen MJ, Sawa A, Mortensen PB (2016) Schizophrenia. Lancet 388(10039):86–97

Palazzo M, Balsari A, Rossini A, Selleri S, Calcaterra C, Gariboldi S, Zanobbio L, Arnaboldi F, Shirai YF, Serrao G, Rumio C (2007) Activation of enteroendocrine cells via TLRs induces hormone, chemokine, and defensin secretion. J Immunol 178(7):4296–4303

Peng L, Bi C, Xia D, Mao L, Qian H (2019) Increased cerebrospinal fluid neurofilament light chain in central nervous system inflammatory demyelinating disease. Mult Scler Relat Disord 30:123–128

Perez-Burgos A, Wang B, Mao YK, Mistry B, Neufeld KA, Bienenstock J, Kunze W (2013) Psychoactive bacteria *Lactobacillus rhamnosus* (JB-1) elicits rapid frequency facilitation in vagal afferents. Am J Physiol Gastrointest Liver Physiol 304(2):G211–G220

Poggioli R, Vergoni AV, Bertolini A (1986) ACTH-(1–24) and α-MSH antagonize feeding behavior stimulated by kappa opiate agonists. Peptides 7:843–848
Prince M, Bryce R, Albanese E, Wimo A, Ribeiro W, Ferri CP (2013) The global prevalence of dementia: a systematic review and metaanalysis. Alzheimers Dement 9(1):63–75
Prins JT, Gazendam-Donofrio SM, Tubben BJ, Van der Heijden FM, Van de Wiel HB, Hoekstra-Weebers JE (2007) Burnout in medical residents: a review. Med Educ 41(8):788–800
Querfurth HW, LaFerla FM (2010) Mechanisms of disease. N Engl J Med 362(4):329–344
Reddy BL, Saier MH (2015) Autism and our intestinal microbiota. Microb Physiol 25(1):51–55
Reigstad CS, Salmonson CE III, Szurszewski JH, Linden DR, Sonnenburg JL, Farrugia G, Kashyap PC (2015) Gut microbes promote colonic serotonin production through an effect of short-chain fatty acids on enterochromaffin cells. FASEB J 29(4):1395–1403
Rice KC, Bayles KW (2008) Molecular control of bacterial death and lysis. Microbiol Mol Biol Rev 72(1):85–109
Richard D (2015) Cognitive and autonomic determinants of energy homeostasis in obesity. Nat Rev Endocrinol 11(8):489–501
Roman ÁC, Carvajal-Gonzalez JM, Merino JM, Mulero-Navarro S, Fernández-Salguero PM (2018) The aryl hydrocarbon receptor in the crossroad of signalling networks with therapeutic value. Pharmacol Ther 185:50–63
Rosenfeld CS (2015) Microbiome disturbances and autism spectrum disorders. Drug Metab Dispos 43(10):1557–1571
Rowan-Nash AD, Korry BJ, Mylonakis E, Belenky P (2019) Cross-domain and viral interactions in the microbiome. Microbiol Mol Biol Rev 83(1):e00044–e00018
Rumah KR, Linden J, Fischetti VA, Vartanian T (2013) Isolation of *Clostridium perfringens* type B in an individual at first clinical presentation of multiple sclerosis provides clues for environmental triggers of the disease. PLoS One 8(10) e76359
Rynn MA, Brawman-Mintzer O (2004) Generalized anxiety disorder: acute and chronic treatment. CNS Spectr 9(10):716–723
Salminen S, von Wright A, Morelli L, Marteau P, Brassart D, de Vos WM, Fondén R, Saxelin M, Collins K, Mogensen G, Birkeland SE (1998) Demonstration of safety of probiotics—a review. Int J Food Microbiol 44(1–2):93–106
Sampson TR, Mazmanian SK (2015) Control of brain development, function, and behavior by the microbiome. Cell Host Microbe 17(5):565–576
Samuel BS, Shaito A, Motoike T, Rey FE, Backhed F, Manchester JK, Hammer RE, Williams SC, Crowley J, Yanagisawa M, Gordon JI (2008) Effects of the gut microbiota on host adiposity are modulated by the short-chain fatty-acid binding G protein-coupled receptor, Gpr41. Proc Natl Acad Sci 105(43):16767–16772
Schwartz MW, Woods SC, Porte D, Seeley RJ, Baskin DG (2000) Central nervous system control of food intake. Nature 404(6778):661–671
Schwarz E, Maukonen J, Hyytiäinen T, Kieseppä T, Orešič M, Sabunciyan S, Mantere O, Saarela M, Yolken R, Suvisaari J (2018) Analysis of microbiota in first episode psychosis identifies preliminary associations with symptom severity and treatment response. Schizophr Res 192:398–403
Scott AP of N and BSA (1992) Sensory neurons: diversity, development, and plasticity. Oxford University Press
Sender R, Fuchs S, Milo R (2016) Are we really vastly outnumbered? Revisiting the ratio of bacterial to host cells in humans. Cell 164(3):337–340
Simon JJ, Skunde M, Walther S, Bendszus M, Herzog W, Friederich HC (2016) Neural signature of food reward processing in bulimic-type eating disorders. Soc Cogn Affect Neurosci 11(9):1393–1401
Sommer F, Rühlemann MC, Bang C, Höppner M, Rehman A, Kaleta C, Schmitt-Kopplin P, Dempfle A, Weidinger S, Ellinghaus E, Krauss-Etschmann S (2017) Microbiomarkers in inflammatory bowel diseases: caveats come with caviar. Gut 66(10):1734–1738

Sonnenburg ED, Smits SA, Tikhonov M, Higginbottom SK, Wingreen NS, Sonnenburg JL (2016) Diet-induced extinctions in the gut microbiota compound over generations. Nature 529(7585): 212–215

Stafstrom CE, Carmant L (2015) Seizures and epilepsy: an overview for neuroscientists. Cold Spring Harb Perspect Med 5(6):a022426

Stecher B, Maier L, Hardt WD (2013) 'Blooming' in the gut: how dysbiosis might contribute to pathogen evolution. Nat Rev Microbiol 11(4):277–284

Suez J, Korem T, Zeevi D, Zilberman-Schapira G, Thaiss CA, Maza O, Israeli D, Zmora N, Gilad S, Weinberger A, Kuperman Y (2014) Artificial sweeteners induce glucose intolerance by altering the gut microbiota. Nature 514(7521):181–186

Sulzer D (2007) Multiple hit hypotheses for dopamine neuron loss in Parkinson's disease. Trends Neurosci 30(5):244–250

Sun L-J, Li J-N, Nie Y-Z (2020) Gut hormones in microbiota-gut-brain cross-talk. J Chin Med Assoc 133:826–833

Tap J, Mondot S, Levenez F, Pelletier E, Caron C, Furet JP, Ugarte E, Muñoz-Tamayo R, Paslier DL, Nalin R, Dore J (2009) Towards the human intestinal microbiota phylogenetic core. Environ Microbiol 11(10):2574–2584

Tatemoto K, Carlquist M, Mutt V (1982) Neuropeptide Y—a novel brain peptide with structural similarities to peptide YY and pancreatic polypeptide. Nature 296:659–660

Tennoune N, Chan P, Breton J, Legrand R, Chabane YN, Akkermann K, Järv A, Ouelaa W, Takagi K, Ghouzali I, Francois M (2014) Bacterial ClpB heat-shock protein, an antigen-mimetic of the anorexigenic peptide α-MSH, at the origin of eating disorders. Transl Psychiatry 4(10) e458-

Thaiss CA, Itav S, Rothschild D, Meijer MT, Levy M, Moresi C et al (2016) Persistent microbiome alterations modulate the rate of post-dieting weight regain. Nature 540:544–551

Tillmann J, Ashwood K, Absoud M, Bölte S, Bonnet-Brilhault F, Buitelaar JK, Calderoni S, Calvo R, Canal-Bedia R, Canitano R, De Bildt A (2018) Evaluating sex and age differences in ADI-R and ADOS scores in a large European multi-site sample of individuals with autism spectrum disorder. J Autism Dev Disord 48(7):2490–2505

Topping DL, Clifton PM (2001) Short-chain fatty acids and human colonic function: roles of resistant starch and nonstarch polysaccharides. Physiol Rev Am J Physiol 81:1031–1064

Tran L, Greenwood-Van Meerveld B (2013) Age-associated remodeling of the intestinal epithelial barrier. J Gerontol A Biol Sci Med Sci 68(9):1045–1056

Tripathi AK, Ray AK, Mishra SK (2022) Molecular and pharmacological aspects of piperine as a potential molecule for disease prevention and management: evidence from clinical trials. Beni-Suef Univ J Basic Appl Sci 11(1):1–24

Turnbaugh PJ, Bäckhed F, Fulton L, Gordon JI (2008) Diet-induced obesity is linked to marked but reversible alterations in the mouse distal gut microbiome. Cell Host Microbe 3:213–223

Ulluwishewa D, Anderson RC, McNabb WC, Moughan PJ, Wells JM, Roy NC (2011) Regulation of tight junction permeability by intestinal bacteria and dietary components. J Nutr 141(5): 769–776

Ulrich-Lai YM, Fulton S, Wilson M, Petrovich G, Rinaman L (2015) Stress exposure, food intake and emotional state. Stress 18(4):381–399

Unger MM, Möller JC, Mankel K, Schmittinger K, Eggert KM, Stamelou M, Stiasny-Kolster K, Bohne K, Bodden M, Mayer G, Oertel WH (2011) Patients with idiopathic rapid-eye-movement sleep behavior disorder show normal gastric motility assessed by the 13C-octanoate breath test. Mov Disord 26(14):2559–2563

Vogt NM, Kerby RL, Dill-McFarland KA, Harding SJ, Merluzzi AP, Johnson SC, Carlsson CM, Asthana S, Zetterberg H, Blennow K, Bendlin BB (2017) Gut microbiome alterations in Alzheimer's disease. Sci Rep 7(1):1

Vuong HE, Yano JM, Fung TC, Hsiao EY (2017) The microbiome and host behavior. Annu Rev Neurosci 40:21–49

Wakefield AJ, Puleston JM, Montgomery SM, Anthony A, O'leary JJ, Murch SH (2002) The concept of entero-colonic encephalopathy, autism and opioid receptor ligands. Aliment Pharmacol Ther 16(4):663–674

Wang JD, Levin PA (2009) Metabolism, cell growth and the bacterial cell cycle. Nat Rev Microbiol 7(11):822–827

Wang D, He X, Zhao Z, Feng Q, Lin R, Sun Y, Ding T, Xu F, Luo M, Zhan C (2015) Whole-brain mapping of the direct inputs and axonal projections of POMC and AgRP neurons. Front Neuroanat 9:40

Waterson MJ, Horvath TL (2015) Neuronal regulation of energy homeostasis: beyond the hypothalamus and feeding. Cell Metab 22(6):962–970

Wexler HM (2007) Bacteroides: the good, the bad, and the nitty-gritty. Clin Microbiol Rev 20(4): 593–621

Williams BL, Hornig M, Parekh T, Lipkin WI (2012) Application of novel PCR-based methods for detection, quantitation, and phylogenetic characterization of *Sutterella* species in intestinal biopsy samples from children with autism and gastrointestinal disturbances. MBio 3(1): e00261–e00211

Wingerchuk DM, Lucchinetti CF, Noseworthy JH (2001) Multiple sclerosis: current pathophysiological concepts. Lab Investig 81(3):263–281

Wolters EC, Braak H (2006) Parkinson's disease: premotor clinico-pathological correlations. Parkinsonism Relat Disord 2006:309–319

World Health Organization (2019) Global status report on alcohol and health 2018. World Health Organization, Geneva

Wren AM, Bloom SR (2007) Gut hormones and appetite control. Gastroenterology 132:2116–2130

Wynne K, Bloom SR (2006) The role of oxyntomodulin and peptide tyrosine–tyrosine (PYY) in appetite control. Nat Clin Pract Endocrinol Metab 2(11):612–620

Xu L, Pu J (2016) Alpha-synuclein in Parkinson's disease: from pathogenetic dysfunction to potential clinical application. Parkinson's disease 2016

Xu R, Wang Q (2016) Towards understanding brain-gut-microbiome connections in Alzheimer's disease. BMC Syst Biol 10(3):277–285

Yano JM, Yu K, Donaldson GP, Shastri GG, Ann P, Ma L et al (2015) Indigenous bacteria from the gut microbiota regulate host serotonin biosynthesis. Cell 161:264–276

Zhang JI, Markovic-Plese S, Lacet B, Raus J, Weiner HL, Hafler DA (1994) Increased frequency of interleukin 2-responsive T cells specific for myelin basic protein and proteolipid protein in peripheral blood and cerebrospinal fluid of patients with multiple sclerosis. J Exp Med 179(3): 973–984

Zhao Y, Lukiw WJ (2018) Microbiome-mediated upregulation of microRNA-146a in sporadic Alzheimer's disease. Front Neurol 9:145

Zhao R, Chen H, Sharp BM (2007) Nicotine-induced norepinephrine release in hypothalamic paraventricular nucleus and amygdala is mediated by N-methyl-D-aspartate receptors and nitric oxide in the nucleus tractus solitarius. J Pharmacol Exp Ther 320(2):837–844

Zhu X, He L, McCluskey LP (2014) Ingestion of bacterial lipopolysaccharide inhibits peripheral taste responses to sucrose in mice. Neuroscience 258:47–61

Zigman JM, Elmquist JK (2003) Minireview: from anorexia to obesity—the yin and yang of body weight control. Endocrinology 144(9):3749–3756

Zinöcker MK, Lindseth IA (2018) The Western diet–microbiome-host interaction and its role in metabolic disease. Nutrients 10(3):365

Chapter 7
Human Diets, Gut Microbiome, and Neuroinflammation

Jyoti Singh, Zoya Khan, and Tripathi Rajavashisth

Abstract A healthy diet provides the fluid, macronutrients, and adequate calories that help to maintain or improve overall health. Our gut harboring a variety of microbiota-a complex network of bacteria, viruses, protozoans, and fungi to sustain a symbiotic association with host body and regulate the immune system, metabolism, and gut physiology. Apart from normal physiology microbiota appears to contribute to behavioral and stress responses. Any modification in composition or absence of any specific species in gut microbiota could lead to the inflammatory response that causes or promotes cardiovascular, renal, neurodegenerative diseases, and many cancers. Alteration in gut microbiota is probably caused by lifestyle, diet, antibiotic treatment, environmental stress, and psychological stress. Growing evidence suggests that diet is a prominent factor in promoting the gut microbiota dysbiosis. In this chapter, we have focused on the critical links among gut microbiota, diet, and inflammation leading to neurodegenerative diseases. We discussed gut microbiome-based studies which include dysbiosis or absence of specific microorganisms induced neurodisorders, introduction of healthy diet-induced modification of gut microbiota to prevent neurological diseases.

Keywords Diet · Gut microbiota · Dysbiosis · Chronic inflammation

Abbreviations

CNS	Central nervous system
CVD	Cardiovascular disease

J. Singh
Academy of Scientific and Innovative Research, CSIR-Indian Institute of Toxicology Research Campus, Lucknow, India

Z. Khan · T. Rajavashisth (✉)
Molecular Biology Unit, Institute of Medical Science, Banaras Hindu University, Varanasi, India

A. K. Tripathi, M. Kotak (eds.), *Gut Microbiome in Neurological Health and Disorders*, Nutritional Neurosciences,
https://doi.org/10.1007/978-981-19-4530-4_7

IBD	Inflammatory bowel disease
LPS	Lipopolysaccharide
MS	Multiple sclerosis
SCFAs	Short-chain fatty acids
TMAO	Trimethylamine-N-oxide

7.1 Introduction

Gut microbiota collectively involves bacteria, viruses, fungi, archaea, and protozoa. Microbial communities reside in the human intestine and stomach count higher than 100 trillion cells (~4 × 10^{13}) preferentially in a symbiotic manner and play an important role in host metabolism (Collins et al. 2012; Sousa et al. 2008). Gut microbiota differs from human to human and appears to be in a complex composition influenced by mode of childbirth, infancy, method of infant feeding, lifestyle, diets, and drugs (Voreades et al. 2014). A number of studies have shown that a healthy human gut possesses a diverse and abundant amount of microbiota (Maukonen and Saarela 2015). Gut microbiota in humans secrete enzymes that contribute to digestion and metabolism releasing diverse metabolites including vitamins, short fatty chain, and amino acids (Koh et al. 2016; Rowland et al. 2018). Alterations in the quantity and quality of microbial communities in human gut often lead to a disturbed homeostasis referred to as dysbiosis that causes or promotes gastroenteric, neurologic, respiratory, metabolic, hepatic, renal, and cardiovascular disorders (Kashtanova et al. 2016).

Diet is an important factor affecting the composition and function of human gut microbiota. In general, the use of dietary fibers, micronutrients like vitamins, minerals, and live beneficial bacteria (probiotics) can be the best way to maintain a balanced intestinal microbiota population. Dietary nutrients could be a possible reason for the distribution of intestinal barrier protective role in such a way that altered host–microbiome interfaces and induced gut microbiota dysbiosis, leading to neuroinflammatory processes and bestowing downstream implications on the host (Maukonen and Saarela 2015).

7.2 Impact of Diet on the Gut Microbiota

Proteins are necessary nutrients for the human body and are important building blocks of tissues that critically contribute to balanced coordination of body function and metabolic reactions. Apart from that, they play vital roles in functioning of human immune system. Dietary proteins also modulate the microbial composition and production of metabolites, with amino acids providing gut microbes with essential carbon and nitrogen (Zhao et al. 2018). Hentges et al. were first reported that proteins affect the gut microbiota by observing that counts of *Bifidobacterium adolescentis* decrease with consuming high beef diet (Hentges et al. 1977) and *Bacteroides* and *Clostridia* counts were increased with a meatless diet. Recent studies have shown that the composition of gut microbiota is associated with the consumption of protein source types, such as animal-based protein (meats, eggs, and cheeses), whey proteins, and plant-based proteins (Singh et al. 2017; Cotillard et al. 2013; Clarke et al. 2014). A large body of evidence has highlighted that intake of pea and whey proteins increased the population of *Bifidobacterium* and *Lactobacillus*, while reducing the numbers of pathogenic *Bacteroides fragilis* and *Clostridium perfringens* (Romond et al. 1998; Meddah et al. 2001; Dominika et al. 2011). Other experimental data showed that mice on high-fat diets with whey-isolated proteins have amplified counts of *Lactobacillaceae* and reduced counts of *Clostridiaceae* compared to mice fed on a regular diet (Guerville and Boudry 2016). Furthermore, dietary protein degradation typically occurs at the lower part of the colon which appears to be optimal for bacteria promoting protein degradation into ammonia, amino acids, short-chain fatty acids (SCFAs), and amine (Yao et al. 2016; Diether and Willing 2019).

Intestinal SCFAs are used as an anti-inflammatory product that contributes to the protection of the mucosal barrier (Voreades et al. 2014). Gut bacteria play a vital role in the regulation of mucosal barrier function (Huang 2013). Human gut microbiome participates in the most energy-efficient process by digesting and degrading peptides to amino acids. Consumption of red meat or dairy products may promote enhanced number of bile-tolerant anaerobic bacteria such as *Alistipes*, *Bacteroides*, and *Bilophila* (Rinninella et al. 2019; David et al. 2014). The alteration of these gut microbiota has been shown to induce proatherogenic compounds like trimethylamine-N-oxide (TMAO) that cause the risk of cardiovascular diseases (Barrea et al. 2019). Moreover, a high intake of animal proteins can increase the count of sulfate-reducing bacteria like *Desulfovibrio* spp. from sulfate amino acids, suggesting the risk of inflammatory bowel diseases (Barrea et al. 2019).

7.3 Dietary Fat and Carbohydrates

Various studies have demonstrated that high-fat diets significantly influence the gut microbiota composition and metabolic activity in the host. Researchers found that a high-fat diet inhibited *Bacteroidetes* growth while increasing the growth of *Firmicutes* and *Proteobacteria* in rodent experiments (Ramos-Romero et al. 2018; Hildebrandt et al. 2009). While a randomized study of human omnivores using either high- or low-fat diets observed slight effects on the gut microbiota composition. Apart from that, a high-fat diet had overall unfavorable effects. High-fat diets cause increased circulating levels of bacteria-derived lipopolysaccharide (LPS) into the bloodstream, that likely results in increased intestinal permeability (Mörkl et al. 2018; Conlon and Bird 2014; Nicholson et al. 2012). LPS, the cell wall component of the Gram-negative bacteria, is a modulator of the immune system and a potent inflammatory agent associated with the development of common metabolic diseases. Increased levels of circulating LPS generate low-grade chronic inflammation (Nicholson et al. 2012). A recent study has concluded that LPS decreased butyrate-producing and anti-inflammatory *Faecalibacterium* and increased *Bacteroides* and *Alistipes*, the former of which was positively correlated with blood lipid markers, including total cholesterol (Wan et al. 2019). The effects of fat on the host microbiota may be passed through bile salts/acids by influencing anaerobic bacteria genera (*Bacteroides*, *Eubacterium*, and *Clostridium*) and involving deconjugation of taurine- and glycine-conjugated bile acids (Tripathi et al. 2022). This deconjugation and 7ab-dehydroxylation of bile salts enhance their hydrophobicity and absorption, which has also been correlated with increased pathological effects on human gut (Conlon and Bird 2014; Nicholson et al. 2012). Further research is required on the relationship among the dietary fats, amount, and type of bile salts, and the quantity, variety, and efficacy of the microbial communities. The phytochemicals are also important to regulate the gut microbiota by modulating the good/bad bacterial diversity (Wan et al. 2019; Tripathi et al. 2022).

Furthermore, gut microbiota promotes carbohydrate digestion and fermentation happened by a process of energy efficient and storage, which illustrate symbiotic relationship between humans' microbiota interface (Murphy et al. 2010). Basically, carbohydrates are classified as digestible and indigestible. Digestible carbohydrates such as sugars, starches, galactose, and fructose can be degraded by intestinal enzymes, and released as glucose in the host circulatory system (Murphy et al. 2010). While, indigestible carbohydrates especially dietary fibbers not easily digested in intestine (Murphy et al. 2010). Dietary fibers involve on digestible oligosaccharides and non-starch polysaccharides are digested by gut microbes. *Ruminococcus*, a species relating to Firmicutes, may act as one of important degraders of nondigestible starch in the human colon (Murphy et al. 2010). Another *Bacteroides* species have ability to degrade diverse plant polysaccharides (Murphy et al. 2010).

Dietary carbohydrates have the potential to modify the gut microbiome. Intake of high levels of dietary carbohydrates can increase the *Bifidobacteria* and reduce *Bacteroides* (Eid et al. 2014). It was also observed that children who feed on plant-based polysaccharides-rich diet had a significantly increased number of *Xylanibacter*, *Firmicutes Prevotella*, and *Bacteroidetes* (De Filippo et al. 2010). Another randomized controlled trial stated that the diversity and function of the host gut microbiota altered immediately when carbohydrates were removed from the diet (David et al. 2014).

Gut microbiota creates an essential capacity for the fermentation of dietary fibers (complex carbohydrates). This fermentation encourages the formation of main metabolites of SCFAs such as *n*-butyrate, propionate, and acetate. SCFAs are mostly produced by *Bacteroidetes* and *Firmicutes*. Studies have shown that several other gut bacteria produced these metabolites by the fermentation of nondigestible carbohydrates (Feng et al. 2018; Levy et al. 2016). SCFAs play a vital role in the maintenance of health and development of diseases by providing a source of energy in the colon (*n*-butyrate) or peripheral tissues (acetate and propionate) as well as by altering the diversity of the microbiota promoting metabolite disorders (Sonnenburg and Bäckhed 2016; Xu et al. 2020). Butyrate has a key role in protecting the tissue barrier, genetic integrity, maintaining immunoregulation, and defending against inflammatory effects (Zmora et al. 2019; Mudgil and Barak 2013). Additionally, SCFAs can directly affect the composition of gut microbiota by serving as a carbon source for intestinal microbiota. High-concentration SCFAs show adverse effects on certain virulence microbe of gut microbiota (Sun and O'Riordan 2013). For instance, at high concentrations and low levels of pH, SCFAs might be inhibiting the *Salmonella* spp. growth, a common foodborne pathogen, by downregulating the expression of invasion genes (El-Gedaily et al. 1997). Apart from that, the types and quantities of SCFAs are primarily determined by the ratio of intestinal microbiota and intake of carbohydrates (El-Gedaily et al. 1997).

7.4 Probiotics and Prebiotics

Probiotics are living microorganisms that provide numerous health benefits. Human probiotic microorganisms belong to various genera such as *Lactobacillus*, *Bifidobacterium*, *Enterococcus*, *Lactococcus*, and *Streptococcus*. Furthermore, *Bacillus* and *Saccharomyces* are widely used in probiotic products (Fijan 2014). Several studies have shown that probiotic strains have several beneficial effects in subjects affected by metabolic syndrome as well as intestinal inflammation (Shaaban et al. 2018). A study based on a dietary trial has shown that probiotics might be a beneficiary agent for the treatment of autistic children (Shaaban et al. 2018). The combination of two strains (*Lactobacilli* and *Bifidobacterium*) even with low bacterial abundance strengthened both behavioral and GI manifestations (Shaaban et al. 2018). Various studies have investigated that probiotics treatment is useful for lactose intolerance, colitis, irritable bowel syndrome, and the prevention of colorectal cancer as well as peptic ulcers (Markowiak and Śliżewska 2017). Furthermore, there is little evidence of the role of probiotics as modulators of human intestinal microbiota. A latest data indicate that even additional amounts of probiotics have limited effects on human intestinal ecology and may even be harmful to the recolonization of microbiota after antibiotic usage (Markowiak and Śliżewska 2017).

Prebiotics are nondigestible food substrates that are actively consumed by intestinal microorganisms conferring a good effect on host's intestinal tract. Carbohydrates sources like seasonal fruits, green vegetables, variety of cereals, and other edible plants are potential prebiotics. Some artificially developed prebiotics include Lactulose, Galactooligosaccharides (GOS), Fructooligosaccharides (FOS), and Lactosaccharose. There are a number of database studies reported on the beneficial effects of prebiotics via stimulating the growth of gut bacteria. The most studied prebiotics are fructans like inulin and different forms of fructo-oligosaccharides that promote the growth of probiotics, such as *Bifidobacteria* or *Lactobacilli* (Markowiak and Śliżewska 2017; Festi et al. 2014). Prebiotics decreased the intestinal inflammation promoted by the enhancing gut barrier integrity and reducing the release of pro-inflammatory cytokines. Likewise, observations have been made by studies conducted on human subjects (Festi et al. 2014). Furthermore, prebiotics leads to the change in gut microbial diversity, stimulate the development of *Bacteroides*, *Bifidobacteria*, *Roseburia*, and *Prevotella* and encourage the relative reduction of *Firmicutes* (Festi et al. 2014; Neyrinck et al. 2011; Parnell and Reimer 2012).

7.5 Micronutrients and Gut Microbiota

Micronutrients such as vitamins and minerals play a significant role in the development of gut microbiota, which, in turn, is an effective mediator of their protective health effects. They regulate metabolism, growth, and differentiation at cellular level, and immune system (Rinninella et al. 2019). Vitamin B family such as riboflavin, thiamine, niacin, pantothenic, biotin, and vitamin K, can also be produced by the fecal or gut microorganisms (Biesalski 2016; Tripathi 2013; Dwivedi et al. 2013; Tripathi et al. 2014). On the other hand, it has been shown that vitamin D can have an effect on the composition of intestinal microbiota. Vitamin D had a positive impact on inflammatory bowel disease (IBD) patients by modulating the gut microbiota and increasing the abundance of potentially beneficial bacterial strains (Sun 2018). In addition, a recent analysis of infant intestinal microbiota found that vitamin D was correlated with increased *Lachnobacterium* but decreased *Lactococcus*. These associations may have potential long-term consequences for immune system regulation and the incidence of asthma/allergic disease (Sordillo et al. 2017). On the other hand, vitamin A had a significant impact on the structure and meta-transcriptome of a gut intestinal bacterial population, such as *Bacteroides vulgatus* (Sordillo et al. 2017).

Moreover, several bacterial physiological processes are processed by metals like vitamins, able to influence the intestinal microbiota. For example, zinc deficiency, a strong risk factor for potentially fatal childhood diarrhea in developing countries, increases the populations of pathogenic bacteria (Sordillo et al. 2017). Iron is also a core component that acts as a cofactor in iron-binding proteins for metabolic pathways, redox reaction, and electron transport chain processes (Frawley and Fang 2014). Human breast milk transmits an iron-binding glycoprotein, i.e., lactoferrin, that protects the undeveloped infant gut from colonization of pathogen (Jaeggi et al. 2015). A study in mice has shown that heme-rich diet decreases microbial diversity, increases the abundance of *Proteobacteria* and reduces the abundance of *Firmicutes* (Constante et al. 2017).

7.6 Gut Microbiota and Neuroinflammatory Diseases

Our immune system is composed of a highly complex network of innate and adaptive immune components showing an immense ability to face and react toward external and internal challenges and maintain the host homeostasis. According to the previously reported study, it has been proved that gut microbiota maintains symbiosis relationship, promote and calibrate all aspect of host immune system (Lobionda et al. 2019). A change in the microbial gut flora is connected to a range of neurological disorders by involving inflammatory proteins. Neuroinflammation involves inflammation found in the nervous system, which is now considered a defining feature in many neurological disorders. Multiple sclerosis (CNS autoimmune disease) is characterized by a prominent infiltration of immune cells from the periphery into the CNS, resulting in demyelination and neuronal damage. It has been observed that the level of inflammatory proteins increases with the infiltration of immune cells in neuro-disorder diseases such as Parkinson's and Alzheimer's, brain injury, and anxiety. A plethora of evidence implicates the close relationship between the neural system and gut microbiota, involving neurological networks, endocrine systems, immunological communication, etc. Gut microbiota metabolites reach the CNS via blood transportation and modulate the CNS and immune cells, which lead to the development of neurological disorders, which we will discuss in the section below.

7.7 Alzheimer's Disease

Amyloid beta and tau proteins are found in extracellular plaques and intracellular neurofibrillary tangles in Alzheimer's disease, respectively, along with synaptic and neuronal loss. Memory loss, personality, and behavioral disorders, and aphasia are phenotypical features of Alzheimer's disease. Inflammation outflow between neurons, tau protein, amyloid beta proteins, and microglia all play important roles in Alzheimer's disease pathology. There is limited research evidence that investigates the microbiota composition in AD patients, and there is little knowledge about any bacterial species that alleviate the pathology of AD. The murine model confirmed that the microbial community of non-AD and AD patients is not quite the same, which is an important step in finding out the correlation between AD and gut microbiota. The microorganism-free AD mouse model showed a lesser magnitude of AD pathology than the conventional mouse model, which showed lower levels of Aβ protein, less activated microglial cells, and so on (Harach et al. 2017). According to a diet pattern study based on patients' diets and risk of developing AD, it was discovered that a healthy diet rich in fruits, vegetables, whole grains, fish, and low-fat dairy was associated with a lower risk of developing AD than an unhealthy diet (Berti et al. 2015). According to an experimental report, it has been found that diet and a few important nutrients can alter the work of the gut microbiota, influencing the synthesis and/or deposition of amyloid proteins. Similarly, pomegranate active alkaloid urolithins have been shown to have neuroprotective effects as well as the ability to prevent Aβ fibrillation, which has been investigated in the in vitro study (Yuan et al. 2016).

Parkinson's disease (PD) is the most common neurodegenerative disease. Growing research studies prove that α-synucleinopathy starts in the enteric nervous system during the early stages of disease, which has a connection with digestive system disorders later in the CNS system (Cussotto et al. 2018). Several research studies (both in vitro and in vivo) have found a link between the onset of Parkinson's disease and the gut microbiota (Sampson et al. 2020). In addition, other studies show the relationship between gut dysbiosis and the severity of PD disease, such as gait abnormality and balancing issues, associated with a change in the composition and population of *Enterobacteriaceae* (Cussotto et al. 2018). For example, an ASO (Alpha-synuclein overexpressing) mouse model with complex gut bacteria showed impaired motor activity, loss of motor control, and striatal function dysfunction. As per a recent study, curli, an amyloid protein secreted by gut microbial *E. coli*, has the ability to trigger the aggregation of α-synuclein I in both the gut and the brain, resulting in GI dysfunction and motor loss, respectively. However, it was found that supplementation with a probiotic cocktail containing *Bifidobacterium animalis* L., *rhamnosus* GG, *lactis*, and *Lactobacillus acidophilus* induced the secretion of butyrate, which consequently reduced nigral dopaminergic neurons' degeneration from MPTP and rotenone-induced mouse mode (Cussotto et al. 2018). Dietary changes are the most widely used method for preventing the onset of Parkinson's disease and restoring healthy gut microbiota. A Mediterranean-style diet high in fruits, green vegetables, and grains was found to have a lower risk of PD pathogenesis than a diet high in oil, butter, sweets, and so on (Alcalay et al. 2012).

7.8 Autism Spectrum Disorder

Autism spectrum disorder (ASD) is another neurologic disorder that manifests abnormal social behavior as well as anxiety and cognitive difficulties. There is mounting evidence that there is a link between the severity of ASD and the resident gut microbiota. This hypothesis is primarily based on clinical studies showing that children with ASD have gastrointestinal problems/symptoms such as constipation and diarrhea (Liu et al. 2019). Further surprising results were observed in ASD children treated with antibiotics, who had lower ASD severity and improved behavioral symptoms, implying that gut bacteria may play a role in ASD pathology. Prevotella, Desulfovibrio, and Bifidobacterium species were more common in ASD patients than in normal patients. Changes in dietary patterns and probiotics can improve the disturbance and modification of gut flora composition in ASD patients (Lee et al. 2018).

7.9 Multiple Sclerosis

It is an inflammatory disease characterized by the immune-mediated demyelination of the neural axon. Furthermore, scientific evidence suggests that changes in specific microbial populations play a role in the onset of Multiple Sclerosis (MS). In support of this, it has been found that MS patients showed an increase in Methanobrevibacter and Akkermansia and a decrease in Butyricimonas. Additionally, probiotic VSL3 treated MS patients show enrichment of specific microbial species in the intestine and inhibit peripheral inflammation mediated by monocytes. A study on relapsing-remitting RR-MS patients revealed that they have a different microbiota population of Haemophilus, Anaerostipes, Coprobacillus, Bacteroides, Prevotella, and Lactobacillus than healthy controls, who have a higher proportion of Acinetobacter Pedobacteria, Akkermansia Dorea, and Blautia. Furthermore, the microbial population recovery in RR MS patients seems to decrease the inflammation process and activate the immune system (Schepici et al. 2019). A recent study found a link between a few specific bacteria populations (Methanobrevibacter (Euryarchaeota phylum) and Akkermansia (Verrucomicrobia phylum) and gene expression of inflammatory proteins like interferon signaling, progenitor cells, NF-κB, and T cells (Jangi et al. 2016). Dietary habits can influence microbial composition. It has been discovered that the level of vitamin D in the gut microbes of MS patients is altered, and that it can be restored by increasing vitamin D consumption. Vitamin D is an important nutrient for maintaining healthy gut microbiota (Chu et al. 2018).

7.10 Conclusion

Every human being has a unique gut microbe that matures throughout early childhood, remains stable, but is prone to change in adulthood. Microbe variations in each person are heavily influenced by genetic and environmental factors such as exposure to hormones, diet, geographical location, and seasonal variations. Dysbiosis in gut microbial communities represents an imbalance in gut microbial diversity and function, promotes distortion of the intestine barrier and an inflammatory response that precipitates a variety of neurological disorders. Fermented foods, high fiber foods, and foods containing oligo- and polysaccharides could all help to improve human gut dysbiosis and helps in reducing the severity of the neuro diseases. Possibilities for improving gut dysbiosis through microbiota-based approaches could be an intriguing way of preventing diseases and maintaining human mental health.

References

Alcalay RN, Gu Y, Mejia-Santana H, Cote L, Marder KS, Scarmeas N (2012) The association between Mediterranean diet adherence and Parkinson's disease. Mov Disord 27(6):771–774

Barrea L, Annunziata G, Muscogiuri G, Laudisio D, Di Somma C, Maisto M, Tenore GC, Colao A, Savastano S (2019) Trimethylamine N-oxide, Mediterranean diet, and nutrition in healthy, normal-weight adults: also a matter of sex? Nutrition 62:7–17

Berti V, Murray J, Davies M, Spector N, Tsui WH, Li Y, Williams S, Pirraglia E, Vallabhajosula S, McHugh P, Pupi A, de Leon MJ, Mosconi L (2015) Nutrient patterns and brain biomarkers of Alzheimer's disease in cognitively normal individuals. J Nutr Health Aging 19(4):413–423

Biesalski HK (2016) Nutrition meets the microbiome: micronutrients and the microbiota. Ann N Y Acad Sci 1372:53–64

Chu F, Shi M, Lang Y, Shen D, Jin T, Zhu J, Cui L (2018) Gut microbiota in multiple sclerosis and experimental autoimmune encephalomyelitis: current applications and future perspectives. Mediat Inflamm 2018:1–17

Clarke SF, Murphy EF, O'Sullivan O et al (2014) Exercise and associated dietary extremes impact on gut microbial diversity. Gut 63:1913–1920

Collins SM, Surette M, Bercik P (2012) The interplay between the intestinal microbiota and the brain. Nat Rev Microbiol 10:735–742

Conlon M, Bird A (2014) The impact of diet and lifestyle on gut microbiota and human health. Nutrients 7:17–44

Constante M, Fragoso G, Lupien-Meilleur J, Calvé A, Santos MM (2017) Iron supplements modulate colon microbiota composition and potentiate the protective effects of probiotics in dextran sodium sulfate-induced colitis. Inflamm Bowel Dis 23:753–766

Cotillard A, Kennedy SP, Kong LC et al (2013) Dietary intervention impact on gut microbial gene richness. Nature 500:585–588

Cussotto S, Sandhu KV, Dinan TG, Cryan JF (2018) The neuroendocrinology of the microbiota-gut-brain axis: a behavioural perspective. Front Neuroendocrinol 51:80–101

David LA, Maurice CF, Carmody RN et al (2014) Diet rapidly and reproducibly alters the human gut microbiome. Nature 505:559–563

De Filippo C, Cavalieri D, Di Paola M, Ramazzotti M, Poullet JB, Massart S, Collini S, Pieraccini G, Lionetti P (2010) Impact of diet in shaping gut microbiota revealed by a comparative study in children from Europe and rural Africa. Proc Natl Acad Sci 107:14691–14696

Diether N, Willing B (2019) Microbial fermentation of dietary protein: an important factor in diet–microbe–host interaction. Microorganisms 7:19

Dominika Ś, Arjan N, Karyn RP, Henryk K (2011) The study on the impact of glycated pea proteins on human intestinal bacteria. Int J Food Microbiol 145:267–272

Dwivedi A, Pal MK, Tripathi AK, Yadav N, Mujtaba SF, Pant MC, Singh SK, Mishra DP, Ray RS, Manjunatha Prabhu BH (2013) Role of type-II pathway in apoptotic cell death induction by photosensitized CDRI-97/78 under ambient exposure of UV-B. Toxicol Lett. https://doi.org/10.1016/j.toxlet.2013.06.210

Eid N, Enani S, Walton G, Corona G, Costabile A, Gibson G, Rowland I, Spencer JPE (2014) The impact of date palm fruits and their component polyphenols, on gut microbial ecology, bacterial metabolites and colon cancer cell proliferation. J Nutr Sci 3:e46

El-Gedaily A, Paesold G, Chen CY, Guiney DG, Krause M (1997) Plasmid virulence gene expression induced by short-chain fatty acids in Salmonella dublin: identification of rpoS-dependent and rpo-S-independent mechanisms. J Bacteriol 179:1409–1412

Feng W, Ao H, Peng C (2018) Gut microbiota, short-chain fatty acids, and herbal medicines. Front Pharmacol. https://doi.org/10.3389/fphar.2018.01354

Festi D, Schiumerini R, Eusebi LH, Marasco G, Taddia M, Colecchia A (2014) Gut microbiota and metabolic syndrome. World J Gastroenterol 20:16079

Fijan S (2014) Microorganisms with Claimed probiotic properties: an overview of recent literature. Int J Environ Res Public Health 11:4745–4767

Frawley ER, Fang FC (2014) The ins and outs of bacterial iron metabolism. Mol Microbiol 93(4): 609–616

Guerville M, Boudry G (2016) Gastrointestinal and hepatic mechanisms limiting entry and dissemination of lipopolysaccharide into the systemic circulation. Am J Physiol Liver Physiol 311: G1–G15

Harach T, Marungruang N, Duthilleul N, Cheatham V et al (2017) Reduction of Abeta amyloid pathology in APPPS1 transgenic mice in the absence of gut microbiota. Sci Rep 7:41802

Hentges DJ, Maier BR, Burton GC, MAF and RKT (1977) Effect of a high-beef diet on the fecal bacterial flora of humans. Cancer Res 37:568–571

Hildebrandt MA, Hoffmann C, Sherrill-Mix SA, Keilbaugh SA, Hamady M, Chen Y, Knight R, Ahima RS, Bushman F, Wu GD (2009) High-fat diet determines the composition of the murine gut microbiome independently of obesity. Gastroenterology 137:1716–1724. e2

Huang X-Z (2013) Bacterial colonization and intestinal mucosal barrier development. World J Clin Pediatr 2:46

Jaeggi T, Kortman GAM, Moretti D et al (2015) Iron fortification adversely affects the gut microbiome, increases pathogen abundance and induces intestinal inflammation in Kenyan infants. Gut 64:731–742

Jangi S, Gandhi R, Cox L et al (2016) Alterations of the human gut microbiome in multiple sclerosis. Nat Commun 7:12015

Kashtanova DA, Popenko AS, Tkacheva ON, Tyakht AB, Alexeev DG, Boytsov SA (2016) Association between the gut microbiota and diet: fetal life, early childhood, and further life. Nutrition 32:620–627

Koh A, De Vadder F, Kovatcheva-Datchary P, Bäckhed F (2016) From dietary fiber to host physiology: short-chain fatty acids as key bacterial metabolites. Cell 165:1332–1345

Lee RWY, Corley MJ, Pang A et al (2018) A modified ketogenic gluten-free diet with MCT improves behavior in children with autism spectrum disorder. Physiol Behav 188:205–211

Levy M, Thaiss CA, Elinav E (2016) Metabolites: messengers between the microbiota and the immune system. Genes Dev 30:1589–1597

Liu S, Li E, Sun Z et al (2019) Altered gut microbiota and short chain fatty acids in Chinese children with autism spectrum disorder. Sci Rep 9(1):287

Lobionda S, Sittipo P, Kwon HY, Lee YK (2019) The role of gut microbiota in intestinal inflammation with respect to diet and extrinsic stressors. Microorganisms 7:271

Markowiak P, Śliżewska K (2017) Effects of probiotics, prebiotics, and synbiotics on human health. Nutrients 9(9):1021. https://doi.org/10.3390/nu9091021. PMCID: PMC5622781

Maukonen J, Saarela M (2015) Human gut microbiota: does diet matter? Proc Nutr Soc 74:23–36

Meddah ATT, Yazourh A, Desmet I, Risbourg B, Verstraete WL, Romond MB (2001) The regulatory effects of whey retentate from Bifidobacteria fermented milk on the microbiota of the simulator of the human intestinal microbial ecosystem. J Appl Microbiol 91:1110–1117

Mörkl S, Lackner S, Meinitzer A et al (2018) Gut microbiota, dietary intakes and intestinal permeability reflected by serum zonulin in women. Eur J Nutr 57:2985–2997

Mudgil D, Barak S (2013) Composition, properties and health benefits of indigestible carbohydrate polymers as dietary fiber: a review. Int J Biol Macromol 61:1–6

Murphy EF, Cotter PD, Healy S et al (2010) Composition and energy harvesting capacity of the gut microbiota: relationship to diet, obesity and time in mouse models. Gut 59:1635–1642

Neyrinck AM, Possemiers S, Druart C, Van de Wiele T, De Backer F, Cani PD, Larondelle Y, Delzenne NM (2011) Prebiotic effects of wheat arabinoxylan related to the increase in bifidobacteria, roseburia and bacteroides/prevotella in diet-induced obese mice. PLoS One 6: e20944

Nicholson JK, Holmes E, Kinross J, Burcelin R, Gibson G, Jia W, Pettersson S (2012) Host-gut microbiota metabolic interactions. Science 80–336:1262–1267

Parnell JA, Reimer RA (2012) Prebiotic fibres dose-dependently increase satiety hormones and alter Bacteroidetes and Firmicutes in lean and obese JCR:LA-cp rats. Br J Nutr 107:601–613

Ramos-Romero S, Hereu M, Atienza L et al (2018) Mechanistically different effects of fat and sugar on insulin resistance, hypertension, and gut microbiota in rats. Am J Physiol Metab 314:E552–E563

Rinninella E, Cintoni M, Raoul P, Lopetuso LR, Scaldaferri F, Pulcini G, GAD M, Gasbarrini A, Mele MC (2019) Food components and dietary habits: keys for a healthy gutmicrobiota composition. Nutrients 11:2393

Romond M-B, Ais A, Guillemot F, Bounouader R, Cortot A, Romond C (1998) Cell-free whey from milk fermented with bifidobacterium breve C50 used to modify the colonic microflora of healthy subjects. J Dairy Sci 81:1229–1235

Rowland I, Gibson G, Heinken A, Scott K, Swann J, Thiele I, Tuohy K (2018) Gut microbiota functions: metabolism of nutrients and other food components. Eur J Nutr 57:1–24

Sampson TR, Challis C, Jain N, Moiseyenko A, Ladinsky MS, Shastri GG, Thron T et al (2020) A gut bacterial amyloid promotes A-synuclein aggregation and motor impairment in mice. elife 9. https://doi.org/10.7554/elife.53111

Schepici G, Silvestro S, Bramanti P, Mazzon E (2019) The gut microbiota in multiple sclerosis: an overview of clinical trials. Cell Transplant 28(12):1507–1527

Shaaban SY, El Gendy YG, Mehanna NS, El-Senousy WM, El-Feki HSA, Saad K, El-Asheer OM (2018) The role of probiotics in children with autism spectrum disorder: a prospective, open-label study. Nutr Neurosci 21:676–681

Singh RK, Chang H-W, Yan D et al (2017) Influence of diet on the gut microbiome and implications for human health. J Transl Med 15:73

Sonnenburg JL, Bäckhed F (2016) Diet–microbiota interactions as moderators of human metabolism. Nature 535:56–64

Sordillo JE, Zhou Y, McGeachie MJ et al (2017) Factors influencing the infant gut microbiome at age 3–6 months: findings from the ethnically diverse vitamin D antenatal asthma reduction trial (VDAART). J Allergy Clin Immunol 139:482–491. e14

Sousa T, Paterson R, Moore V, Carlsson A, Abrahamsson B, Basit AW (2008) The gastrointestinal microbiota as a site for the biotransformation of drugs. Int J Pharm 363:1–25

Sun J (2018) Dietary vitamin D, vitamin D receptor, and microbiome. Curr Opin Clin Nutr Metab Care 21(6):471–474. https://doi.org/10.1097/MCO.0000000000000516

Sun Y, O'Riordan MX (2013) Regulation of bacterial pathogenesis by intestinal short-chain fatty acids. Adv Appl Microbiol 85:93–118. https://doi.org/10.1016/B978-0-12-407672-3.00003-4. PMCID: PMC4029053

Tripathi AK (2013) Attenuated antioxidant property of UV-B irradiated piperine in ischemia-reperfusion injury in SD rat. G-JEST 1:29–36

Tripathi AK, Dwivedi A, Pal MK et al (2014) Attenuated neuroprotective effect of riboflavin under UV-B irradiation via miR-203/c-Jun signaling pathway in vivo and in vitro. J Biomed Sci. https://doi.org/10.1186/1423-0127-21-39

Tripathi AK, Ray AK, Mishra SK (2022) Molecular and pharmacological aspects of piperine as a potential molecule for disease prevention and management: evidence from clinical trials. Beni Suef Univ J Basic Appl Sci 11(1):1–24

Voreades N, Kozil A, Weir TL (2014) Diet and the development of the human intestinal microbiome. Front Microbiol. https://doi.org/10.3389/fmicb.2014.00494

Wan Y, Wang F, Yuan J et al (2019) Effects of dietary fat on gut microbiota and faecal metabolites, and their relationship with cardiometabolic risk factors: a 6-month randomised controlled-feeding trial. Gut 68:1417–1429

Xu H, Wang X, Feng W, Liu Q, Zhou S, Liu Q, Cai L (2020) The gut microbiota and its interactions with cardiovascular disease. Microb Biotechnol 13:637–656

Yao CK, Muir JG, Gibson PR (2016) Review article: insights into colonic protein fermentation, its modulation and potential health implications. Aliment Pharmacol Ther 43:181–196
Yuan T, Ma H, Liu W, Niesen DB, Shah N, Crews R, Rose KN, Vattem DA, Seeram NP (2016) Pomegranate's neuroprotective effects against Alzheimer's disease are mediated by urolithins, its ellagitannin-gut microbial derived metabolites. ACS Chem Neurosci 7(1):26–33
Zhao J, Zhang X, Liu H, Brown MA, Qiao S (2018) Dietary protein and gut microbiota composition and function. Curr Protein Pept Sci 20:145–154
Zmora N, Suez J, Elinav E (2019) You are what you eat: diet, health and the gut microbiota. Nat Rev Gastroenterol Hepatol 16:35–56

Chapter 8
Dietary Fatty Acids, Gut Microbiome, and Gut–Brain Communication: A Current Perspective

Santosh Anand, S. K. Sukrutha, B. R. Shilpa, and A. Nagarathna

Abstract Globally, the significance of gut microbial composition in alleviating several intestinal complications has gained tremendous clinical consideration. Increasing lines support the evidence that, healthy lipids such as omega fatty acids and short-chain fatty acids play a pivotal role in orchestrating the composition and shaping the host microbial population in the gastrointestinal system by producing protective derivatives, restoring intestinal gut barrier, modulating gut microbial community, nervous, and immune system, attenuating inflammation, insulin resistance, and fat accumulation. In the above context, the present chapter uncovers the recent advancements carried out in the recent past on the impact of long- and short-chain fatty acids on gut microbiome–brain communication.

Keywords Gut microbiome · Fatty acids · Gut dysbiosis · Derivatives · Clinical implications

Abbreviations

ALA	Alpha-linolenic acid
DHA	Docosahexaenoic acid
EPA	Eicosapentaenoic acid
FFA	Free fatty acid
GIS	Gastrointestinal system

S. Anand (✉) · B. R. Shilpa
Department of Biotechnology, School of Applied Sciences, REVA University, Bengaluru, India
e-mail: santosh.anand@reva.edu.in

S. K. Sukrutha
Department of Microbiology, Bangalore University, Bengaluru, India

A. Nagarathna
Department of Biotechnology, Ramaiah College of Arts, Science and Commerce, Bengaluru, India

A. K. Tripathi, M. Kotak (eds.), *Gut Microbiome in Neurological Health and Disorders*, Nutritional Neurosciences,
https://doi.org/10.1007/978-981-19-4530-4_8

GM	Gut microbiota
HDACs	Histone deacetylases
HFD	High-fat diet
HYA	10-Hydroxy-*cis*-12-octadecenoic acid
LCFAs	Long-chain fatty acids
LPS	Lipopolysaccharide
MCFA	Medium-chain fatty acid
MUFA	Monounsaturated fatty acid
OFAs	Omega fatty acids
PUFAs	Polyunsaturated fatty acids
SFA	Saturated fatty acid
TLR	Toll-like receptor

8.1 Introduction

A plethora of microorganisms such as bacteria and eukarya inhabiting the gastrointestinal tract are collectively designated "Gut microbiome" (GM) and has been evolved and adapted within the human body for many years (Milisavljevic et al. 2013). Human GM encompasses about 100 trillion microorganisms, with 35,000 bacterial species. Oral cavity, genital organs, respiratory tract, skin, and gastrointestinal system (GIS) harbor a wide array of microorganisms.

Until recent decades, knowledge about the microbiota composition, identification, characterization, distribution, and host–microbiota interactions was understudied due to limited technologies. Of late, due to the advancement in high throughput robust sequencing technologies and Omics approaches, identification, and profiling of microorganisms through culture-independent techniques accompanied by large-scale funded projects such as the Human Microbiome Project consortium and Metagenomics of the Human Intestinal Tract consortium have paved a breakthrough in deciphering the commensal microbial gut community and its interaction with the host (Human Microbiome Project Consortium 2012).

The association between microbes and healthy aging was first documented by Elie Metchnikoff (Bischoff 2016). With advancement in age, the GM is modulated with regard to its constitution and functionality (Tiihonen et al. 2010). A large body of evidence supports the significance of a favorable GM behind healthy aging, and any physiological function alterations upon aging has the ability to influence the composition and activities of microbial species inhabiting GM (Kumar et al. 2016). Maintaining a healthy living style and intake of nutritional diet has been considered beneficial for healthy aging (Choi and Lee 2016). Diet plays a chief role in articulating the chemostatic niche of the gut. Recently, various processes undertaken during food processing and preservation have manifested microbiome alteration imparting harmful health consequences, particularly in elderly population (Cotillard et al. 2013; Keenan et al. 2015). Alterations in physiological parameters like appetite reduction, masticatory abnormalities, and constipation have been associated with advancing age (Candela et al. 2014).

Intake of a healthy diet is the major imperative factor influencing host nutrition and metabolism. A well-balanced diet is a prerequisite to maintaining healthy aging. Diet enriched with high fat alters the homeostasis, metabolism, and gut commensal microflora. Evidences suggest that, high-fat diet (HFD) may lead to the development of a wide array of health impediments such as obesity (Black et al. 2013), inflammatory diseases such as colitis (Ananthakrishnan et al. 2014), gut dysbiosis (Agus et al. 2016), and intestinal gut barrier impairment (Rohr et al. 2020). The present chapter highlights the impact of long- and short-chain fatty acids on GM in ameliorating various gastrointestinal complications/gut dysbiosis.

8.2 Role of Long- and Short-Chain Fatty Acids

Long-chain fatty acids (LCFAs), omega fatty acids (OFAs), also known as polyunsaturated fatty acids (PUFAs), and essential fatty acids through fish oil have been documented to have a wide spectrum of health benefits such as in cardiovascular disease (Mozaffarian and Wu 2011; Schunck et al. 2018), neurological disorders (Stavrinou et al. 2020), pregnancy (Shrestha et al. 2020; Carlson et al. 2013), infant growth (Derbyshire 2018; Lauritzen et al. 2016), aging (Stavrinou et al. 2020), and inflammatory diseases (Simopoulos 2002; Innes and Calder 2018; Calder 2017). In the recent past, evidences support that, PUFAs play a pivotal role in maintaining the harmony of GM and gut homeostasis. Diet enriched with PUFAs promotes the growth and maintenance of a healthy bacterial community in the intestine and influences the GM to synthesize biologically important metabolites, which helps in amelioration of the complications in both in vitro and in vivo conditions. The above evidences suggest that, GM with supplementation of PUFA enriched diet embodies a new target for personalized therapeutic management for attenuating various age-linked gastrointestinal ailments. Furthermore, it may also serve as an ancillary platform for potential clinical implications in the development of targeted treatments associated with GIS.

Unlike LCFAs, short-chain fatty acids (SCFAs) specifically propionate, acetate, and butyrate, are synthesized by the gut microbial community which participates in multifaceted functions like maintaining low pH environment, augmenting nutrient absorption, nurturing the development of advantageous bacterial ecosystem, migration of mucosal cells, managing gut barrier integrity, thereby aiding in gut homeostasis with advancing age (Nagpal et al. 2018). Influences of SCFAs are confined not only to the intestines but also to distant tissues (Milligan et al. 2017). They are also gifted with anti-inflammatory and antineoplastic functions. SCFAs can maintain the epithelial barrier integrity by regulating the levels of tight junction proteins, the reduction of which would assist bacterial translocation and lipopolysaccharides to stimulate inflammation (Wang et al. 2012; Feng et al. 2018). In addition, SCFAs inhibit histone deacetylases (HDACs) that are associated with the regulation of genes responsible for promoting pathogenesis in various ailments. HDACs upregulation leads to decline in histone acetylation and finally results in gene silencing (Ratajczak et al. 2019). Higher levels of SCFAs, on the other hand, have been reported to display toxic effects on certain beneficial species of the GM. Among the SCFAs, butyrate is the chief energy source for the colonocytes and thereby responsible for maintaining colon health (van der Beek et al. 2017). Endowed with this vast range of activities, SCFAs combat against intestinal ailments including colorectal cancer, Crohn's disease, and ulcerative colitis (Shao et al. 2019; Neurath 2020).

8.3 Alterations in the Gut Ecosystem

Commensal microorganisms are believed to colonize in humans from birth and their composition is influenced by course of birth. Infants delivered through vaginal route harbors a large population of *Lactobacilli* (Avershina et al. 2014; Aagaard et al. 2012). Conversely, birth by C-section results in fewer and late colonization of *Bacteroides* genus and *Clostridium* species, respectively (Jakobsson et al. 2014). Within a year, the microbial community gets diversified and the compositions are unique to each infant (Palmer et al. 2007). The microbial population pattern resembles the adult microbiota by 2.5 years of age (Rodriguez et al. 2015; Koenig et al. 2011). As age advances, shaping of microbial population in general and gut in particular is influenced by various factors such as gender, age, diet, stress, antibiotic therapy, and environmental conditions (Donaldson et al. 2016).

Microflora in the small intestine are predominantly anaerobic in nature and small intestine consists of 10^7–10^8 bacterial cells, whereas, the large intestine constitutes 10^{10}–10^{11} cells per mL (Lloyd-Price et al. 2016; Sekirov et al. 2010; Lecocq et al. 2013). It comprises three predominant phyla, namely Bacteroidetes (*Prevotella* and *Porpyromonas*) (9–42%), Firmicutes (*Eubacteria*, *Clostridium*, and *Ruminococcus*) (30–52%), and Actinobacteria (*Bifidobacterium*) (1–13%) and in which 2% is composed of phylum Lactobacillae (*Enterobacteria* and *Streptococci*) (Bourlioux et al. 2003; Rajoka et al. 2017). In older adults, a significant increment in Bacteroidetes phyla and *Clostridium* cluster IV is more prevalent (Vemuri et al. 2019). Owing to the enormous gut microbial diversity, these commensals deliver protection against pathogenic microorganisms, maintain mucosal barrier integrity, and supply nutrients such as vitamins during the course of aging. In addition, the interface between commensal microbiota and the mucosal immune system is a prerequisite for smooth working of the immune system (Claesson et al. 2011).

8.4 Impact of Fatty Acids on Gut Microbiome

8.4.1 Effect on Immune System

Intestinal immunity is influenced by dietary intake and gut microbial community. Healthy lipids produce immunoprotective metabolites and enzymes which support the GM and the host to regulate various ailments such as allergy, infection, and inflammation. Intake of a diet enriched with n-3 fatty acid in BALB/c mice (5 months old) modulated GM by resulting in a remarkable increment in the microbial composition of *Blautia*, *Oscillibacter*, *Clostridales*, *Robinsoniella*, *Lactococcus*, and *Eubacterium* and significant alleviation in the number of Lachnospiraceae, Anaerotruncus, and Roseburia members. In colon and spleen tissue, altered microbial composition augmented the production of IL-10, an anti-inflammatory cytokine and protected against peanut oral allergy with minor alterations in the serological parameters (Myles et al. 2014). Excessive multiplication of invariant natural killer T cells is inhibited in the colon by means of intestinal *Bacteroides fragilis*-derived from glycosphingolipids and confers protection from cell-mediated, oxazolone-induced colitis in neonatal mice (An et al. 2014). 17,18-epoxyeicostetraenoic acid (17,18-EpETE), an anti-allergic metabolite, synthesized from alpha-linolenic acid (ALA) metabolism mitigated allergic diarrhea in the mice gut supplemented with a ALA-enriched diet (Kunisawa et al. 2015).

In 2018, Ohue-Kitano et al. (2018) for the first time reported that, G protein-coupled receptor 40 and PPAR-g play a pivotal role in anti-inflammatory M2 macrophage differentiation by the influence of gut lactic acid bacteria-derived ALA metabolites such as 13-oxo-9(Z),15(Z)-octadecadienoic acid (13-oxo) and 13-hydroxy9(Z),15(Z)-octadecadienoic acid (13-OH) in male C57BL/6 mice. In another study, intestinal gut microbe-derived butyrate, remarkably increased anti-inflammatory M2 macrophage, M2 macrophage-associated protein, Arg1 protein expression, and promoted M2 macrophage polarization in mice with dextran sulfate sodium (DSS)-induced colitis. M2 macrophage significantly reduced intestinal inflammation and encouraged wound healing. After treatment with the butyrate, colons expressed increased levels of Arg1 protein, reduced levels of mucosal injury, pro-inflammatory cytokines including IL-6, IL-1β, and TNF-α in the serum compared to control mice (Ji et al. 2016). In a similar study, 10-hydroxy-*cis*-12-octadecenoic acid (HYA), a derivative of linoleic acid and synthesized by the gut microbes reduced epithelial barrier impairment in Caco-2 cells and ameliorated gut inflammation in DSS-induced colitis in mice (Miyamoto et al. 2015).

Experiments conducted by Souza et al. (2020) on male Nile tilapia fish fed with alga (1.2% *Schizochytrium* sp.) for 105 days led to a significant enhancement in the microbial community of *Bacillus*, *Alphaproteobacteria*, *Betaproteobacteria*, and *Gammaproteobacteria* (Proteobacteria), whereas, *Erysipelotrichia*, *Negativicutes*, *Bacteroidia*, and Caldilineae were dominant in the GM of the control group. Further, a superficial decrease in lymphocyte count and white blood corpuscles was evident. Overall, these results suggest that, algal diet apparently modulates the gut immunosystem in male Nile tilapia fish without disturbing the overall structure and integrity of the intestinal villi. A bioactive PUFA, 10-oxo-trans-11-octadecenoic acid (KetoC) significantly inhibited NF-κB signalling pathway, pro-inflammatory markers such as TNFα, IL-6, and IL-1β via NF-κB p65 through GPR120 receptor and demonstrated anti-inflammatory property in RAW 264.7 cells stimulated with *Porphyromonas gingivalis* lipopolysaccharide (Sulijaya et al. 2018).

8.4.2 Effect on Gut Ecosystem

Supplementation of 600 mg of omega-3 PUFA for 2 weeks in healthy 45-year-old individuals exhibited a reduction in *Faecalibacterium* genus, increase in *Coprococcus*, *Ruminococcus*, *Roseburia*, *Blautia*, *Roseburia eubacterium rectale*, and *Subdoligranulum* genera (Noriega et al. 2016). In 2012, Liu et al. (2012)) reported that, diet enriched with SCFAs and Omega-6 PUFAs for 14 weeks significantly reduced the proportion of Bacteroidetes phylum, Porphyromonadaceae, and Lachnospiraceae in wild-type rats compared to n-3 PUFA diet. Results from the studies with milk, lard fat, or PUFAs (38% fat) diet for 3 weeks in C57BL/6 germ-free mice suggested that, milk fat favors the growth of sulfite-reducing bacteria, Bilophila genus of Proteobacteria phylum. However, PUFAs augmented Bacteroidetes phylum, and lowered the abundance of Firmicutes (Devkota et al. 2012).

In a double-blinded randomized crossover clinical study, named as COMIT (Canola Oil Multicenter Intervention Trial), the effect of supplementation of 60 g of one of the following five different unsaturated dietary oils: DHA-enriched high oleic canola oil (37.95 g oleic acid and 3.48 g DHA/60 g oil), conventional canola oil (35.17 g oleic acid/60 g oil), high oleic canola oil (42.88 g oleic acid/60 g oil), a blend of 60:40 flax/safflower (22.48 g linolenic acid and 19.19 g ALA/60 g oil), and a blend of 25:75 corn/safflower oil (41.61 g linolenic acid/60 g oil) for 30 days on GM were evaluated in 25 volunteers with a risk of metabolic syndrome. Findings displayed variations in GM at the genus level rather than the phylum level. High oleic canola oil revealed maximum level of Faecalibacterium among all other oils that were tested. Conversely, the lowest level was evident in high oleic canola oil enriched with DHA. Evaluation between canola and canola/DHA indicated that canola was linked to *Coprobacillus* and *Blautia*, whereas DHA fortified with canola oil was connected with the family Lachnospiraceae of the phylum Firmicutes. Instead, the comparison between all the canola oils and the PUFA-rich oils (i.e., corn/safflower and flax/safflower) exposed a correlation between the genera Parabacteroidetes, Turicibacter, Enterobacteriaceae, and Prevotella family with the first group versus Isobaculum genus, associated with the second group. The authors further speculated that, differences in the distribution of GM between canola and DHA enriched with canola oil could be attributed to an interaction between the GM- and DHA-derived metabolites (Pu et al. 2016).

Kaliannan et al. (2015) demonstrated that, ratio of n-6/n-3 PUFA in the tissues influenced GM profile. A high tissue n-6/n-3 PUFA ratio has the ability to augment the proportions of lipopolysaccharides (LPS)-producing and/or pro-inflammatory bacteria such as of the phylum Proteobacteria and decrease those of LPS-suppressing and/or anti-inflammatory bacteria including *Bifidobacterium*, *Lactobacillus* (primarily *L gasseri*), *Akkermansia muciniphila*, *Clostridium* clusters IV, XIVa, and *Enterococcus faecium*. In contrast, low n-6/n-3 PUFA ratio exhibited the opposite effect. Impact of dietary ALA on the gut microbiome was studied in adult C57BL/6 J mice for 4 weeks. Microbial composition in the intestine indicated that, diet enriched with ALA remarkably increased the abundance of Prevotella and Parabacteroides. Conversely, a significant reduction in Firmicutes phylum such as *Lactobacillus*, *Clostridium* cluster XIVa, *Lachnospiraceae*, and *Streptococcus* were observed compared to the control diet (Todorov et al. 2020).

Dietary intake of *Cucumis melo* var. *agrestis* seed oil for six weeks has significantly amplified the synthesis of fecal SCFAs and reduced plasma cholesterol levels in hypercholesterolemic hamsters. Further, it enhanced the abundance of *Streptococcaceae*, *Eubacteriaceae*, *Ruminococcaceae*, *Clostridiales_vadinBB60_group*, and *Desulfovibrionaceae* family in the intestine (Hao et al. 2020). Impact of HFD-PUFA (mixture of fish oil and safflower oil) and standard nutritional therapy was evaluated in a randomized control study on 16 premature infants with enterostomy. Results revealed differences in the composition of intestinal microbiota in treated and control groups. Reduction in the population of *Streptococcus*, *Clostridium*, and many pathogenic genera within the Enterobacteriaceae family in the treated group was evident. Presence of *E. coli*, *Pantoea*, *Serratia*, and *Citrobacter* was documented in the control group. Nevertheless, these pathogenic bacteria were not reported in the treated group (Younge et al. 2017).

Sea buckthorn seed oil alleviated the population of SCFA-producing bacteria such as *Bacteroidales_S24-7_group*, *Ruminococcaceae*, and *Eubacteriaceae* and also enhanced blood cholesterol levels in hypercholesterolemia hamsters (Hao et al. 2019). Influence of high-saturated fat diet (HFD) and its supplementation with a commercial fish oil (HFO) on lipid profile, GM and blood glucose level was studied in zebrafish adults. In all the experimental groups (control, HFD, HFO), relative copiousness in the phyla of Fusobacteria, Proteobacteria, Cyanobacteria, and Tenericutes was observed. In HFD group, the body lipid profile, population of *Pseudomonas* and *Acinetobacter* were elevated while *Tenericutes* exhibited decrement. Contrastingly, a remarkable decrease in body lipid profile, blood glucose level, *Pseudomonas* and *Acinetobacter* population and increase in *Tenericutes* ck1c4-19 was noticed. Overall these results suggest that, fish oil supplementation alleviates the effect of high-saturated fat diet in zebrafish adults (Arias-Jayo et al. 2019).

8.4.3 Effect on Gut Inflammatory Diseases

Disruption of intestinal tight junction barrier function, intense infiltration of inflammatory cells, and intestinal mucosal injury are influenced by increase in the community of *E. coli, Clostridium* spp., and decrease in *Lactobacillus* and *Bacteroides* spp., in intestinal allografted CR rat. However, dietary intake of fish oil remarkably enhanced the population of *Lactobacillales* spp. and reduced the abundance of *Clostridium, E. coli*, and *Bacteroides* spp., respectively. Furthermore, fish oil ameliorated inflammation, reduced mucosal injury, and restored epithelial integrity and intestinal tight junction barrier function (Li et al. 2011). In a two rat cohort study, Acorn-fed ham animals promoted the growth of a plethora of gut bacteria having anti-inflammatory properties such as *Blautia, Dorea, Parasutterella, Bilophila, Parabacteroides, Alistipes, and others such as Bacteroides, Butyricimonas, Staphylococcus, Enterococcus, Absiella,* and *Phascolarctobacterium*. It reduced the proportion of *Prevotella, Mucispirillum, Lactobacillus, Clostridium, Lachnoanaerobaculum, Ruminococcus, Oscillibacter,* and *Desulfovibrio*. Furthermore, oleic acid in acorn-fed ham diet displayed anti-inflammatory property and presented a protective role by effectively mitigating the symptoms of ulcerative colitis, compared to the rats kept on diet enriched with conventional vegetable rat feed (Fernandez et al. 2020).

In a 6-week randomized control study, the impact of omega-3 fatty acid, probiotic (VSL#3), and a mixture of omega-3 and probiotic on blood lipids, insulin sensitivity, and inflammation was investigated in 60 overweight (BMI > 25), 40–60 years old healthy adults. Results indicated that, groups supplemented with omega-3 fatty acid and VSL#3 had a profound impact on HDL, insulin sensitivity, and C-reactive protein. Decrease in the level of inflammatory markers such as pro-inflammatory cytokines, IL-1β, TNF-α, and IL-6 was observed. Further, it increased the population of *Bifidobacteria*, *Lactobacillus*, and *Streptococcus* in the intestine (Rajkumar et al. 2014). In another study, diet enriched with omega-6 fatty acids resulted in elevated metabolic endotoxemia and systemic low-grade inflammation. Nevertheless, transgenic conversion of tissue from omega-6 to omega-3 fatty acids significantly increased intestinal alkaline phosphatase enzyme. This resulted in alteration in the GM composition resulting in alleviated growth of lipopolysaccharide-producing *E. coli* and Gram-negative bacteria and increase in lipopolysaccharide-suppressing and/or anti-inflammatory bacteria such as *Bifidobacterium*. A significant reduction in decreased metabolic endotoxemia and inflammation was evident (Kaliannan et al. 2015).

In a pilot study, supplementation of eicosapentaenoic acid (EPA) as free fatty acid (FFA) at 2 g/daily for 90 days increased the abundance of *Porphyromonadaceae* and decreased *Ruminococcaceae* in the feces. Further, it attenuated the population of mucosal-adherent members of the *Bacteroidaceae,* inflammation, and increased the goblet cell population in ulcerative colitis patients (Prossomariti et al. 2017). In 2015, Caesar et al. (2015)) reported that, dietary intake of lard oil for 11 weeks induced the activation of Toll-like Receptor (TLR) such as TLR2 and TLR4, promoted gut microbiota-induced white adipose tissue (WAT) inflammation by recruitment of CD45+ cells and reduced insulin sensitivity in mice. In addition, increase in the expression of genes for CCL2, a chemokine mediator for accumulation of macrophage during WAT inflammation was observed in lard fed than fish oil-fed mice. Furthermore, lard oil increased the microbial population of the genera Bacteroides, Turicibacter, and Bilophila. Contrastingly, fish oil reduced inflammation and significantly increased the microbial diversity of Actinobacteria (*Bifidobacterium* and *Adlercreutzia*), lactic acid bacteria (*Lactobacillus* and *Streptococcus*), alpha-, and delta-proteobacteria. Overall these results indicate that, lard oil, enriched with saturated fatty acids promotes inflammation through gut microbial interaction, whereas fish oil, rich in PUFAs alleviates the inflammation and modulates the GM composition.

8.4.4 Effect on Obesity

Growing body of evidences on animal models suggest that, diet enriched with medium chain fatty acids (MCFA) and PUFA has an impact on GM which further influences obesity and associated etiologies. Zhou et al. (2017)) studied the effect of high-fat diet (HFD) containing 20% (w/w) rapeseed oil in combination with MCFAs (10–30%) in healthy male C57BL/6 J mice (7 weeks old) for a duration of 6 weeks. Findings reported an increase in the population of Bacteroidetes, Allobaculum, Lachnospiraceae, and decrease in *Helicobacter* spp. Proteobacteria. Furthermore, 30% (w/w) MCFA decreased the body weight and optimized the serum lipid profile and liver triacylglycerol concentration. Gene expression levels for enzymes encoding fatty acid degradation increased noticeably, whereas the enzymes for fatty acid synthesis reduced significantly in the liver. Many studies have demonstrated that, SCFA production favored the growth of Bifidobacterium, Ruminococcaceae, and *Lactobacillus* which are linked with and are correlated in reducing effects of obesity, type 2 diabetes mellitus, inflammatory bowel disease, and cardiovascular diseases in the hosts (Zhang et al. 2015; Sanmiguel et al. 2015; Ossa et al. 2018; Yang et al. 2019).

In a similar study, supplementation of HFD with different PUFA ratios (P/S), on development of metabolic syndrome was studied in C57Bl/6 J mice for 8 weeks. Results demonstrated that, saturated fat consisting of palm oil increased weight gain and hepatic lipid accumulation than unsaturated fat diet consisting of olive oil and safflower oil (Wit et al. 2012). An increase in the number of *Bacteroidetes*, *Bifidobacterium* spp. (Actinobacteria), *Enterobacteriales* (Proteobacteria), *Lactobacillus* spp. (Firmicutes), and reduction in *Clostridial* cluster XIVa (Firmicutes) followed by decrease in weight gain was observed in 8-week-old healthy female ICR Swiss mice, with HFD of n–3 PUFAs (EPA + DHA) for 19 weeks (Mujico et al. 2013). Ghosh et al. (2013)) in their studies highlighted that, n-6 PUFA-rich diets caused dysbiosis, infiltration of inflammatory mediators, and weight gain. In contrast, fish oil enriched with n-3 PUFA reversed effects in aged mice and favored regulatory T-cell recruitment.

Patterson et al. (2014) showed that, dietary saturated fatty acid (SFA) intake (palm oil) for 16 weeks in C57BL/6 J mice resulted in increased body weight, body fat mass specifically subcutaneous fat mass with a notable reduction in the populations of *Bacteroidetes* at the phylum level compared to monounsaturated fatty acid (MUFA, olive oil) and control (high maize starch) fed diet. Further, fortification of PUFA-enriched diet (flaxseed/fish oil) remarkably enhanced the intestinal population of *Bifidobacterium* at the genus level, EPA and DHA concentrations in the tissues. Overall, these results imply that, consumption of diet rich in MUFA and PUFA is generally beneficial when compared to SFA. Maternal PUFA production during gestation or lactation markedly reduced weight gain and markers of metabolic disruption in male transgenic murine C57BL/6 WT offspring (4 weeks old) fed with HFD. However, in female offsprings, maternal fatty acid status revealed no alteration in terms of body weight. Decreased maternal n-3 PUFA exposure led to reduced abundance of *Epsilonproteobacteria*, *Bacteroides*, *Akkermansia*, and higher relative abundance of *Clostridia* (Robertson et al. 2018).

Dietary metabolism of PUFA by gut microflora resulted in 10-hydroxy-*cis*-12-octadecenoic acid (HYA) production and attenuates HFD-induced obesity in mice and improves glucose homeostasis through conversion of linoleic acid to HYA and thus reducing the adipose inflammation. Further, HYA activates GPR40 and GPR120 receptors and stimulates GLP-1 secretion promoting intestinal peristalsis, thereby, suppressing lipid absorption (Miyamoto et al. 2019). Transplantation of n-3 PUFA altered GM of fat-1 mice to wild-type littermates protecting against obesity, glucose intolerance, and hepatic steatosis (Bidu et al. 2018). In a parallel study, Bacteroides population was significantly higher in male Sprague–Dawley rats, supplemented with the combination of D-fagomine and n-3 PUFA or both for 23 weeks. However, Lactobacillales and Bifidobacteriales population were higher in independent n-3 group and D-fagomine, respectively. Nevertheless, in control group, populations of both the bacterial genera were significantly lower. Furthermore, complementary effects of D-fagomine and n-3 PUFAs in the rat intestinal tract resulted in reduced weight gain, lowered the amount of acetic acid and total SCFAs in the feces (Hereu et al. 2019).

8.4.5 Impact on Type 2 Diabetes Mellitus

Type 2 diabetes mellitus (T2DM) is a type of metabolic disorder, characterized by deficiency of insulin hormone due to the non-functioning of beta cells of Langerhans in the pancreas (Stumvoll et al. 2005). In a pilot randomized trial, 6 months intake of 100 g of sardines, 5 days/week in 35 drug naive type 2 diabetes patients resulted in decrease in the phylum Firmicutes and increase in the abundance of *E. coli* in both the control and sardine groups. However, in the sardine group, a decrease in Firmicutes/Bacteroidetes ratio and high population of Bacteroides-Prevotella phyla

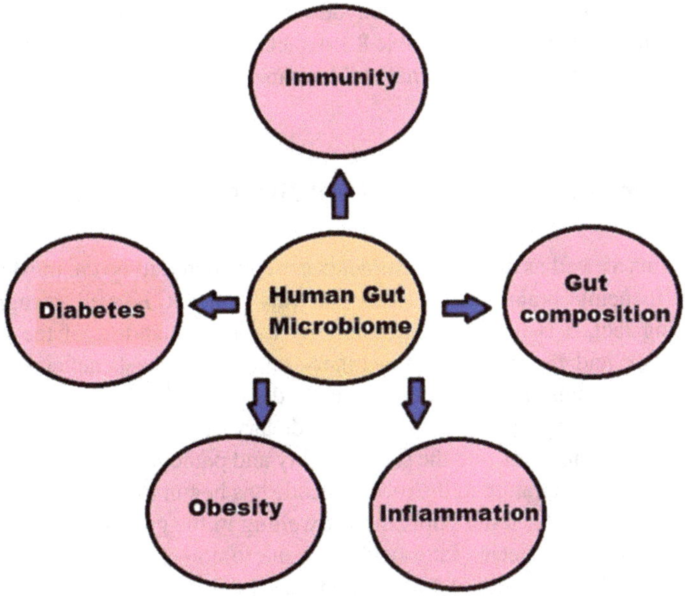

Fig. 8.1 Diagrammatic representation of human gut microbiome and its therapeutic relation relation with diabetes, obesity and inflammation

was evident. Furthermore, in the control group, Firmicutes/Bacteroidetes ratio was found to be reduced. In both the groups, glycemic control was not significant and sardine group exhibited high omega-3 index, in relation to controls. Nevertheless, sardine group revealed an increase in adiponectin levels in the plasma (Balfego et al. 2016). Overall Fig. 8.1 illustrates the influence of GM on various factors leading to pathologies.

Supplementation of 36 g/kg EPA for 11 weeks in male C57BL/6 J mice prevented saturated fat-induced insulin resistance and reversed inflammatory conditions in the adipose tissue (Kalupahana et al. 2010). In a randomized controlled trial, a study group with supplementation of omega-3 fatty acid increased insulin sensitivity and reduced inflammation, whereas, GM was not affected. However, the population of total aerobes, total anaerobes, *Lactobacillus*, bifidobacteria, and *Streptococcus* was abundantly present in the probiotic group and probiotic supplemented with omega-3 supplemented groups. Significant improvement in high-density lipoprotein, insulin sensitivity, and inflammation attenuation was observed in probiotic fortified with omega-3 fatty acid group than probiotic alone (Rajkumar et al. 2014). Intake of diet enriched with flaxseed oil attenuated T2DM, inflammation in male Sprague-Dawley rats through a remarkable reduction in fasting blood glucose level, glycated hemoglobin, blood lipid, and plasma lipopolysaccharide than control and corn oil-treated groups. Metabolites synthesized by GM namely acetic acid, propionic acid, butyric acid, and populations of Bacteroidetes and Alistipes were significantly elevated in flaxseed oil-treated rats. Furthermore, pro-inflammatory markers such as IL-1β,

IL-6, IL-17A, TNF-α, and malondialdehyde were reduced compared to other experimental groups (Zhu et al. 2020). Table 8.1 depicts the updated findings in relation to the influence of fatty acids on animal and human GM.

8.5 Dietary Fats–Gut Microbiota: Brain Communication

Dietary habits as well as dietary constituents are considered to be the most powerful factors influencing brain health. Although this concept is well recognized as longstanding fact, it is only made possible by the understanding of the gut–brain reciprocal talk and the molecular machinery underlying the beneficial impact of several dietary nutrients on neurological disorders. Several findings have documented a strong correlation between dietary lipids, their derivatives and improper lipid metabolism, and the susceptibility and pathogenesis to brain diseases. A growing body of literature in the recent decade has begun to explore the chief role of these in numerous bacterial populations dwelling in the gut, and also to decipher the interplay between dietary fatty acids, gut microbiome, and neurodegenerative diseases. Several findings have documented that the dynamic gut–brain communication is orchestrated through multifarious communication pathways that include autonomic nervous system, enteric nervous system, immune system, and the hypothalamic–pituitary–adrenal axis of the endocrine system (Lynch and Pedersen 2016). Secretion of various hormones by the enteroendocrine cells namely cholecystokinin, peptide-YY, serotonin, and glucagon-like peptide have been found to influence appetite, metabolism, and nutrient absorption (Marrone and Coccurello 2019). Therefore, the gut–brain axis assimilates neuronal, hormonal, and immune signals wherein the microbial population along with the dietary fatty acids impacts the brain.

8.6 Conclusion

Interaction between the gut microbiome and humans has evolved for ages. Assembly of diversed gut microbiome at early stage of life is now considered crucial for a healthy life. In the recent past, several attempts have been made to delineate the gut microbiome composition and its influence on gut immunity and various pathophysiological disease conditions. Intake of diet enriched with fatty acids such as SCFA and PUFAs help in the synthesis of mediators and derivatives to counteract the dysbiosis and aids in the modulation of obesity, insulin resistance, and inflammation in the gut. Further, it embraces an ancillary therapeutic platform for resolving and restoring the gut immunity and altered gut microbial community. Although comprehensive research on dosage of fatty acids has been conducted so far, animal studies and clinical trials are much warranted to establish these evidences leading to

Table 8.1 Influence of Omega fatty acids on animal and human gut microbiota

Host	Diet	Main outcome	References
Imprinting Control Region mice	Natural saline group, high-dose fish oil group (10 mg/kg), and low dose fish oil group (5 mg/kg) for 2 weeks	Decrease in Firmicutes phylum, *Helicobacter*, Uncultured bacterium clone WD2_aaf07d12 (GenBank: EU511712.1), Clostridiales, *Sphingomonadales*, *Pseudomonas*	Yu et al. (2014)
C57BL/6 mice	Corn oil diet or corn oil + fish oil diet for 5 weeks	Corn oil increased the population of Enterobacteriaceae family Fish oil diet enriched microbiota with *Lactobacillus* and *Bifidobacteria* genera of *Firmicutes* phylum	Ghosh et al. (2013)
C57Bl/6 Wild-type germ free mice	High fat diet (45%) for fish oil or lard	Increased levels of *Lactobacillus* genera and *Akkermansia muciniphila* species with fish oil and lard diet increased levels of *Bilophila* genus of *Proteobacteria* phylum	Caesar et al. (2015)
C57BL/6 and Resisting-like molecule knockout mice	High-fat diet (45% fat) for 21 weeks	Decrease in Bacteroidetes phylum and an increase in both Firmicutes and Proteobacteria phyla	Hildebrandt et al. (2009)
Pig	6.0% soybean oil (SBO), 6.0% palm oil (PO), and 7.5% encapsulated palm oil (EPO, contains 80% palm oil)	Diet with PO, increased abundance of Proteobacteria and decreased Firmicutes Diet with EPO and SBO decreased abundances of Proteobacteria and increased abundance of Firmicutes	Yang et al. (2019)
Pregnant women	high intake of omega-3 PUFAs	Increased level of *F. prausnitzii* species and a lower abundance of *Bacteroides* genera	Mokkala et al. (2016)
Male C57BL/6NCrl mice/6 mice	HFD (60% of fat) or high-carbohydrate (11% of fat) for 12 weeks	HFD diet increased Firmicutes and decreased Bacteroidetes	Daniel et al. (2014)
Drug-naive patients with type 2 diabetes	Standard diet (control group: CG), or a standard diet enriched with 100 g of sardines 5 days a week (sardine group: SG) for 6 months	Decreased phylum Firmicutes and increased *E. coli* population observed in both the groups SG decreased Firmicutes/Bacteroidetes and increased *Bacteroides-Prevotella*	Balfego et al. (2016)
Healthy volunteers aged ≥50 years, male and female	4 g daily mixed EPA/DHA for 8 weeks	Reversible increase in *Bifidobacterium*, *Oscillospira*, *Roseburia* and *Lachnospira*, *Lactobacillus*	Watson et al. (2018)

(continued)

Table 8.1 (continued)

Host	Diet	Main outcome	References
		species Decrease in abundance of *Coprococcus* and *Faecalibacterium*	

a therapeutic strategy in relation to short- and long-chain fatty acids maintaining a healthy gut ecosystem.

Acknowledgment The authors acknowledge the Department of Biotechnology, REVA University, Bengaluru for providing the necessary facility.

Conflict of Interest The authors declare no conflict of interest.

References

Aagaard K, Riehle K, Ma J, Segata N, Mistretta TA, Coarfa C et al (2012) A metagenomic approach to characterization of the vaginal microbiome signature in pregnancy. PLoS One 7:e36466

Agus A, Denizot J, Thevenot J, Martinez-Medina M, Massier S, Sauvanet P et al (2016) Western diet induces a shift in microbiota composition enhancing susceptibility to adherent-invasive *E. coli* infection and intestinal inflammation. Sci Rep 6:19032

An D, Oh SF, Olszak T, Neves JF, Avci FY, Erturk-Hasdemir D et al (2014) Sphingolipids from a symbiotic microbe regulate homeostasis of host intestinal natural killer T cells. Cell 156(1–2): 123–133

Ananthakrishnan AN, Khalili H, Konijeti GG, Higuchi LM, Silva PD, Fuchs CS et al (2014) Long-term intake of dietary fat and risk of ulcerative colitis and Crohn's disease. Gut 63(5):776–784

Arias-Jayo N, Abecia L, Lavín JL, Tueros I, Arranz S, Ramirez-García A et al (2019) Host-microbiome interactions in response to a high-saturated fat diet and fish-oil supplementation in zebrafish adult. J Funct Foods 60:103416

Avershina E, Storro O, Oien T, Johnsen R, Pope P, Rudi K (2014) Major faecal microbiota shifts in composition and diversity with age in a geographically restricted cohort of mothers and their children. FEMS Microbiol Ecol 87:280–290

Balfego M, Canivell S, Hanzu FA, Sala-Vila A, Martinez-Medina M, Murillo S et al (2016) Effects of sardine-enriched diet on metabolic control, inflammation and gut microbiota in drug-naive patients with type 2 diabetes: a pilot randomized trial. Lipids Health Dis 15:78

Bidu C, Escoula Q, Bellenger S, Spor A, Galan M, Geissler A et al (2018) The transplantation of ω3 PUFA–Altered Gut microbiota of FAT-1 mice to wild-type littermates prevents obesity and associated metabolic disorders. Diabetes 67(8):1512–1523

Bischoff SC (2016) Microbiota and aging. Curr Opin Clin Nutr Metab Care 19(1):26–30

Black MH, Watanabe RM, Trigo E, Takayanagi M, Lawrence JM, Buchanan TA, Xiang AH (2013) High-fat diet is associated with obesity-mediated insulin resistance and β-cell dysfunction in Mexican Americans. J Nutr 143(4):479–485

Bourlioux P, Koletzko B, Guarner F, Braesco V (2003) The intestine and its microflora are partners for the protection of the host: report on the Danone Symposium "The Intelligent Intestine," held in Paris, June 14, 2002. Am J Clin Nutr 78(4):675–683

Caesar R, Tremaroli V, Kovatcheva-Datchary P, Cani PD, Backhed F (2015) Crosstalk between gut microbiota and dietary lipids aggravates WAT inflammation through TLR signaling. Cell Metab 22(4):658–668

Calder PC (2017) Omega-3 fatty acids and inflammatory processes: from molecules to man. Biochem Soc Trans 45:1105–1115

Candela M, Biagi E, Brigidi P, O'Toole PW, De Vos WM (2014) Maintenance of a healthy trajectory of the intestinal microbiome during aging: a dietary approach. Mech Ageing Dev 136–137:70–75

Carlson SE, Colombo J, Gajewski BJ, Gustafson KM, Mundy D, Yeast J et al (2013) DHA supplementation and pregnancy outcomes. Am J Clin Nutr 97(4):808–815

Choi YJ, Lee DH (2016) Microbiota: a key for healthy aging. Ann Geriatr Med Res 20(4):168–176

Claesson MJ, Cusack S, O'Sullivan O, Greene-Diniz R, de Weerd H, Flannery E et al (2011) Composition, variability, and temporal stability of the intestinal microbiota of the elderly. Proc Natl Acad Sci U S A 108(1):4586–4591

Cotillard A, Kennedy SP, Kong LC, Prifti E, Pons N, Le Chatelier E et al (2013) Dietary intervention impact on gut microbial gene richness. Nature 500:585–588

Daniel H, Gholami AM, Berry D, Desmarchelier C, Hahne H, Loh G, Mondot S, Lepage P, Rothballer M, Walker A, Böhm C, Wenning M, Wagner M, Blaut M, Schmitt-Kopplin P, Kuster B, Haller D, Clavel T (2014) High-fat diet alters gut microbiota physiology in mice. ISME J 8(2):295–308. https://doi.org/10.1038/ismej.2013.155

Derbyshire E (2018) Brain health across the lifespan: a systematic review on the role of omega-3 fatty acid supplements. Nutrients 10(8):1094

Devkota S, Wang Y, Musch MW et al (2012) Dietary-fat-induced taurocholic acid promotes pathobiont expansion and colitis in Il10−/− mice. Nature 487(7405):104–108

Donaldson GP, Lee SM, Mazmanian SK (2016) Gut biogeography of the bacterial microbiota. Nat Rev Microbiol 14:20–32

Feng W, Ao H, Peng C (2018) Gut microbiota, short-chain fatty acids, and herbal medicines. Front Pharmacol 9:1354

Fernandez J, de la Fuente VG, García MTF, Sanchez JG, Redondo BI, Villar CJ et al (2020) A diet based on cured acorn-fed ham with oleic acid content promotes anti-inflammatory gut microbiota and prevents ulcerative colitis in an animal model. Lipids Health Dis 19(1):28

Ghosh S, Molcan E, DeCoffe D, Dai C, Gibson DL (2013) Diets rich in n-6 PUFA induce intestinal microbial dysbiosis in aged mice. Br J Nutr 110(3):515–523

Hao W, He Z, Zhu H, Liu J, Kwek E, Zhao Y et al (2019) Sea buckthorn seed oil reduces blood cholesterol and modulates gut microbiota. Food Funct 10(9):5669–5681

Hao W, Zhu H, Chen J, Kwek E, He Z, Liu J et al (2020) Wild melon seed oil reduces plasma cholesterol and modulates gut microbiota in hypercholesterolemic hamsters. J Agri Food Chem 68:2071–2081

Hereu M, Ramos-Romero S, Marín-Valls R, Amezqueta S, Miralles-Perez B, Romeu M et al (2019) Combined buckwheat d-fagomine and fish omega-3 PUFAs stabilize the populations of gut Prevotella and Bacteroides while reducing weight gain in rats. Nutrients 11(11):2606

Hildebrandt MA, Hoffmann C, Sherrill-Mix SA, Keilbaugh SA, Hamady M, Chen Y-Y, Knight R, Ahima RS, Bushman F, Wu GD (2009) High-fat diet determines the composition of the murine gut microbiome independently of obesity. Gastroenterology 137(5):1716–1724.e2. https://doi.org/10.1053/j.gastro.2009.08.042

Human Microbiome Project Consortium (2012) Structure, function and diversity of the healthy human microbiome. Nature 486(7402):207–214

Innes JK, Calder PC (2018) Omega-6 fatty acids and inflammation. Prostaglandins Leukot Essent Fatty Acids 132:41–48

Jakobsson HE, Abrahamsson TR, Jenmalm MC, Harris K, Quince C, Jernberg C et al (2014) Decreased gut microbiota diversity, delayed Bacteroidetes colonisation and reduced Th1 responses in infants delivered by caesarean section. Gut 63:559–566

Ji J, Shu D, Zheng M, Wang J, Luo C, Wang Y et al (2016) Microbial metabolite butyrate facilitates M2 macrophage polarization and function. Sci Rep 6:24838

Kaliannan K, Wang B, Li XY, Kim KJ, Kang JX (2015) A host-microbiome interaction mediates the opposing effects of omega-6 and omega-3 fatty acids on metabolic endotoxemia. Sci Rep 5: 11276

Kalupahana NS, Claycombe K, Newman SJ, Stewart T, Siriwardhana N, Matthan N et al (2010) Eicosapentaenoic acid prevents and reverses insulin resistance in high-fat diet-induced obese mice via modulation of adipose tissue inflammation. J Nutr 140(11):1915–1922

Keenan MJ, Marco ML, Ingram DK, Martin RJ (2015) Improving healthspan via changes in gut microbiota and fermentation. Age 37(5):98

Koenig JE, Spor A, Scalfone N, Fricker AD, Stombaugh J, Knight R et al (2011) Succession of microbial consortia in the developing infant gut microbiome. Proc Natl Acad Sci U S A 108: 4578–4585

Kumar M, Babaei P, Ji B, Nielsen J (2016) Human gut microbiota and healthy aging: recent developments and future prospective. Nutr Healthy Aging 4(1):3–16

Kunisawa J, Arita M, Hayasaka T, Harada T, Iwamoto R, Nagasawa R et al (2015) Dietary ω3 fatty acid exerts anti-allergic effect through the conversion to 17, 18-epoxyeicosatetraenoic acid in the gut. Sci Rep 5:9750

Lauritzen L, Brambilla P, Mazzocchi A, Harslof LB, Ciappolino V, Agostoni C (2016) DHA effects in brain development and function. Nutrients 8(1):6

Lecocq M, Detry B, Guisset A, Pilette C (2013) FcαRI-mediated inhibition of IL-12 production and priming by IFN-γ of human monocytes and dendritic cells. J Immunol 190(5):2362–2371

Li Q, Zhang Q, Wang C, Tang C, Zhang Y, Li N et al (2011) Fish oil enhances recovery of intestinal microbiota and epithelial integrity in chronic rejection of intestinal transplant. PLoS One 6(6): e20460

Liu T, Hougen H, Vollmer AC, Hiebert SM (2012) Gut bacteria profiles of Mus musculus at the phylum and family levels are influenced by saturation of dietary fatty acids. Anaerobe 18(3): 331–337

Lloyd-Price J, Abu-Ali G, Huttenhower C (2016) The healthy human microbiome. Genome Med 8(1):51

Lynch SV, Pedersen O (2016) The human intestinal microbiome in health and disease. N Engl J Med 375(24):2369–2379

Marrone MC, Coccurello R (2019) Dietary fatty acids and microbiota-brain communication in neuropsychiatric diseases. Biomol Ther 10(1):12

Milisavljevic V, Garg M, Vuletic I, Miller JF, Kim L, Cunningham TD et al (2013) Prospective assessment of the gastroesophageal microbiome in VLBW neonates. BMC Pediatr 13:49

Milligan G, Shimpukade B, Ulven T, Hudson BD (2017) Complex pharmacology of free fatty acid receptors. Chem Rev 117:67–110

Miyamoto J, Mizukure T, Park SB, Kishino S, Kimura I, Hirano K et al (2015) A gut microbial metabolite of linoleic acid, 10-hydroxy-cis-12-octadecenoic acid, ameliorates intestinal epithelial barrier impairment partially via GPR40-MEK-ERK pathway. J Biol Chem 290(5): 2902–2918

Miyamoto J, Igarashi M, Watanabe K, Karaki SI, Mukouyama H, Kishino S et al (2019) Gut microbiota confers host resistance to obesity by metabolizing dietary polyunsaturated fatty acids. Nat Commun 10(1):4007

Mokkala K, Röytiö H, Munukka E, Pietilä S, Ekblad U, Rönnemaa T, Eerola E, Laiho A, Laitinen K (2016) Gut microbiota richness and composition and dietary intake of overweight pregnant women are related to serum zonulin concentration a marker for intestinal permeability. J Nutr 146(9):1694–1700. https://doi.org/10.3945/jn.116.235358

Mozaffarian D, Wu JH (2011) Omega-3 fatty acids and cardiovascular disease. J Am Coll Cardiol 58(20):2047–2067

Mujico JR, Baccan GC, Gheorghe A, Diaz LE, Marcos A (2013) Changes in gut microbiota due to supplemented fatty acids in diet-induced obese mice. Br J Nutr 110(4):711–720

Myles IA, Pincus NB, Fontecilla NM, Datta SK (2014) Effects of parental omega-3 fatty acid intake on offspring microbiome and immunity. PLoS One 9. e87181

Nagpal R, Mainali R, Ahmadi S, Wang S, Singh R, Kavanagh K et al (2018) Gut microbiome and aging: physiological and mechanistic insights. Nutr Healthy Aging 4(4):267–285

Neurath MF (2020) Host–microbiota interactions in inflammatory bowel disease. Nat Rev Gastroenterol Hepatol 17(2):76–77

Noriega BS, Sanchez-Gonzalez MA, Salyakina D, Coffman J (2016) Understanding the impact of omega-3 rich diet on the gut microbiota. Case Rep Med 2016:3089303

Ohue-Kitano R, Yasuoka Y, Goto T, Kitamura N, Park SB, Kishino S et al (2018) α-Linolenic acid-derived metabolites from gut lactic acid bacteria induce differentiation of anti-inflammatory M2 macrophages through G protein-coupled receptor 40. FASEB J 32(1):304–318

Ossa JC, Yanez D, Valenzuela R, Gallardo P, Lucero Y, Farfan MJ (2018) Intestinal inflammation in Chilean infants fed with bovine formula vs. breast milk and its association with their gut microbiota. Front Cell Infect Microbiol 8:190

Palmer C, Bik EM, DiGiulio DB, Relman DA, Brown PO (2007) Development of the human infant intestinal microbiota. PLoS Biol 5(7) e177

Patterson E, RM O'D, Murphy EF, Wall R, Sullivan OO, Nilaweera K et al (2014) Impact of dietary fatty acids on metabolic activity and host intestinal microbiota composition in C57BL/6J mice. Br J Nutr 111(11):1905–1917

Prossomariti A, Scaioli E, Piazzi G, Fazio C, Bellanova M, Biagi E et al (2017) Short-term treatment with eicosapentaenoic acid improves inflammation and affects colonic differentiation markers and microbiota in patients with ulcerative colitis. Sci Rep 7(1):7458

Pu S, Khazanehei H, Jones PJ, Khafipour E (2016) Interactions between obesity status and dietary intake of monounsaturated and polyunsaturated oils on human gut microbiome profiles in the canola oil multicenter intervention trial (COMIT). Front Microbiol 7:1612

Rajkumar H, Mahmood N, Kumar M, Varikuti SR, Challa HR, Myakala SP (2014) Effect of probiotic (VSL# 3) and omega-3 on lipid profile, insulin sensitivity, inflammatory markers, and gut colonization in overweight adults: a randomized, controlled trial. Mediat Inflamm 2014. 348959

Rajoka MSR, Shi J, Mehwish HM, Zhu J, Li Q, Shao D et al (2017) Interaction between diet composition and gut microbiota and its impact on gastrointestinal tract health. Food Sci Hum Wellness 6:121–130

Ratajczak W, Rył A, Mizerski A, Walczakiewicz K, Sipak O, Laszczyńska M (2019) Immunomodulatory potential of gut microbiome-derived short-chain fatty acids (SCFAs). Acta Biochim Pol 66(1):1–12

Robertson RC, Kaliannan K, Strain CR, Ross RP, Stanton C, Kang JX (2018) Maternal omega-3 fatty acids regulate offspring obesity through persistent modulation of gut microbiota. Microbiome 6(1):95

Rodriguez JM, Murphy K, Stanton C, Ross RP, Kober OI, Juge N et al (2015) The composition of the gut microbiota throughout life, with an emphasis on early life. Microb Ecol Health Dis 26: 26050

Rohr MW, Narasimhulu CA, Rudeski-Rohr TA, Parthasarathy S (2020) Negative effects of a high-fat diet on intestinal permeability: a review. Adv Nutr 11(1):77–91

Sanmiguel C, Gupta A, Mayer EA (2015) Gut microbiome and obesity: a plausible explanation for obesity. Curr Obes Rep 4(2):250–261

Schunck WH, Konkel A, Fischer R, Weylandt KH (2018) Therapeutic potential of omega-3 fatty acid-derived epoxyeicosanoids in cardiovascular and inflammatory diseases. Pharmacol Ther 183:177–204

Sekirov I, Russell SL, Antunes LC, Finlay BB (2010) Gut microbiota in health and disease. Physiol Rev 90(3):859–904

Shao S, Wang D, Zheng W, Li X, Zhang H, Zhao D et al (2019) A unique polysaccharide from *Hericium erinaceus* mycelium ameliorates acetic acid-induced ulcerative colitis rats by modulating the composition of the gut microbiota, short chain fatty acids levels and GPR41/43 respectors. Int Immunopharmacol 71:411–422

Shrestha N, Sleep SL, Cuffe JSM, Holland OJ, Perkins AV, Yau SY et al (2020) Role of omega-6 and omega-3 fatty acids in fetal programming. Clin Exp Pharmacol Physiol 47(5):907–915

Simopoulos AP (2002) Omega-3 Fatty Acids in Inflammation and Autoimmune Diseases. J Am Coll Nutr 21(6):495–505

Souza FP, Lima ECS, Urrea-Rojas AM, Suphoronski SA, Facimoto CT, Bezerra Junior JD et al (2020) Effects of dietary supplementation with a microalga (*Schizochytrium* sp.) on the hemato-immunological, and intestinal histological parameters and gut microbiota of Nile tilapia in net cages. PLoS One 15(1) e0226977

Stavrinou PS, Andreou E, Aphamis G, Pantzaris M, Ioannou M, Patrikios IS et al (2020) The effects of a 6-month high dose omega-3 and omega-6 polyunsaturated fatty acids and antioxidant vitamins supplementation on cognitive function and functional capacity in older adults with mild cognitive impairment. Nutrients 12(2):325

Stumvoll M, Goldstein BJ, van Haeften TW (2005) Type 2 diabetes: principles of pathogenesis and therapy. Lancet 365(9467):1333–1346

Sulijaya B, Takahashi N, Yamada M, Yokoji M, Sato K, Aoki-Nonaka Y et al (2018) The anti-inflammatory effect of 10-oxo-trans-11-octadecenoic acid (KetoC) on RAW 264.7 cells stimulated with *Porphyromonas gingivalis* lipopolysaccharide. J Periodontal Res 53(5):777–784

Tiihonen K, Ouwehand AC, Rautonen N (2010) Human intestinal microbiota and healthy ageing. Ageing Res Rev 9(2):107–116

Todorov H, Kollar B, Bayer F, Brandao I, Mann A, Mohr J et al (2020) α-Linolenic acid-rich diet influences microbiota composition and villus morphology of the mouse small intestine. Nutrients 12(3):732

van der Beek CM, Dejong CHC, Troost FJ, Masclee AAM, Lenaerts K (2017) Role of short-chain fatty acids in colonic inflammation, carcinogenesis, and mucosal protection and healing. Nutr Rev 75(4):286–305

Vemuri R, Gundamaraju R, Shinde T et al (2019) *Lactobacillus acidophilus* DDS-1 modulates intestinal-specific microbiota, short-chain fatty acid and immunological profiles in aging mice. Nutrients 11(6):1297

Wang D-C, Sun C-H, Liu L-Y, Sun X-H, Jin X-W, Song W-L et al (2012) Serum fatty acid profiles using GC-MS and multivariate statistical analysis: potential biomarkers of Alzheimer's disease. Neurobiol Aging 33:1057–1066

Watson H, Mitra S, Croden FC, Taylor M, Wood HM, Perry SL, Spencer JA, Quirke P, Toogood GJ, Lawton CL, Dye L, Loadman PM, Hull MA (2018) A randomised trial of the effect of omega-3 polyunsaturated fatty acid supplements on the human intestinal microbiota. Gut 67 (11):1974–1983. https://doi.org/10.1136/gutjnl-2017-314968

Wit ND, Derrien M, Bosch-Vermeulen H, Oosterink E, Keshtkar S, Duval C et al (2012) Saturated fat stimulates obesity and hepatic steatosis and affects gut microbiota composition by an enhanced overflow of dietary fat to the distal intestine. Am J Physiol Gastrointest Liver Physiol 303(5):G589–G599

Yang WY, Lee Y, Lu H, Chou CH, Wang C (2019) Analysis of gut microbiota and the effect of lauric acid against necrotic enteritis in Clostridium perfringens and Eimeria side-by-side challenge model. PLoS One 14(5) e0205784

Younge N, Yang Q, Seed PC (2017) Enteral high fat-polyunsaturated fatty acid blend alters the pathogen composition of the intestinal microbiome in premature infants with an enterostomy. J Pediatr 181:93–101. e6

Yu H-N, Zhu J, Pan W-s, Shen S-R, Shan W-G, Das UN (2014) Effects of fish oil with a high content of n-3 polyunsaturated fatty acids on mouse gut microbiota. Arch Med Res 45(3):195–202. https://doi.org/10.1016/j.arcmed.2014.03.008

Zhang YJ, Li S, Gan RY, Zhou T, Xu DP, Li HB (2015) Impacts of gut bacteria on human health and diseases. Int J Mol Sci 16(4):7493–7519

Zhou S, Wang Y, Jacoby JJ, Jiang Y, Zhang Y, Yu LL (2017) Effects of medium-and long-chain triacylglycerols on lipid metabolism and gut microbiota composition in C57BL/6J mice. J Agric Food Chem 65(31):6599–6607

Zhu L, Sha L, Li K, Wang Z, Wang T, Li Y et al (2020) Dietary flaxseed oil rich in omega-3 suppresses severity of type 2 diabetes mellitus via anti-inflammation and modulating gut microbiota in rats. Lipids Health Dis 19(1):20

Chapter 9
Role of Short-Chain Fatty Acids from Gut Microbiota in Neuroendocrine Pathogenesis Management

Neha Sahu, Prabhat Upadhyay, and Sunil Kumar Mishra

Abstract Short-chain fatty acids (SCFAs) are important beneficial molecules produced as a result of microbial fermentation and are currently researched for their action against various diseases such as cancer and diabetes. Dietary fibers (DF) are a major source of SCFA. DF affects gut microbial composition which eventually stimulates the microbial metabolite production, i.e., SCFAs. Soluble DFs (pectin, inulin, arabinoxylan, and hemicellulose) are rich sources of SCFAs. 90–95% of SCFAs present in the colon constitutes acetic acid, propionic acid, and butyric acid. Among these, butyrate has been the most potent showing effects against colon cancer. These data suggest that the antiproliferative, apoptotic, and differentiating properties of the various SCFAs are linked to the degree of induced histone hyperacetylation, miRNA-106b regulation, histone deacetylation, and p21 gene regulation. SCFA also modulates hunger, inflammation, and is potentially linked with the maintenance of type 2 diabetes. The mode of action of SCFAs involves various mechanisms such as regulation of GPCR receptors, inhibition of histone deacetylase (HDAC), and secretion of gut hormones such as peptide YY (PYY) and glucagon-like peptide 1 (GLP-1). The metabolites also cross the blood–brain barrier, involve in neurogenesis, brain homeostasis, and synthesis of neurotransmitters (serotonin). The chapter highlights the role of SCFAs as a critical factor in the maintenance of gut and immune homeostasis. SCFAs with potential activity and vast effects could prove to be the future target for therapy of diseases.

N. Sahu
Department of Botany, Govt Post Graduate College, Bina, India

P. Upadhyay (✉)
Molecular Science Lab, National Institute of Immunology, New Delhi, India
e-mail: upadhyayprabhat@nii.ac.in

S. K. Mishra
Department of Pharmaceutical Engineering and Technology, Indian Institute of Technology (Banaras Hindu University), Varanasi, India
e-mail: skmishra.phe@iitbhu.ac.in

A. K. Tripathi, M. Kotak (eds.), *Gut Microbiome in Neurological Health and Disorders*, Nutritional Neurosciences,
https://doi.org/10.1007/978-981-19-4530-4_9

Keywords Dietary fibers · Gut hormones · Cancer · GLP-1 · Neurogenesis · ROS · MicroRNA

9.1 Introduction

Fats are categorized into different classes of compounds such as mono, di and triglycerides, phosphatides, cerebrosides, sterols, terpenes, fatty alcohols, and fatty acids (Fig. 9.1a). Structurally fatty acids (FA) are made up of long aliphatic chains constituting carbon, hydrogen, and oxygen, arranged in a linear manner. There are two types of fatty acids (a) saturated FA containing no double bond and (b) unsaturated FA containing one or more double bonds (Chow 2007). Fatty acids are the building blocks of fats in the human body. They are a source of energy, and are part of cellular machineries such as plasma membrane and receptors. These fatty acids have been used by mankind since ancient times as soap, detergents candles.

On the basis of the presence of number of double bonds, FA has been further divided into two types, i.e., monounsaturated type where one double bond is found and polyunsaturated type FA containing 2 or more double bonds. Both are abbreviated as MUFA and PUFA, respectively. Fatty acids are further categorized on the basis of length of the carbon chain namely short, medium, long, and very long FAs containing carbon chains C2–6, C8–14, C14–28, and C $>$ 28, respectively (Fig. 9.1b).

Among these, short-chain fatty acids (SCFAs) are basically saturated fatty acids. These are formed by gut microbiota as a metabolic end-product of fermentation. Their production in the gut microbiota is dependent on the nature of food intake. SCFAs play a major role in gut health maintenance through different mechanisms (Arango et al. 2003). In recent years, the importance of the SCFA in human well-being has been discovered and many publications have highlighted its role against different diseases such as cancer, diabetes, and inflammation (Martin-Gallausiaux et al. 2021). Therefore, SCFAs have been the subject of numerous investigations

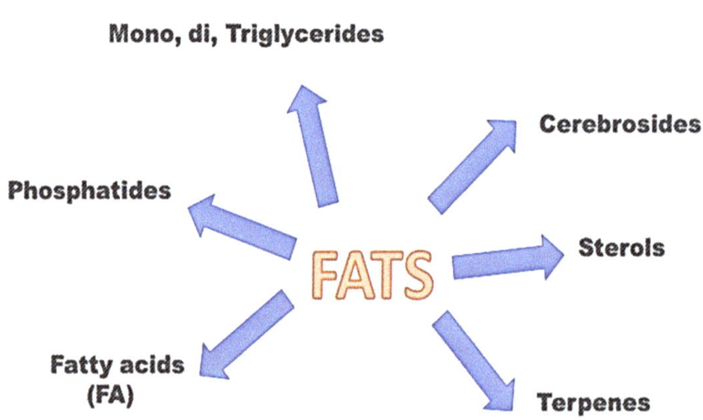

Fig. 9.1 (**a**) Diagrammatic representation of different types of fats. (**b**) Classification of fatty acids

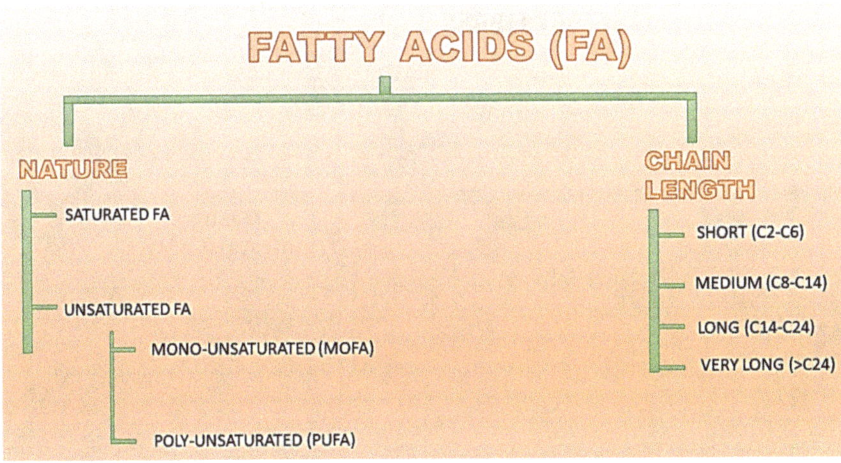

Fig. 9.1 (continued)

(Den Besten et al. 2013). In this book chapter, the effects and mode of action of SCFAs on human health are discussed in detail.

9.2 Occurrence

The gut microbiota represents a rich population of microorganisms in the gastrointestinal tract of humans and other mammals. Most of the microbiota constitute a bacterial population and a few strains of archaea, fungi, and protozoa, in a symbiotic relationship with their host species. Gut microbiota protects the gut against infections, builds strong immunity, and plays a critical part in nutrient utilization and metabolism (Brestoff and Artis 2013; Belkaid and Hand 2014; Bindels et al. 2015). These have the ability to ferment residual carbohydrates and proteins of small intestine during digestion. SCFAs are mainly produced by these microbiota (Ríos-Covián et al. 2016). The proximal part of the colon is the main region where microbial fermentation of residual oligo- and polysaccharides takes place. Gut microbial composition fluctuates as a result of dietary changes which eventually affects the microbial metabolite production (Den Besten et al. 2013). The major source of SCFAs is dietary fiber (DF) as host enzymes in the upper gastrointestinal tract are unable to perform degradation of DF. Therefore, DF reaches intact to the colon, where gut microbiota through fermentation process (Fig. 9.2), produce a number of metabolites, SCFAs being a major product (Bugaut 1987). Soluble dietary fibers (pectin, inulin, arabinoxylan, and hemicellulose) are rich sources of SCFAs.

Probiotics, as the name suggests, are live beneficial microorganisms specifically bacteria and are administered in limited amounts to host for health benefits. The most common beneficial bacteria used in probiotics are namely Lactobacilli and

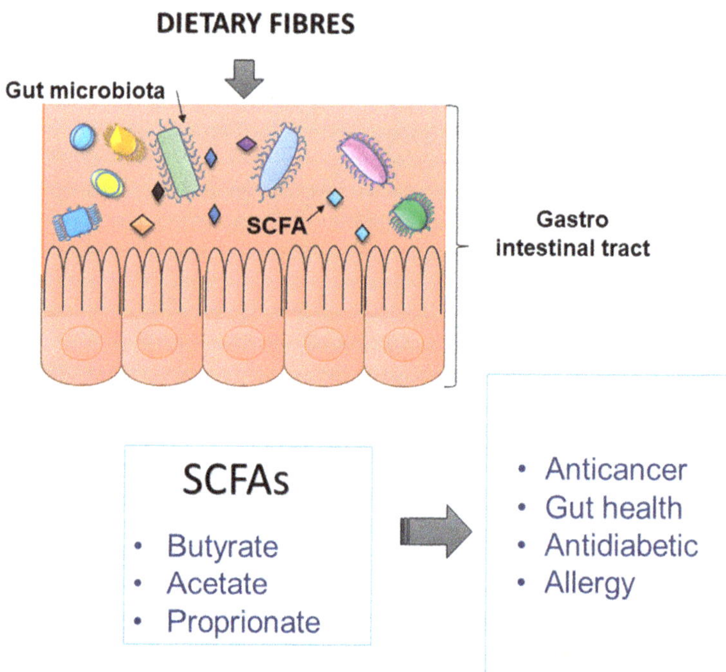

Fig. 9.2 Overview of SCFAs and their role

Bifidobacteria. Yogurt and lassi are Indian probiotics used by common people in India. Probiotic food has the ability to increase SCFAs and hydration in the colon (Sanders et al. 2019). Another term prebiotics are a class of nondigestible food ingredient, which have the ability to promote the host health through nutrient metabolism in gastrointestinal tract by stimulating the activity of beneficial bacteria in the colon (Gibson and Roberfroid 1995). Prebiotics have been used to manipulate microorganisms in the host to improve measurable health outcomes. The concept of prebiotics has attracted a lot of attention in the field of gastrointestinal microbiology.

Both Probiotics and Prebiotics serve to increase the community of beneficial microorganisms and their byproducts such as SCFAs in the host.

9.3 Chemistry of SCFAs

SCFAs (C2–C6) can be steam distilled at atmospheric pressure due to presence of a simple short aliphatic chain hence called volatile fatty acids (VFAs). Majority of the SCFAs are saturated but some of them are unsaturated too. These are water soluble compared to other fatty acids with longer aliphatic chains (Chow 2007). SCFAs backbone contain a carboxyl group (COOH) attached to chains of 1–5 additional carbons and is designated C2 (acetic acid), C3 (propionic acid), C4 (butyric acid), C5 (valeric acid), and C6 (caproic acid) (Enciu et al. 2018). SCFAs present in the colon constitute acetic acid, propionic acid, and butyric acid. These are the most abundant, representing 90–95% of SCFAs (Ríos-Covián et al. 2016).

9.4 Role of SCFA and Its Mode of Action in prognosis of diseases

9.4.1 Anticancer Activity

SCFAs have been shown to be protective in regard to colon carcinogenesis. A study focused on anticancer role of SCFA against human colon carcinoma (Hinnebusch et al. 2002) showed that among all the SCFAs, butyrate (C4) is the most potent followed by Propionate (C3) and valerate (C5) as potential anticancer agent. Butyrate exerts its cellular effects through induction of histone hyperacetylation, eventually activating the transcription of genes which caused growth arrest and differentiation in human colon carcinoma. The link between SCFA and histones was further supported by SCFA-induced transactivation of the differentiation marker gene, intestinal alkaline phosphatase (IAP), which was blocked by histone deacetylase (HDAC). Butyrate also enhanced cellular apoptosis. These data suggest that the antiproliferative, apoptotic, and differentiating properties of the various SCFAs are linked to the degree of induced histone hyperacetylation.

Yusuf et al. explored the potential relation between SCFA and the incidence of colorectal cancer in patients. Results showed that the SCFA level is decreased significantly in colorectal cancer patients compared to noncancerous patients (Yusuf et al. 2019). A research by Khan et al. (2019)) revealed that SCFA induced from edible mushroom profoundly improved the colon health by downregulating oncogenic signalling molecules. The mechanism involved stimulation of G-protein-coupled receptors accompanied by modulation of histone deacetylases. Additionally, in another study by the same group, the extract of flower buds showed antiproliferative activities against colon cancer by changes in gut microbiome. The 16S Sequencing data revealed a marked decrease in pathogenic bacteria, however, a significant increase in SCFA-producing bacteria. These findings provide strong evidence of the vital role of SCFA against cancer (Yang et al. 2020).

Majority of the study indicates that butyrate plays an important role in the cellular differentiation of mucosal epithelial cells in colon further decreasing the chances of colorectal cancer (Stein et al. 2000).

Butyrate had differential effects in colon cells at different stages of cancer development (Fig. 9.3). The role of SCFA specifically butyrate in inducing glutathione-S-transferase activity and reducing the risk of colon cancer has been reviewed by (Scharlau et al. 2009). The chemopreventative effects of the SCFA butyrate are, in part, mediated through induction of tumor suppressor genes, i.e., p21

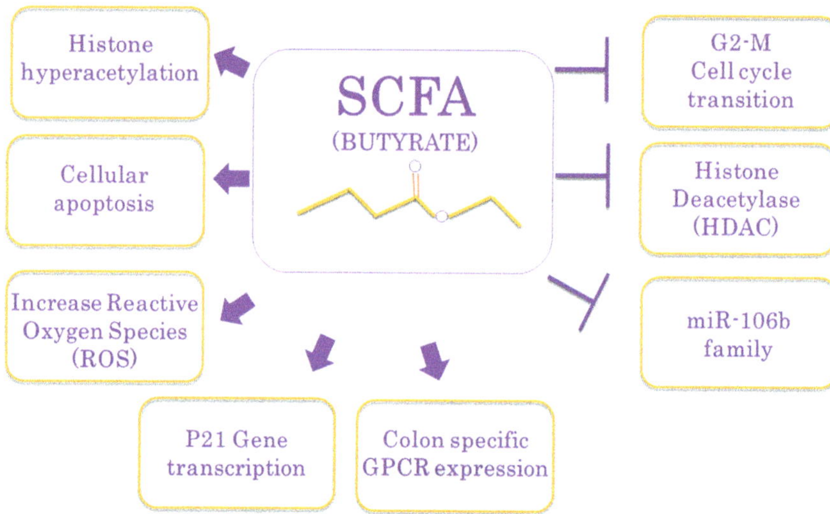

Fig. 9.3 Mode of action of SCFAs as an anticancer agent

expression. Butyrate modulated expressions of MicroRNAs of cancerous cells. The miRNAs are short noncoding RNAs that function in regulating the expression of tumor suppressors genes or oncogenes and their expression alters significantly in several tumor tissues and cancer cell lines, e.g., miR-106b (Sagar 2021). The miR-106b directly targets genes involved in tumorigenesis, proliferation, invasion, migration, and metastases. Butyrate-induced p21 protein expression was dampened by treatment with a miR-106b mimic. It was concluded that SCFAs play part in the prevention of carcinogenesis through modulation of miRNAs (Lee et al. 2011).

Further in a different study role of SCFA in modulating apoptosis, cell cycle, intracellular redox state, and glucose metabolism was investigated in the Caco-2 human colon cancer cell line. Butyrate-induced apoptosis and G2-M arrest. SCFA treatment led to reductions in glutathione availability and increase in levels of reactive oxygen species (Matthews et al. 2012).

GPR43 is a GPCR receptor of SCFA and found in abundance in normal colon tissues but was markedly reduced in most colon cancer tissues. Tang et al. (2011)) investigated the effect of SCFA against colon cancer by monitoring the expression of GPR43. Result revealed loss of GPR43 in colon cancer cells, however, when treated with SCFAs, restoration of GPR43 expression was observed. The increased GPR43 expression further increased apoptosis. In conclusion, GPR43 suppress tumor through SCFA-induced apoptosis and cell growth arrest in colon cancer.

Research shows that the level of SCFA-producing bacteria was lower in the mice suffering from colon cancer compared to healthy mice which suggest a decline in SCFA level as well. After giving a high-fiber diet to mice with colon cancer, SCFA-producing bacteria as well as SCFA levels were increased. Remarkably, an increase in SCFA butyrate receptor and decrease in cancer tissue was observed (Bishehsari et al. 2018).

Some carbohydrates, e.g., Inulin-type fructans promote microbial production of SCFA, which further affects cancer cell proliferation in other organs of host. When these inulin-type fructans were given as treatment to mice suffering from liver cancer, it showed a reduction in inflammation and cancer cell growth possibly through a cAMP level-dependent pathway. Furthermore, the activation of propionate receptor GPR43 was also observed which reduces cancer growth. This was the first report of the effect of SCFA transformed from nutrients against liver cancer (Bindels et al. 2012).

Another role of SCFAs is in maintenance of the intestinal barrier by regulating intestinal P-glycoprotein (P-gp) and breast cancer resistance protein (BCRP). Rats treated with SCFAs showed reduced expression and function of intestinal P-gp, but increased expression of intestinal BCRP. Butyrate exhibited the strongest induction or inhibitory effect. In conclusion, SCFAs show contrasting results on intestinal P-gp and BCRP (Qiu-shi et al. 2021). Further in breast cancer research, significant changes in the interaction of intestinal flora were observed in premenopausal breast cancer women when compared to healthy premenopausal women. In women suffering from breast cancer, very low level of SCFA-producing bacteria and SCFA-producing enzymes was observed. Moreover, the in vitro testing of propionate and butyrate showed positive result to inhibit breast cancer cell growth (He et al. 2021).

9.4.2 Gut and Brain

Multiple modes decoded now (immunological, endocrine, neuronal, and so on) through gut are involved in the communication of the brain and this happened due to production of SCFAs by gut microbes. These pathways are now being explored by researchers all over the world for pathophysiology of different neurological diseases (Silva et al. 2020).

In a few years, through many studies it was found that the gut bacteria mediate the level of neuronal transmitters (for example, γ-aminobutyric acid (GABA), glutamate, oxytocin, dopamine, acetylcholine, and serotonin (5-hydroxytryptamine (5-HT))). It has also shown that SCFAs cross the blood–brain barrier, increase neurogenesis, improve neuronal homeostasis, and function and contribute to serotonin biosynthesis. The level of these chemicals is mediated by the microbes through the production of various short-chain fatty acids. These fatty acids influence symptoms of various neurological disorders like Alzheimer's disease, Parkinson disease, Autism spectrum and others (Chen et al. 2021).

9.4.3 Diabetes

Diabetes is a metabolic disorder with abnormal glucose metabolism generating various long-term complications (Cooper et al. 2001). Research shows that one of the most prominent reasons for onset of diabetes is dietary changes. Recent research has confirmed that, in diabetic patients, dietary fiber plays a positive role in managing plasma glucose and insulin and also shows hypocholesterolemic effect (Giacco et al. 2002; Chandalia et al. 2000; Plazonić et al. 2009). A recent study by Reynolds et al. (2020) examined the effect of dietary fiber intake on glucose levels in diabetes patients and found a dose response relationships between dietary fiber and glucose level. Often a high dietary fiber intake is responsible for changing the pH of gut and eventually enhances the level of SCFA-producing microbiota in diabetic humans (Zhao et al. 2018). The SCFAs produced by the microbiota are being studied with respect to their effect against diabetic complications. Results revealed that SCFA modulates hunger, inflammation, and is potentially linked with maintenance of type 2 diabetes. When a disease occurs, it affects the gut microbiota of the particular patient, which further affects the SCFA levels.

The mode of action of SCFAs involve various mechanisms such as regulation of GPCR receptors, inhibition of histone deacetylase (HDAC), and secretion of gut hormones. Extensive research led to identification of four GPCR receptors of SCFA namely GPR43, GPR41, GPR109A, and Olfr78 (Hwang et al. 2014). Among these SCFA receptors, GPCR 41 and GPCR43, are specifically expressed by immune cells, gut epithelia, and adipose tissue (Castro-Barquero et al. 2018). SCFA activates these GPCRs, which further activate the production of immunity-related cytokines such as IL-1, IL-6, and IL-12 (Kim et al. 2013). SCFA also follows another mechanism that is independent of GPCR which means it can work on cells devoid of GPCR. These molecules can pass through the cell using simple diffusion or carrier proteins (Kim et al. 2015). Once inside the cell, these intracellular SCFAs inhibit the HDACs as their main function is to regulate proteins, particularly histones. This suggests that dietary fiber can produce a SCFA profile that could have anti-inflammatory effects on the body. Moreover, SCFAs enhance important physiological processes such as glycolysis and oxidative phosphorylation in actively dividing lymphocytes throughout immune responses. In addition, SCFAs boost fatty acid biogenesis, which is important for cell proliferation and differentiation (Lee et al. 2011; Kim 2018).

The maximum SCFA production takes place in the cells of distal region of colon. Research by Larraufie et al. (2018) shows that propionate and butyrate strongly increased the expression of appetite-suppressing molecules viz. gut hormones peptide YY (PYY) and glucagon-like peptide 1 (GLP-1) in the same region suggesting a strong correlation between PYY and GLP-1 production and SCFA. Both GLP-1 and PYY have potential antidiabetic and anti-obesity effects (Fig. 9.4). GLP-1 stimulates SCFA receptors eventually increasing insulin level and lowering glucagon level. Psichas et al. (2015) examined the correlation of the PYY and GLP-1 with SCFA receptors and free fatty acid receptor 2 (FFA2 also known as GPR43) present on

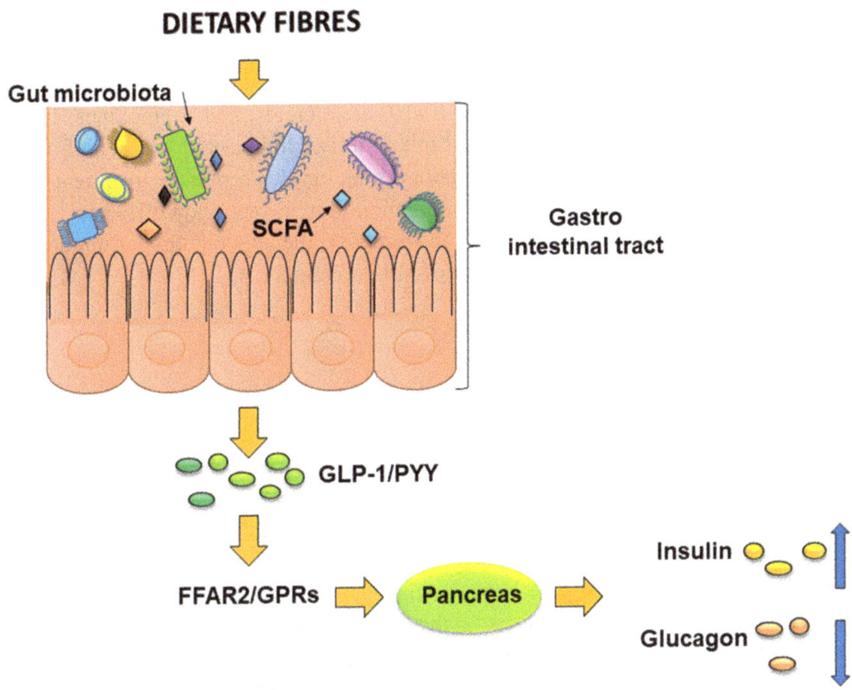

Fig. 9.4 Mode of action of SCFAs against diabetes

cells of the colon using in vitro and in vivo mice models. Results revealed that propionate remarkably enhanced the secretion of both PYY and GLP-1 in wild type. However, in FFA2-deficient (FFA2−/−) mice no significant increase was observed. Intra-colonic infusion of propionate elevated PYY and GLP-1 levels in jugular vein plasma in rats and in portal vein plasma in both rats and mice. However, propionate did not significantly stimulates gut hormone release in FFA2−/− mice indicating that FFA2−/− deficiency in primary colonic cultures impairs GLP-1 secretion to SCFAs (Tolhurst et al. 2012; Karaki et al. 2006; Christiansen et al. 2018). The above findings were supported by the research of (Wang and Xie 2020) where probiotics showed antidiabetic effect mediated by an increase in the production of SCFA-producing bacteria and GLP-1. In conclusion, the available data clearly shows that SCFAs are effective in treating diabetes through various mechanisms, e.g., stimulating pancreas via a receptor on pancreatic cells.

9.4.4 *Inflammatory Regulation*

During inflammation, leukocytes from the bloodstream reach the location of injury or inflammation. The recruitment of leukocytes involve a multistep process of transcription and translation of several proteins (Luster et al. 2005). SCFAs help in the process by promoting the recruitment of circulating leukocytes to the inflammatory site (Maslowski et al. 2009; Vinolo et al. 2011). SCFAs also promote cytokines production such as TNF-α, IL-2, IL-6, IL-10, eicosanoids, and chemokines. In an experiment where mice (exposed to dust mites) were administered with a high-fiber diet showed lowered levels of inflammation in the lungs. Propionate treatment also reduced dust mites-induced hepatic inflammation in wild-type and GPR43 receptor-deficient mice (Maslowski et al. 2009). In summary, the above findings support the positive role of SCFAs in the reduction of inflammation.

9.4.5 Gut Health

SCFA production by fermentation shows a considerable positive effect on gut health of host. It could be due to a significant reduction in the pH value leading to a mildly acidic condition in the proximal colon compared to distal region. Colonization of pathogenic bacteria is inhibited in such mild acidic conditions, however, favors the flourishing of SCFAs producing bacteria (Fattahi et al. 2020). SCFAs produced by gut microbes have prominent effects on the metabolic activities of tissues outside of the intestines, e.g., liver and adipose tissue through modulation of fat and carbohydrate homeostasis (Delzenne et al. 2011). SCFAs specifically acetate and butyrate protect the gut epithelia by favoring mucus production and decreasing the interaction with luminal microorganisms and toxins in the gastrointestinal tract (Allen et al. 2015).

9.5 Conclusion

SCFAs are the key metabolites that connect dietary fibers and gut microbiota to the intestinal health. Fats or lipids are a class of macromolecules containing carboxylic acids as a functional group at one end of the aliphatic chain. An alteration in both the abundance and function of gut microorganisms (named dysbiosis) has now been shown to be a trait that is common in multiple pathogenesis. Among SCFAs, butyrate has been studied extensively owing to its potent activities against critical diseases, e.g., cancer and diabetes. The report namely "Food, Nutrition and the Prevention of Cancer: A Global Perspective" a new dimension of the field came into view because the accumulated evidence was showing that cancer and diet were strongly interrelated. SCFAs act as a ligand for its specific cell surface receptors such as GPR41, GPR43, and GPR109A, and further activate and inhibit the intracellular target proteins such as HDAC. The chapter highlights the role of SCFAs as a critical factor in the maintenance of gut and immune homeostasis. SCFAs with potential activity and vast effects, could prove to be the future target for therapy of diseases.

References

Allen T, Mba GMNV, Shoja G, Razavi E (2015) Complementary medicines and cancer. J Cancer Sci Res 1:17

Arango D, Mariadason JM, Wilson AJ et al (2003) c-Myc overexpression sensitises colon cancer cells to camptothecin-induced apoptosis. Br J Cancer 89:1757–1765. https://doi.org/10.1038/sj.bjc.6601338

Belkaid Y, Hand TW (2014) Role of the microbiota in immunity and inflammation. Cell 157:121–141. https://doi.org/10.1016/j.cell.2014.03.011

Bindels LB, Porporato P, Dewulf EM et al (2012) Gut microbiota-derived propionate reduces cancer cell proliferation in the liver. Br J Cancer 107:1337–1344. https://doi.org/10.1038/bjc.2012.409

Bindels LB, Delzenne NM, Cani PD, Walter J (2015) Towards a more comprehensive concept for prebiotics. Nat Rev Gastroenterol Hepatol 12:303–310. https://doi.org/10.1038/nrgastro.2015.47

Bishehsari F, Engen PA, Preite NZ et al (2018) Dietary fiber treatment corrects the composition of gut microbiota, promotes SCFA production, and suppresses colon carcinogenesis. Genes (Basel) 9. https://doi.org/10.3390/genes9020102

Brestoff JR, Artis D (2013) Commensal bacteria at the interface of host metabolism and the immune system. Nat Immunol 14:676–684. https://doi.org/10.1038/ni.2640

Bugaut M (1987) Occurrence, absorption and metabolism of short chain fatty acids in the digestive tract of mammals. Comp Biochem Physiol – Part B Biochem 86:439–472. https://doi.org/10.1016/0305-0491(87)90433-0

Castro-Barquero S, Lamuela-Raventós RM, Doménech M, Estruch R (2018) Relationship between mediterranean dietary polyphenol intake and obesity. Nutrients 10:1523. https://doi.org/10.3390/nu10101523

Chandalia M, Garg A, Lutjohann D et al (2000) Beneficial effects of high dietary fiber intake in patients with type 2 diabetes mellitus. N Engl J Med 342:1392–1398. https://doi.org/10.1056/nejm200005113421903

Chen Y, Xu J, Chen Y (2021) Regulation of neurotransmitters by the gut microbiota and effects on cognition in neurological disorders. Nutrients 13. https://doi.org/10.3390/NU13062099

Chow C (2007) Fatty acids in foods and their health implications, 3rd edn. CRC Press, Boca Raton

Christiansen CB, Gabe MBN, Svendsen B et al (2018) The impact of short-chain fatty acids on glp-1 and pyy secretion from the isolated perfused rat colon. Am J Physiol – Gastrointest Liver Physiol 315:G53–G65. https://doi.org/10.1152/ajpgi.00346.2017

Cooper ME, Bonnet F, Oldfield M, Jandeleit-Dahm K (2001) Mechanisms of diabetic vasculopathy: an overview. Am J Hypertens 14:475–486. https://doi.org/10.1016/S0895-7061(00)01323-6

Delzenne NM, Neyrinck AM, Bäckhed F, Cani PD (2011) Targeting gut microbiota in obesity: effects of prebiotics and probiotics. Nat Rev Endocrinol 7:639–646. https://doi.org/10.1038/nrendo.2011.126

Den Besten G, Van Eunen K, Groen AK et al (2013) The role of short-chain fatty acids in the interplay between diet, gut microbiota, and host energy metabolism. J Lipid Res 54:2325–2340. https://doi.org/10.1194/jlr.R036012

Enciu A-M, Codrici E, Mihai S et al (2018) Role of nutraceuticals in modulation of gut-brain axis in elderly persons. In: Gerontology. IntechOpen. 5 Princes Gate Court, London, SW7 2QJ, UNITED KINGDOM https://www.intechopen.com/chapters/58594

Fattahi Y, Heidari HR, Khosroushahi AY (2020) Review of short-chain fatty acids effects on the immune system and cancer. Food Biosci 38:100793. https://doi.org/10.1016/j.fbio.2020.100793

Giacco R, Clemente G, Riccardi G (2002) Dietary fibre in treatment of diabetes: myth or reality? Dig Liver Dis 34:S140–S144. https://doi.org/10.1016/S1590-8658(02)80182-7

Gibson GR, Roberfroid MB (1995) Dietary modulation of the human colonic microbiota: introducing the concept of prebiotics. J Nutr 125:1401–1412. https://doi.org/10.1093/jn/125.6.1401
He C, Liu Y, Ye S et al (2021) Changes of intestinal microflora of breast cancer in premenopausal women. Eur J Clin Microbiol Infect Dis 40:503–513. https://doi.org/10.1007/s10096-020-04036-x
Hinnebusch BF, Meng S, Wu JT et al (2002) The effects of short-chain fatty acids on human colon cancer cell phenotype are associated with histone hyperacetylation. J Nutr 132:1012–1017. https://doi.org/10.1093/jn/132.5.1012
Hwang SJ, Yoon WB, Lee OH et al (2014) Radical-scavenging-linked antioxidant activities of extracts from black chokeberry and blueberry cultivated in Korea. Food Chem 146:71–77
Karaki SI, Mitsui R, Hayashi H et al (2006) Short-chain fatty acid receptor, GPR43, is expressed by enteroendocrine cells and mucosal mast cells in rat intestine. Cell Tissue Res 324:353–360. https://doi.org/10.1007/s00441-005-0140-x
Khan I, Huang G, Li X et al (2019) Mushroom polysaccharides and *Jiaogulan saponins* exert cancer preventive effects by shaping the gut microbiota and microenvironment in Apc Min/+ mice. Pharmacol Res 148. https://doi.org/10.1016/j.phrs.2019.104448
Kim CH (2018) Microbiota or short-chain fatty acids: which regulates diabetes? Cell Mol Immunol 15:88–91. https://doi.org/10.1038/cmi.2017.57
Kim DH, Park MH, Choi YJ et al (2013) Molecular study of dietary heptadecane for the anti-Inflammatory modulation of NF-kB in the aged kidney. PLoS One 8. e59316
Kim MS, Jeon WK, Lee KW et al (2015) Ameliorating effects of ethanol extract of fructus mume on scopolamine-induced memory impairment in mice. Evidence-based Complement. Altern Med 2015:1–8
Larraufie P, Martin-Gallausiaux C, Lapaque N et al (2018) SCFAs strongly stimulate PYY production in human enteroendocrine cells. Sci Rep 8. https://doi.org/10.1038/s41598-017-18259-0
Lee HJ, Lee G, Jang NR et al (2011) Biological synthesis of copper nanoparticles using plant extract. In: Technical proceedings of the 2011 NSTI nanotechnology conference and expo, NSTI-Nanotech 2011, pp 371–374
Luster AD, Alon R, von Andrian UH (2005) Immune cell migration in inflammation: Present and future therapeutic targets. Nat Immunol 6:1182–1190. https://doi.org/10.1038/ni1275
Martin-Gallausiaux C, Marinelli L, Blottière HM et al (2021) SCFA: mechanisms and functional importance in the gut. In: Proceedings of the nutrition society. Cambridge University Press, pp 37–49
Maslowski KM, Vieira AT, Ng A et al (2009) Regulation of inflammatory responses by gut microbiota and chemoattractant receptor GPR43. Nature 461:1282–1286. https://doi.org/10.1038/nature08530
Matthews GM, Howarth GS, Butler RN (2012) Short-chain fatty acids induce apoptosis in colon cancer cells associated with changes to intracellular redox state and glucose metabolism. Chemotherapy 58:102–109. https://doi.org/10.1159/000335672
Plazonić A, Bucar F, Maleš Ž et al (2009) Identification and quantification of flavonoids and phenolic acids in burr parsley (*Caucalis platycarpos* L.), Using high-performance liquid chromatography with diode array detection and electrospray ionization mass spectrometry. Molecules 14:2466–2490
Psichas A, Sleeth ML, Murphy KG et al (2015) The short chain fatty acid propionate stimulates GLP-1 and PYY secretion via free fatty acid receptor 2 in rodents. Int J Obes 39:424–429. https://doi.org/10.1038/ijo.2014.153
Reynolds AN, Akerman AP, Mann J (2020) Dietary fibre and whole grains in diabetes management: systematic review and meta-analyses. PLoS Med 17.:e1003053. https://doi.org/10.1371/journal.pmed.1003053
Ríos-Covián D, Ruas-Madiedo P, Margolles A et al (2016) Intestinal short chain fatty acids and their link with diet and human health. Front Microbiol 7:185. https://doi.org/10.3389/fmicb.2016.00185

Sagar SK (2021) mi R-106b as an emerging therapeutic target in cancer. Genes Dis. https://doi.org/10.1016/j.gendis.2021.02.002

Sanders ME, Merenstein DJ, Reid G et al (2019) Probiotics and prebiotics in intestinal health and disease: from biology to the clinic. Nat Rev Gastroenterol Hepatol 16:605–616

Scharlau D, Borowicki A, Habermann N et al (2009) Mechanisms of primary cancer prevention by butyrate and other products formed during gut flora-mediated fermentation of dietary fibre. Mutat Res 682:39–53. https://doi.org/10.1016/j.mrrev.2009.04.001

Silva YP, Bernardi A, Frozza RL (2020) The role of short-chain fatty acids from gut microbiota in gut-brain communication. Front Endocrinol (Lausanne) 11:25. https://doi.org/10.3389/FENDO.2020.00025/BIBTEX

Stein J, Zores M, Schröder O (2000) Short-chain fatty acid (SCFA) uptake into Caco-2 cells, by a pH-dependent and carrier mediated transport mechanism. Eur J Nutr 39:121–125. https://doi.org/10.1007/s003940070028

Tang Y, Chen Y, Jiang H et al (2011) G-protein-coupled receptor for short-chain fatty acids suppresses colon cancer. Int J Cancer 128:847–856. https://doi.org/10.1002/ijc.25638

Tolhurst G, Heffron H, Lam YS et al (2012) Short-chain fatty acids stimulate glucagon-like peptide-1 secretion via the G-protein-coupled receptor FFAR2. Diabetes 61:364–371. https://doi.org/10.2337/db11-1019

Vinolo MAR, Rodrigues HG, Nachbar RT, Curi R (2011) Regulation of inflammation by short chain fatty acids. Nutrients 3:858–876. https://doi.org/10.3390/nu3100858

Wang J, Xie Y (2020) Review on microbial degradation of zearalenone and aflatoxins. Grain Oil Sci Technol 3:117–125. https://doi.org/10.1016/j.gaost.2020.05.002

Qiu-shi X, Jia-xin Z, Liu M et al (2021) Short-chain fatty acids exert opposite effects on the expression and function of p-glycoprotein and breast cancer resistance protein in rat intestine. Acta Pharmacol Sin 42:470–481. https://doi.org/10.1038/s41401-020-0402-x

Yang K, Zhang L, Liao P et al (2020) Impact of gallic acid on gut health: focus on the gut microbiome, immune response, and mechanisms of action. Front Immunol 11:580208. https://doi.org/10.3389/fimmu.2020.580208

Yusuf F, Adewiah S, Syam AF, Fatchiyah F (2019) Altered profile of gut microbiota and the level short chain fatty acids in colorectal cancer patients. J Phys Conf Ser 1146:012037. https://doi.org/10.1088/1742-6596/1146/1/012037

Zhao L, Zhang F, Ding X et al (2018) Gut bacteria selectively promoted by dietary fibers alleviate type 2 diabetes. Science 80–359:1151–1156. https://doi.org/10.1126/science.aao5774

Chapter 10
Potential Role of Probiotics on Gut Microbiota in Neurological Disease

Jovel Varghese Jose and S. Aliya

Abstract Increasing research evidence cites that the gut microbiota and its composition in the gastrointestinal tract greatly influence the host's physiological and neuronal functions, immune and circulatory systems. A bidirectional communication or cross-talk comprising the microbiome and host exists between the brain and intestinal microbiota called microbiota-gut-brain (MGB) axis and neurological diseases have been thought to progress via the same. It serves various functions such as the production of neurons and neurotransmitters, maintenance of neuroendocrine system and also controls our response to stress and memorization. Any dysfunction in the axis impacts an individual's behavior and disease pathogenesis eventually affecting the central nervous system. Pre-clinical and clinical data of microbiota-directed therapies using probiotics which are non-pathogenic live microorganisms, showed a significant association of gut microbiome dysbiosis and its increasing potential on the development of mental disorders in a host. Animal model studies have elucidated the significance of MGB axis though in humans, more substantial evidence is required. This chapter primarily outlines the associated benefits of the gut-microbiota axis and how a dysbiosis affecting the same is capable of triggering neurological disorders, often causing a defective development in brain function and molecular mechanisms of the gut-brain axis. A paradigmatic shift with the focus on microbiota-targeted therapy involving the modification of an individual's gut microbiome with the aid of probiotics offers promising future prospects.

Keywords Gut microbiota · Gut-brain axis · Neurological diseases · Probiotics · Psychobiotics · Dysbiosis

---

J. V. Jose (✉)
Amity University, Mumbai, India

Blue Marble Space Institute of Science, Seattle, WA, USA

S. Aliya
Sree Sankara College, Kalady, India

A. K. Tripathi, M. Kotak (eds.), *Gut Microbiome in Neurological Health and Disorders*, Nutritional Neurosciences,
https://doi.org/10.1007/978-981-19-4530-4_10

10.1 Introduction

The gastrointestinal (GI) tract's surface epithelium covers 32 m^2 and stretches over 5 m in length (Helander and Fändriks 2014), anatomically GI organs are mesentery sectioned with each organ being composed of the mucosa, submucosa, muscular layer, and serosa. A pathogenic microorganism is an infectious, disease-causing agent which in the case of neurological disorders, negatively impacts the functioning of brain inducing both neurological and mental defects. Infectious diseases like HIV (caused by human immuno-deficiency virus) and Lyme disease (caused by *Borrelia burgdorferi*) can cause serious psychiatric and neurodegenerative effects on the host. Also in late stages of the parasitic infection, brain toxoplasmosis (triggered by *Toxoplasma gondii*) has indicated a possible link leading to a suicidal nature (Coccaro et al. 2016). These expressions of neurological infectious diseases are caused by the direct hindrance of neurotransmitter signaling by pathogens. Characterization of the gut microbiome has revolutionized our perception of gastrointestinal and metabolic processes. All bacteria are not pathogenic invaders, instead may have a potential function in maintaining immunity and homeostasis, this notion has impacted a major shift in neuroscience and neuropsychiatric research. The gastrointestinal (GI) microbiome is a distinctive assortment of commensal microorganisms inhabiting in diverse niches in the GI tract which substantially help both in the advancement and improvement of psychiatric and neurological disorders.

Probiotic microorganisms belong to diverse groups and play multiple roles for example short-chain fatty acid (SCFA) production, ferment undigested carbohydrates, synthesize vitamins and metabolites (Quigley 2013). Gestational age, delivery mode, diet, and antibiotic exposure regulate the influence the composition in gastrointestinal colonization by microorganisms (Fouhy et al. 2012). The important potential phyla present in human gut are Bacteroides and firmicutes (Bäckhed 2011). The other beneficial and opportunistic bacterial gut flora including *Lactobacillus* sp., *Bifidobacterium* sp., *Propionobacteria*, *Enterococci*, *Peptostreptococci*, etc. are a beneficial group but those such as *Actinobacteria*, *Bacteroides* sp., *Clostridia*, *Peptococci*, *Enterobacteria*, *Streptococci*, *Staphylococci*, and yeasts are of the opportunistic group (Joshi et al. 2018). The MGB axis is referred to as a bidirectional system of communication by the gut microbiome with the central nervous system (CNS) sending signals to the brain and vice versa (Fung et al. 2017). Behavioral factors like stress accelerate the corticotrophin-releasing hormone production in the hypothalamus, further activating the hypothalamic–pituitary–adrenal (HPA) axis (Belmaker and Agam 2008). The synthesis of neuroactive metabolic molecules by gut microbiota modulates the pathogenesis of several neurological diseases such as amyotrophic lateral sclerosis, Parkinson's disease, multiple sclerosis, etc. (Girolamo et al. 2017). This chapter focuses on the cross-talk-mediated signaling of gut microbiota to the brain that impacts its functional development, summarizes the involvement of the gut-intestinal microbiota in the progression of neuropsychiatric disorders while making remarks on the potential therapeutic benefits of probiotics particularly in their actions against neurological disorders.

10.2 Microbiome-Gut-Brain Axis: A Bi-directional Communication System

10.2.1 Role and Developmental Role and Mechanism of Action of Gut-Brain Axis

The early life microbial colonization in an individual right after birth proceeds through a rapid development of unique site-specific microbial niches, and change in the microbiota composition from primitive in earlier life to mature form (Dominguez-Bello et al. 2016). "Barker Hypothesis" (1993) stated that the development of a fetus and its sensitivity to neurological and metabolic diseases is substantially affected by its intrauterine habitat in life. The GI tract instantaneously develops and facilitates for the colonization of microbiota. The pathogen *Campylobacter jejuni* stimulates behavioral abnormalities in early life, decreased motor function and increased anxiety (Forsythe et al. 2010). DNA methylation due to prenatal stress affects functional advancement of the HPA axis leading to hypersensitivity and glucocorticoids hypersecretion also decreased the binding capacity of the hippocampal glucocorticoid receptor (Murgatroyd et al. 2009). The cholinergic signaling synchronized with the HPA axis blocking nicotinic and muscarinic receptors causes the defect in barrier function which prevents an increase in macromolecular permeability (Gareau et al. 2007). Development in the womb is initially established and continued after the birth period but the emotions and storage of memories are controlled by the limbic system.

The process of new neuron formation known as neurogenesis takes place in specific regions of the brain throughout life (Fig. 10.1). Neurogenesis and memory endurance level have a complementary relationship between plasticity and the ability

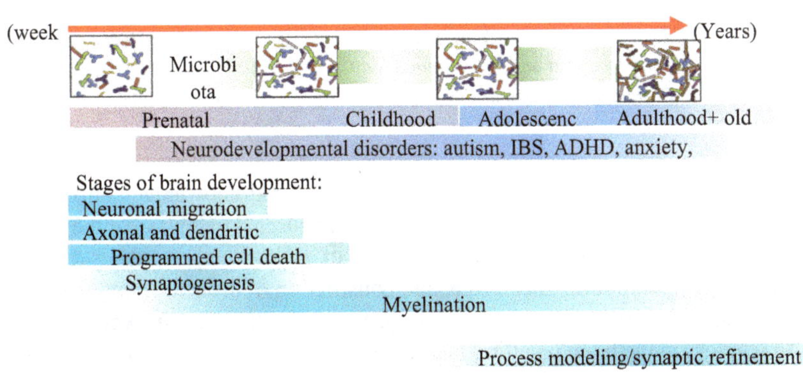

Fig. 10.1 The development of brain and gut microbiome. During the prenatal stage brain and gut microbiota begins, in first 3 years is critical developmental stage. Disturbance in development can impact communication between these systems and it can also facilitate pathogenesis of neural development disorders for instance autism, IBS, attention-deficit hyperactivity disorders (ADHD), anxiety, and obesity (Ogbonnaya et al. 2015).

Table 10.1 Summary of microbiome observed in human studies of different neurological disorders

Neurological disorders	Gut microbiota	References
Parkinson's disease	*Lactobacillus, Bifidobacterium, Akkermansia; Prevotella, Lachnospiraceae, Faecalibacterium*	Qian et al. (2018), Petrov et al. (2017)
Multiple system atrophy	*Bacteroides, Proteobacteria, Prevotella*	Tan et al. (2018)
Alzheimer's diseases	*Ruminococcaceae, Enterococcaceae, Lactobacillaceae; Firmicutes, Bifidobacterium, Lachnospiraceae*	Zhuang et al. (2018)
Amyotrophic lateral sclerosis	*Dorea, Oscillibacter, Anaerostipes, Lachnospira*	(Mazzini et al. 2018)
Multiple sclerosis	*Akkermansia, Clostridial clusters*	Cekanaviciute et al. (2017)
Autism spectrum disorders	*Betaproteobacteria, Sutterella, Bifidobacterium, Firmicutes, Lactobacillus, Clostridium, Candida, Bacteroidetes*	Tap et al. (2017), Zhang et al. (2018)
Depression and anxiety	*Proteobacteria, Bacteroides, Oscillibacter, Alistipes, Lachnospiraceae, Faecalibacterium, Bifidobacterium*	Lin et al. (2017b)
Chronic fatigue	*Firmicutes, Faecalibacterium*	Nagy-Szakal et al. (2017)
Schizophrenia	*Helicobacter pylori, Lachnospiraceae, Ruminococcaceae, Lactobacillaceae, Firmicutes, Halothiobacillaceae, Brucellaceae, Micrococcineae, Bacteroidetes, Actinobacteria, Veillonellaceae*	Breban et al. (2017)
Eating disorders (anorexia nervosa)	*Firmicutes, Bacteroidetes, and Actinobacteria*	Kleiman et al. (2015)
Dementia	*L. helveticus, L. pentosus, Saccharomyces cerevisiae*	Yeon et al. (2010), Jung et al. (2012), Lee et al. (2007)
ADHD	*Firmicutes, Bacteroidetes, Proteobacteria, Actinobacteria and Faecalibacterium*	Lynch and Pedersen (2016)
Epilepsy	*Firmicutes and Bacteroidetes*	Thursby and Juge (2017)

to encompass new information without deteriorating stored knowledge (Zhao et al. 2008). Microbiota communicates with CNS through neuronal pathways of the vagus nerve belonging to the parasympathetic division of ANS which regulates various functions such as gut motility, heart rate, and constriction of bronchi (Forsythe et al. 2010). Probiotic *Lactobacillus rhamnosus* sp. induces alteration in expression of GABA receptors in the brain resulting in reduced corticosterone and in an anxiolytic effect through signaling from the vagus nerve (Bravo et al. 2011). Probiotic mixture comprising *B. animalis* subsp. and *Propionibacterium jensenii* restored the imbalance in gut microbiota and attenuated activation of neonatal stress pathways in HPA axis. Presence of fatty acid composition also influences neurophysiologic conditions

(Forsythe et al. 2010) which was observed in the immune conditions of stressed animals while experiencing neonatal maternal separation (Wall et al. 2012).

Antibiotics fundamentally show detrimental effects on neurodegenerative and neurodevelopment diseases disrupting gut microbiota involved in neuro-modulatory signaling. SCFAs derived from microbiota have an essential role in the functioning and promotion of microglial maturation (Barouei et al. 2012). When the levels of neurogenesis are low after birth, memories become more resistant to remodeling and a rise in the stability of memories dependent on the hippocampus (Erny et al. 2015). Decrease in neurogenesis due to cortex proliferative subventricular zone and microglia phagocytose neural precursor cells (NPC) arising a change in neural development (Akers et al. 2014) impacting memory formation and cognitive function.

These microglial cells are involved in CNS development at an early stage, and also in phagocytosis, antigen presentation, and regulation of inflammation during their lifetime. The amalgamation of bromodeoxyuridine (BrdU) in the hippocampus accentuated neurogenesis in adult germ-free mice (Cunningham et al. 2013).

The increased volume and modified morphology of dendrites indicated that, for amygdala and hippocampus, normal morphology and ultrastructure requirement of microbiota is significant (Borre et al. 2014). Probiotic *Bifidobacterium longum* and *Lactobacillus helveticus* combination can prevent a decrease in neurogenesis in the hippocampus region when influenced by stress (Luczynski et al. 2016). Astrocytes are various operational groups of glial cells, having functions such as neurotransmitter clearance, ion homeostasis, glycogen storage, maintenance of blood–brain barrier, and neuronal signaling additionally to their distinguished neuroinflammatory function (Ait-Belgnaoui et al. 2014). Toll-like receptor (TLR) signaling controls hippocampal neurogenesis in mice signifying the potential regulatory role of microbial components in neurogenesis. TLR4 KO mice demonstrated the formation of spatial reference memory and fear of learning and enhanced neurogenesis (Okun et al. 2012), whereas TLR2 inadequacy diminished both hippocampal volume and neurogenesis (Rolls et al. 2007).

Prenatal oral administration of probiotics to pregnant maternal indicated that lactation normalized high glucocorticoid (cortisol) secretion and restored corticotrophin-releasing hormone (CRH); for example, oral intake of genetically modified *Enterococcus faecium* to pregnant mice indicated offspring possessing the specific bacterial species in meconium and amniotic fluid (Jiménez et al. 2008) confirming that maternal microbial transmission in mammals is possible and maternal tryptophan along with serotonin (5-HT) neurotransmitter hormone is essential for neurodevelopment conversion led by the placenta that influences fetal brain development. Serotonin [5-hydroxytryptamine (5-HT)] is a biogenic amine neurotransmitter in the CNS and in gut where synthesis depends on the availability of its precursor, and tryptophan synthesis maintains cognitive activity and also regulates gastrointestinal secretion (Costedio et al. 2007). Tryptophan metabolism and availability depend on enteric microbiota and consequently impact the central serotonin concentrations as well as kynurenine, whereas the activity of tryptophan and indoleamine-2,3-dioxygenase enzymes induced altered enzyme activity in irritable bowel syndrome (IBS) (Schwarcz et al. 2012) and downstream neuroactive metabolites in the brain. The probiotic *Bifidobacterium infantis* sp. affects tryptophan metabolism along this pathway (Desbonnet et al. 2010). The gut communicates through hormonal signaling pathways to the brain resulting in the release of peptides in gut from entero-endocrine cells (such as orexin, ghrelin, and leptin), circadian pattern, sexual behavior, and anxiety (Cameron and Doucet 2007). Saresella M and colleagues (2017) (Saresella et al. 2017) reported that on MS patients where data supported the chance that diet would probably be utilized as a tool to regulate the immune system in anti-inflammatory way as a significance of changes in the gut microbiota.

Present results of this trial study show that a skewing of the constitution of the microbiota characterized by the plenty of *Lachnospiraceae* family, a diminish of IL-17-producing T CD4+ lymphocytes and PD-1 expressing T CD4+ lymphocytes, and an enhance of PD-L1 stating monocytes was considered in those individuals following a HV/LP diet. In these same patients, positive correlations between *Lachnospiraceae* and anti-inflammatory IL-10- and TGFβ-producing CD14+ monocytes, as well as between *Lachnospiraceae* and CD4+/CD25+/FoxP3+ T-reg lymphocytes were also observed. The concurrent development during initial postnatal life of the microbiota, gastrointestinal tract maturation, and neurogenesis of the hippocampus simultaneously forms the MGB axis. Identification of pathways and mechanisms of this complex interconnectedness has substantial therapeutic effects for several diseases.

10.2.2 Effects of Human Microbiome and Probiotics on ENS, ANS, and CNS

10.2.2.1 Effect of Human Microbiome and Probiotics on ENS

The enteric nervous system (ENS) or the "second brain" is a part of the peripheral and autonomic nervous system which tends to control GI tract functioning (Turner 2009). It is located within the GI tract wall, hence it remains protected from the intestinal luminal contents, and is composed of ganglia and millions of neurons. It serves several diverse functions such as the secretions from the gut and pancreas, reflexes, blood flow, gastrointestinal motility and physiology, GI-endocrine modulation, and protective reactions (Yan and Polk 2011). The gut microbiome helps in the overall development and maintenance of the intestinal barrier, prevents pathogenic production of emetic toxins, and also protects the intestinal sensory nerves from pathogenic invasion (Borthakur et al. 2008; Kamm et al. 2004). The excitatory irregularity of the ENS often caused as a result of gut dysbiosis has also been considerably reduced with the aid of probiotics (Bercik et al. 2011).

10.2.2.2 Effects of Human Microbiome and Probiotics on ANS

The autonomic nervous system (ANS) comprises the sympathetic and parasympathetic systems and is also composed of both motor and sensory neurons that transverse between the various internal organs and the central nervous system (CNS) (Azpiroz 2005). The sympathetic and parasympathetic systems together constitute the autonomic nervous system (ANS) which is also composed of both motor and sensory neurons that transverse between the various internal organs and the central nervous system (CNS) (Azpiroz 2005). The sympathetic nervous system prepares and aids the body in triggering the fight or flight response and subsequent reflexes, whereas the parasympathetic system on the other hand aids in returning the body functions adjusted by the former, from the excited/activated state to its normal stature (O'Mahony et al. 2011). Both prebiotics and probiotics regulate the generation of pro-inflammatory cytokines, thereby sustaining the intestinal barrier, promoting the ideal functioning of the brain-gut axis (although the supporting evidence is still under scrutiny), and exerting anti-inflammatory effects when administered in combination with certain fatty acid supplements (Desbonnet et al. 2010; Clarke et al. 2010; Wall et al. 2010).

10.2.2.3 Effects of Human Microbiome and Probiotics on Central Nervous System

The central nervous system (CNS) is majorly composed of the brain and spinal cord. It has a network of millions of neurons that transverse the comprehensive length of the body (Neufeld et al. 2011). One of the best examples for a CNS output includes human behavior and few studies have indicated the possible correlation between an individual's neurochemical characteristics with that of his/her behavioral constitution (Heijtz et al. 2011). The crucial role of the gut microbiome on an individual's CNS and corresponding mental status is already quite evident from several studies based on animal and human models (Galland 2014). The administration of probiotics has not only been found to positively impact the functioning of the human brain but has also shown to modify its neurochemistry both directly and indirectly. It has also markedly helped to decrease anxiety and depression with prolonged treatment (Bravo et al. 2011).

The mechanisms involved in the effect of probiotics on CNS include (Sharma and Kaur 2020):

(a) Restoration of the functioning of the hypothalamic-pituitary-adrenal (HPA) axis that plays an essential role in our response to stress (Varghese and Brown 2001). Indirectly, it can act as a mediator in a number of neurological diseases such as anxiety, depression, IBS, etc. (Smith and Vale 2006). Increased activity of the HPA may be induced in conditions of prolonged stress and anxiety as reported in patients suffering from bipolar or depression associated disorders (Guilliams and Edwards 2010). Probiotic interventions that focus on the normalization of the HPA axis in patients experiencing psychotic disorders via regulating the levels of cortisol (CORT) or adrenocorticotropic hormone (ACTH) have given encouraging results (Savignac et al. 2015).

(b) Regulation and production of brain-derived neurotrophic factor (BDNF) and SCFAs; neuronal factors such as BDNF are mostly proteins that aid in controlling the functioning of neurons including their differentiation, integrity, and survival (Edelmann et al. 2014). Neurodegenerative diseases are frequently accompanied by an impaired or abnormal expression of neurotrophic factors (Rao et al. 2007). A study by Ranuh et al. (2019)) demonstrated the direct linkage between the increased BDNF levels in the brain after probiotic administration and their corresponding stimulation of the gut-brain axis. When it comes to SCFAs, their increased production by the gut microbiome was shown to alleviate the oxidative stress induced during neurological disorders (Hamer et al. 2008).

(c) Proto-oncogene activation and expression are likely to be enhanced in patients suffering from psychiatric disorders in particular Alzheimer's disease and dementia (Lu et al. 1998). According to Smith et al. (2014)), probiotic administration can directly reinstate the expression of proto-oncogenes.

(d) Stimulation of the vagus nerve, a major constituent of the parasympathetic nervous system by probiotic intervention, can help in the treatment of several neurological disorders like depression, anxiety, schizophrenia, IBS, etc. (Johnson and Wilson 2018). Enhanced activity of the vagus nerve helps to decrease the liberate pro-inflammatory cytokines such as TNF-α in stress-related disorders (Herman et al. 2016) (Fig. 10.2).

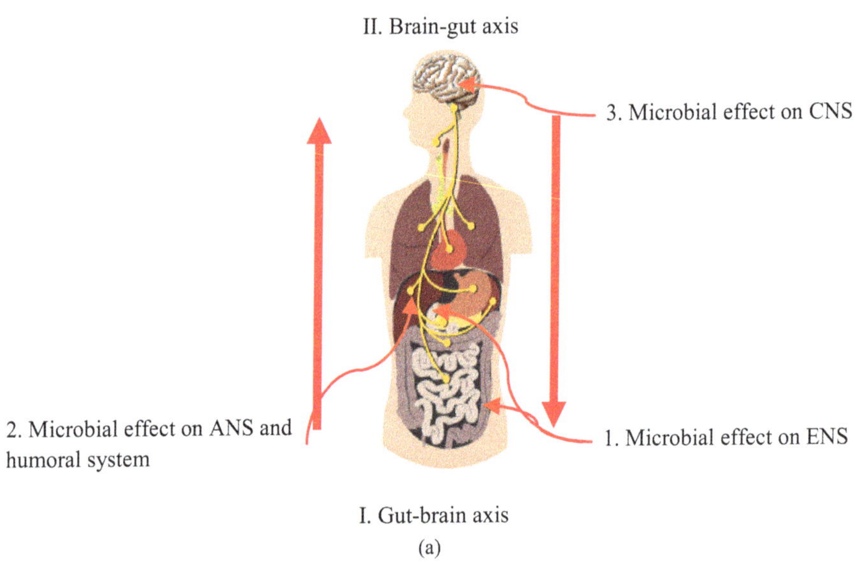

Fig. 10.2 (**a**) Effect of neurogastroenterology due to the gut microbial impact in gastrointestinal tract. (**b**) Representative image of the effect of probiotics on CNS, ENS, and ANS

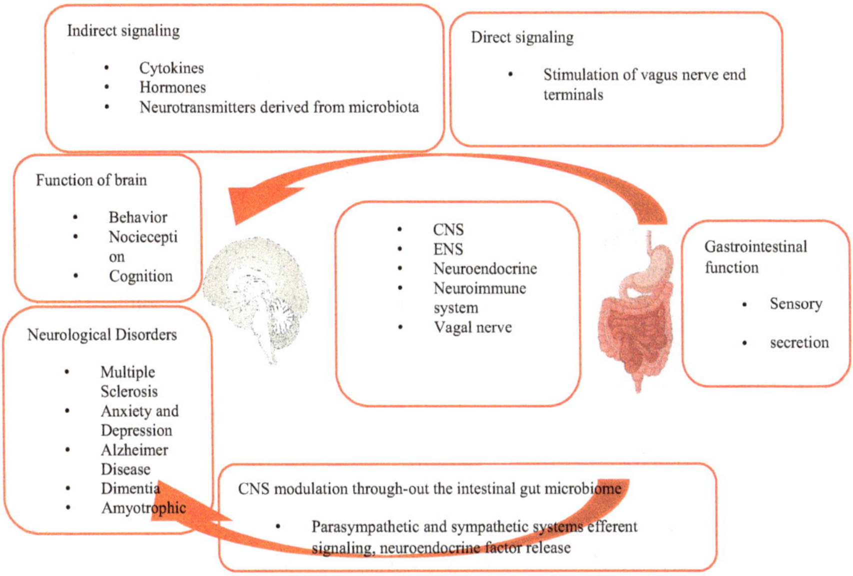

(b)

Fig. 10.2 (continued)

10.3 Neurological Diseases Influenced by Imbalance of Gut-Brain Axis

10.3.1 Amyotrophic Lateral Sclerosis

Amyotrophic lateral sclerosis (ALS) is a sequential neurodegenerative disorder. The symptoms include cramping, muscle spasm, weakness, muscle twitching, and stiffness, problems with coordination, speaking, breathing, and swallowing difficulties, weight loss, gastroparesis, and an increase in metabolic rate, relating to the death of motor neurons, spinal cord, and eventually the brain. Neuroinflammation is recognized as a disease driver for the development of ALS disorder (Skaper et al. 2018). Immune pathway de-regulation is a principal feature in the brain and spinal cord tissue affected with ALS (D'Erchia et al. 2017). ALS is of two types: (i) Sporadic ALS, signifying the most general form although causation is unknown, (ii) Familial ALS, occurring due to genetic changes (Steyn et al. 2018; Toepfer et al. 2000).

Modifications in monocytes, neutrophils, CD4+, CD8+, and natural killer T cells have been observed in patients with ALS causing progression in the disease rate (Perner et al. 2018; Zhou et al. 2017). Association of the gut microbiome and ALS in a transgenic mouse model with G93A genetic mutation of the superoxide dismutase gene (SOD1/SOD1 G93A) disclose disrupted BBB and increased permeability of the gut, reduced amount of butyrate-producing bacteria *Peptostreptococcus* and

Butyrivibrio fibrisolvens resulting in an elevated level of serum and IL-17 (intestinal pro-inflammatory cytokine) as well as damaged intestinal wall due to decreased expression of E-cadherin and zonula occludens (ZO-1) (Wu et al. 2015). Furthermore, the mutant mice SOD1G93A treated with 2% butyrate in drinking water, improved intestinal barrier function by bacteria *B. fibrisolvens* as well as delayed symptoms like weight loss and even death, in comparison to control mice (Zhang et al. 2017). Butyrate-producing bacteria like *Lachnospira*, *Anaerostipes*, and *Oscillibacter* were in reduced levels in feces of ALS patients and indicated increased levels of *Dorea* spp. which synthesizes harmful end product ethanol in glucose metabolism (Fang et al. 2016). The therapeutic potential of probiotics for the improvement of the ALS condition is yet to be fully explored.

10.3.2 Epilepsy

It is a neurological globally prevalent disease. Symptoms like epileptic seizures are often experienced by the affected individuals and lack of proper medical care during severe seizures can even prove to be fatal (Gómez-Eguílaz et al. 2018). Although multiple therapeutic alternatives are in practice, their benefits remain limited or short-termed. The use of probiotics for the treating of pharmaco-resistant or refractory epilepsy (that often involve highly convulsive seizures) is already undergoing trials (Iannone et al. 2020). Such studies have revealed promising results by reducing the symptomatic seizures by over 50%, thus eventually contributing to a better quality of life among the patients (Krauss and Sperling 2011).

10.3.3 Autistic Spectrum Disorder

Autism or autistic spectrum disorder (ASD) comprises an extensive range of neuropsychiatric disorder exhibited mainly during infancy wherein affected children tend to experience numerous neurodevelopmental disabilities along with stereotypical or repetitive actions, communicational, and social interaction difficulties and intestinal problems (Lord et al. 2000; Horvath et al. 1999). Real-time PCR meta-analysis studies indicated autistic children have a lower copiousness of bacterial phyla comprising *Proteobacteria*, *Bacteroidetes*, *Firmicutes*, *Bifidobacteria* spp. and especially *A. muciniphila* mucolytic bacterium in their feces with comparison to controls (Cao et al. 2013; Wang et al. 2011). A probiotic consortium of *Lactobacillus*, *Streptococcus*, and *Bifidobacterium* showed considerable changes in fecal cytokine levels of ASD patients (Tomova et al. 2015).

Gut microbiota produces metabolites such as acetate, valerate, and propionate in lower levels in autistic patients. Also, ASD animal model MIA mouse resembles characteristics of ASD in mouse offspring which are caused due to gut-intestinal barrier defects and exhibited changes in significant microbial metabolite 4-ethylphenyl-sulfate (4EPS) associated with anxiety behaviors (Adams et al. 2011) and *Bacteroides fragilis* reverses gut-intestinal barrier non-regulation in the MIA mouse. Alternative ASD mouse model strain BTBR T+ Itpr3tf/J is directed by multiple genetic modifications inducing autism-like behavior leading to change in gut microbiota with a reduction in SCFAs, tryptophan, and bile acid metabolism (Meyza and Blanchard 2017). Simultaneously, these presymptomatic and symptomatic data substantiate the conception of a MGB axis dysfunction in ASD.

10.3.4 Dementia

It is a term collectively used to describe the various signs and symptoms of both cognitive and psychological impairment, often characterized by difficulties associated with communication, coordination, visual perception, memory loss, etc. Many conditions can contribute to dementia such as Alzheimer's disease, Parkinson's disease, brain injury, and traumas (Bachstetter et al. 2015). The administration of probiotics and prebiotics helps in stimulation of the functioning of the gut microbiome, in turn affecting the MGB axis (Cryan and Dinan 2012). With the advancement in age, the composition of gut microbiome changes denoted by the loss of a few beneficial microbes reduced heterogeneity and a possible increase of pathogenic microbes. Aging is probably one of the major risk factors that lead to the development of dementia and hence the microbiome-gut-brain axis is also likely to be a crucial factor in the development of dementia (as per the ongoing study references from Medical University of Graz, probiotics in dementia-undergoing clinical trials).

10.3.5 Multiple Sclerosis (MS)

Multiple sclerosis (MS) is a chronic autoimmune neurological disorder of the central nervous system. The neuropathological characteristics of MS include axonal damage, demyelination, progressive neurological disability, neurodegeneration, and abnormal T-cell-facilitated immune responses triggered against myelin antigens (Ota et al. 1990). The clinical disclosures include dizziness, vision loss, vertigo, motor dysfunction, pain, numbness, impaired coordination, fatigue, and depression. MS is mainly divided into four: (i) Progressive-relapsing MS (PRMS), (ii) Primary Progressive MS (PPMS), (iii) Secondary Progressive MS (SPMS), (iv) Relapsing-Remitting MS (RRMS) (Noseworthy et al. 2000).

Neuropathogenesis of MS is dependent upon important environmental factors like gut dysbiosis (Mowry and Glenn 2018). RRMS is distinguished by a reduction in bacteria responsible for generating T regulatory cells (Tregs) which are immune cells accountable to anti-inflammatory reactions, CD4+ T cells that generate tolerogenic dendritic cells, regulatory B cells, IL-10, and macrophage suppression.

CD4+ T cells, dendritic cells, monocytes, or B cells produce pro-inflammatory reactions due to the increase in specific bacteria (Shahi et al. 2017). Germ-free mice having autoimmune encephalomyelitis (EAE) develop a substantially weakened pathology while the commensal microbiota in experimental mice stimulates CD4+ T cells which are myelin-specific and also obtained autoantibodies produced to myelin oligodendrocyte glycoprotein by B cells (Berer et al. 2011).

MS patient's fecal gut microbiota has a different bacterial composition in comparison with controls (Shahi et al. 2017). Presence in increased concentration of Proteobacteria species like *Acinetobacter calcoaceticus* (Cekanaviciute et al. 2017), *Pseudomonas*, Mycoplasma, (Chen et al. 2016), Bilophila (Miyake et al. 2015) in white matter lesions; *Akkermansia*, *Acinetobacter*, *Prevotella*, *Clostridium*, *Bacteroidetes*, and *Lactobacillus* bacterial genera in MS patients gut microbiota induces production of SCFAs (Berer et al. 2011; Branton et al. 2016) and helps in the maintenance of immune cell for producing an anti-inflammatory reaction. The importance of bacteria *Faecalibacterium* in MS patients is its production of butyrate which increases the production of T^{regs} (Arpaia et al. 2013). Colonization of *Clostridium perfringens* type B in the human GI tract increases epsilon toxin level which deteriorates the blood–brain barrier (BBB), dominating oligodendrocyte and neuronal damage and also activating autoimmune demyelinating action (Rumah et al. 2013).

Oral administration of various combinations of probiotics like *Lactobacillus* species increased regulatory T cells and increase in IL-10 production and decreased IL-17 and (Takata et al. 2011), *Streptococcus thermophiles*, *Bifidobacterium bifidum* in both rat and mouse models improved clinical score of EAE and showed a decrease in Th1 and Th17 cells together with T^{regs} development (Kwon et al. 2013). Clinically, it was indicated that laboratory mice colonizing with MS patients gut microbiota increased the severity of experimental autoimmune encephalomyelitis (EAE) in MS animal models which can be reduced by the oral administration of *Bifidobacterium animalis* (Ezendam et al. 2008). An anti-inflammatory bacterium *Prevotella histicola* from human celiac disease patients has immunomodulatory potential which suppresses disease in EAE models through the initiation of tolerogenic dendritic cells, and FoxP3+ regulatory T cells and a decrease in pro-inflammatory Th1 and Th17 responses (Mangalam et al. 2017). *Lactobacillus* sp., *Bifidobacterium bifidum*, and *Streptococcus thermophiles* enhanced regulatory T cells, producing increased IL-10 and decreased TNF-α, IFN-c, IL-17, and reduced Th1 and Th17 cells along with the development of T^{regs} (Ezendam et al. 2008). These results produce evidence that the microbiome present in the human GI tract has a substantial impact on CNS-specific autoimmunity.

10.3.6 Alzheimer's Disease

Alterations occurring to the gut microbiome have the ability to influence the advancement of neurological diseases such as Alzheimer's Disease (AD), often recognized with a gradual decrease in cognitive abilities and memory and eventually in most cases, cause dementia (Mangalam et al. 2017). Although age is regarded as a notable risk factor for AD, the causative factors that majorly trigger it are still unclear (Kohler et al. 2016). There are several crucial factors that accompany the degenerative process of AD such as an impaired immune response, markers or signaling processes (Bhattacharjee and Lukiw 2013). Currently, no cure or treatment protocol is available for AD and the ways of disease progression by plaque and tangle formation or its further propagation in the brain still remain unknown (Aisen and Davis 1994). Probiotics possibly modulate and prevent the cognitive impairment in AD by the production of metabolites and neurotransmitters such as SCFAs and GABA. They help maintain a state of eubiosis, promote inflammatory responses and reduce cell damage caused by oxidative stress (Balin and Hudson 2014).

10.3.7 Anxiety and Depression

Anxiety is a psychological state of an individual characterized by apprehension or intense fear, associated with a lack of response of adaptation by the subject in a particular situation often commonly experience psychiatric disorders. Depression is a psychological state characterized by unhappiness or irritability and joint from various psycho-physiological alterations such as appetite distortions, sleep, constipation, and inability to experience work pleasure. Due to the intervention of the ANS, the two disorders generate alteration in the stability and constitution of the gut microbiota causing changes in colon mobility.

These patients exhibit increased inflammatory levels, dysfunction of the HPA axis, and neurotransmitter signaling. MDD patients showed significantly reduced *Firmicutes* and an increase in *Bacteroidetes*, *Proteobacteria*, and *Actinobacteria* but compared with controls confirmed the constitution of the altered gut microbiome. MDD patients also had increased *Enterobacteriaceae*, *Prevotella*, *Klebsiella*, *Streptococcus*, *Clostridium*, and *Alistipes*, decreased levels of *Faecalibacterium*, *Bifidobacterium*, *Lactobacillus*, *Prevotellaceae* and increase in *Thermoanaerobacteraceae* revealed that gut microbiota dysbiosis is significantly linked with metabolic changes of bile acids and tryptophan against the controls (Lin et al. 2017a). Desbonnet *et al.* 2008 (Desbonnet et al. 2010) observed peripheral HPA levels and concentrations of the serotonin precursor, tryptophan altered due to *B. infantis* for development of protective mechanisms preceding stress disclosure.

Germ-Free (GF) mice present an excessive liberation of adrenocorticotropic hormone (ACTH) and corticosterone resulting in stress. GF mice exhibit low anxiety suggesting that the intestinal microbiota affects the developmental behavior (Neufeld et al. 2011). Supplementing GF mice with *Bifidobacterium infantis* enhanced HPA stress response, comprising a reverse in an increase of plasma (ACTH) and corticosterone (Sudo et al. 2004). *L. rhamnosus* reduced stress-induced corticosterone and changed GABA receptor gene expression levels in the brain also signifying modulatory communication pathway as vagus nerve between the gut microbiome and the brain (Bravo et al. 2011). When fecal microbiota of MDD patients was transplanted to GF mice, induced metabolic disturbances in host and induced depression indicating dysfunctionality of gut microbiota playing a significant role in MDD.

Individuals with depression indicated higher levels of cytokines and immunoglobulins IgA and IgM, resulting in inflammatory processes and gastrointestinal disorders (Lima-Ojeda et al. 2017). Enteropathogens disturb mood through the immune system in humans. Reichenberg et al. (2001) indicated that following intravenous infusion with endotoxins from *Salmonella abortus equi* to healthy volunteers underwent an increase in anxiety, and depression levels. Alterations in the neuroendocrine pathways and several neurobiological mechanisms involving serotonin neurotransmitter reduction, reduction of dopamine in anxiety and noradrenaline in depression. The low doses of *Campylobacter jejuni* through oral administration can stimulate anxiogenic effects in mice (Lyte et al. 2006). The intake of probiotics has anti-inflammatory and antioxidant effects on depressed patients and their ability to regulate BDNF growth factor levels (Logan and Katzman 2005).

Lactobacillus rhamnosus JB-1 and combination of *Lactobacillus helveticus* R0052 and *Bifidobacterium longum* R0175 improved anxiolytic activity in specific-pathogen-free (*SPF*) rat and GF mice, reduced extreme HPA, changing GABA receptor level in particular brain regions, restored serotonin and norepinephrine levels in inflammatory stress response, and also promoted potential psychological properties in humans (Liang et al. 2015). The enteric microbiota shows a significant impact on potential therapeutic approach, modifying the host's gut microbiota affecting neurochemical, behavioral, and immunological parameters appropriate to the brain-gut axis disorders provided health benefits to the host with psychobiotics as emerging treatment (Petschow et al. 2013).

10.3.8 Schizophrenia

Schizophrenia is a mental illness that disturbs the human mind's functioning with severe occurrences of psychosis and passivity, followed by periods of normal mental activity (Grover et al. 2019). The prevalence of schizophrenia and autism has been correspondingly higher in patients suffering from *C. difficile* infection, possibly because the organism is known to synthesize a phenylalanine derivative in the gut that in turn regulates the levels of catecholamine in the brain (Argou-Cardozo and Zeidán-Chuliá 2018).

Further evidence from genetic studies focusing on twins and their adoption also suggests a linkage between the gut microbiome and schizophrenia. Patients suffering from schizophrenia often experience both mental and inflammatory stress, poor nutrition, and lactose sensitivity. These symptoms can be relieved with sufficient and proper probiotic administration (Ledochowski et al. 1998).

10.4 Psychobiotics

Probiotic microorganisms that may positively impact the psychological condition of a host upon adequate administration are referred to as psychobiotics (Dinan and Cryan 2013). The correlation of the gut microbiome with an individual's psychological condition via the gut-brain axis is increasingly gaining importance. Hereby, gut microbiota has found to exert an anti-inflammatory effect along with a modulatory reaction on the functioning of neurotransmitters (Beck et al. 2019). It has often been implied that a healthy mind means a healthy body vice versa possibly indicating their close relationship and also the fact that the gut microbiome can have both a direct and indirect connection with the mental status of a host (Dinan et al. 2013).

The gut microbiome has the ability to shape and alter one's thinking and mental abilities proving why it has been termed as the second brain. Dinan et al. (2013)) referred psychobiotics to some psychotropic bacteria colonizing the gut which could either positively or negatively influence the mental status of a host. Gut microbiota dysbiosis has also been associated with an unstable mental status of the host often leading to neurological disorders such as depression (Luczynski et al. 2016), autism disorders (Critchfield et al. 2011), schizophrenia (Grover et al. 2019), Parkinson's disease (Holmqvist et al. 2014), Alzheimer's disease, dementia (Kohler et al. 2016), etc. Psychobiotics are capable of releasing hormones and modulating the functioning of neurotransmitters which in turn act upon the gut-brain axis in Table 10.2 below (Ho et al. 2015). Studies reveal that the cytokines such as INFα and TNF-α have the ability to cause depression depending on their circulation and production levels, therefore a therapeutic dosage of psychobiotics is likely to alleviate such conditions and even improve mood swings (Petschow et al. 2013). Psychobiotics offer promising results in the treatment of various neurological diseases; targeting the gut microbiome and alteration of the same in a positive manner would greatly benefit the host (Rooks and Garrett 2016).

10.5 Therapeutic Manipulation, Implications, and Future Prospects

The significant functional and structural alteration of central nervous system can be concomitant with gut dysbiosis directing to the assumption that manipulation of gut microbiota is potentially a sensible method to confine clinical complications in neurological disorders. Certain treatments with antibiotics active against specific species are used to control intestinal flora of children with ASD and their GI tract showed a considerable negative impact on CNS with a reduction of pathogenic bacteria (Critchfield et al. 2011), but this method has only been partially successful and was found to follow a significant development in neurological disorder only during the administration of adequate drug dosage (Sandman et al. 2012). However,

Table 10.2 Representative table of gut microbiota producing neurochemicals within the human gut

Genus	Neurochemical	References
Bacillus, Lactobacillus	Acetylcholine	Kawashima et al. (2007)
Bacillus, Escherichia, Lactobacillus, Lactococcus, Streptococcus	Dopamine	Shishov et al. (2009)
Bifidobacterium, Lactobacillus	GABA	Barrett et al. (2012)
Enterococcus, Lactobacillus, Lactococcus, Streptococcus	Histamine	Thomas et al. (2012) Landete et al. (2008)
Bacillus, Escherichia	Norepinephrine	Tsavkelova et al. (2000)
Enterococcus, Escherichia, Lactobacillus, Lactococcus, Streptococcus	Serotonin	Shishov et al. (2009)

soon after treatment with certain phytochemicals reverses the neurological impairment in its initial phase (Tripathi et al. 2022).

A probiotic is defined as living microorganisms that aid in recovering the gut microbiota balance, with improvement in the integrity of the gut mucosal barrier and immunomodulation, ensure health benefits to the host, and IgA mucosal response to the benefit of the host when governed in acceptable amounts (D'Mello et al. 2015). Administration of probiotic bacteria modifies the bacterial composition of the gut, with an increase in protected positive strains and a reduction in negative strains. Furthermore, the concentration of specific bacterial products which causes immune system alterations, anti-anxiety effects (Curran et al. 2016), memory and learning improvements (Adler and Wong-Kee-You 2015), inflammation, modification of the CNS function and structure, and modulation of gene expression is reduced by probiotics when it exceeds the intestinal wall (Murgatroyd et al. 2009). This beneficial effect is different and strain-dependent.

Many presymptomatic and animal research investigations indicated the probiotic potential for treatment and prevention of numerous diseases, comprising of CNS and GI diseases, mainly using genera *Lactobacillus* and *Bifidobacterium* bacteria (Sánchez et al. 2017). Across the MGB axis, probiotics initiate brain function, for example anxiety and depression normalization (Wallace and Milev 2017); the brain neurological processing is influenced by the mechanism of probiotic supplements, for depression and anxiety in particular. For instance, chronic therapy with the probiotic *Lactobacillus rhamnosus* diminishes anxiety, depression, and stress responses merely in the existence of an intact vagus nerve, by reducing mRNA expression of GABAAα2 in the amygdala and prefrontal cortex, however, enhanced GABAAα2 mRNA expression in the hippocampus, therefore, vagus nerve identification plays a significant role in modulatory communication pathway between the gut, which is susceptible to probiotics or bacteria, and the brain (Bravo et al. 2011).

Bercik et al. (2011)) demonstrated a chronic administration of *Bifidobacterium infantis* in rats which was restricted by early age from maternal contact and possessed stress-related mood and gastrointestinal disorders resulting in immune normalization, reversal of behavioral deficiencies, and rehabilitation of basal noradrenaline concentrations in the brainstem. Moreover, *Bifidobacterium longum* NCC3001 and *Bifidobacterium longum* 1714 normalize anxiety-like behavior, with beneficial effect on cognition, *Bifidobacteria infantis* raised serotoninergic precursors and weakened inflammatory immune response, further indicated an antidepressant effect (Quigley et al. 2012) and hippocampal brain-derived neurotrophic factor (BDNF) in mice with infectious colitis (Bercik et al. 2011). In rats, the administration of *Lactobacillus helveticus*, *Lactobacillus farciminis*, and *Lactobacillus rhamnosus* can prevent chronic-stress-induced intestinal abnormalities and decreases psychological distress (Zareie et al. 2006). Lavasani et al. (2010)) observed that instead of administrating monostrain probiotic strains, three *Lactobacillus* strains such as *L. paracasei* DSM 13434, *L. plantarum* DSM 15312, and L. *plantarum* DSM 15313 in mice showed therapeutic efficiency on development of experimental autoimmune encephalomyelitis (EAE), by inhibiting disease progression (Boksa and El-Khodor 2003), gut microbiota modulation and improved destructive antisocial behavioral pattern, communication and anxiety problems in patients (Cai et al. 2015). Hsiao et al. (2013) described oral administration of *Bacteroides fragilis* which is a human commensal by a maternal immune activation (MIA) model of ASD in mouse offspring; modified the gut permeability, changed the composition of gut microbiota, and ameliorated the defects in communicative, stereotypic, with improved anxiety-like and sensorimotor communicative behaviors also *Bacteroides thetaiotaomicron* administration has also substantially improved abnormal behaviors and signifies that modulating therapies utilizing probiotics is a safer and efficient treatment for autism spectrum disorder (Cai et al. 2015).

Fecal microbiota transplantation (FMT) is used for the treatment of gastrointestinal diseases by introducing the enteric bacteria from healthy donor's feces to affected/diseased individuals with the sole purpose of restoring gut microbiota balance or to effect in eubiosis. This is done through oral administration of capsules comprising of donor microbiota sample in freeze-dried form. So far the results have been promising without any adverse effects and hence maybe considered as a safe alternative strategy for the treatment in organ transplant recipients (Bajaj et al. 2017) and cancer patients (Hefazi et al. 2017). Although this mode of treatment would be most ideal for gut dysbiosis-related diseases for instance inflammatory bowel diseases, metabolic syndromes, etc. (Sayar and Cetin 2016).

Clustered and regularly interspaced short palindrome repeats (CRISPR) technology can be utilized to make genetic changes in probiotic microorganisms via gene editing. This technology in association with the bacterial neuroimmune and metabolic systems can be used to engineer genes through upregulation and downregulation of their expression via therapeutic signaling molecules and blocking exopolysaccharides produced by the bacteria (Petimar et al. 2019). CRISPR-CAS targets the bidirectional gut-brain axis changing the neurological and intestinal effects leading to neurodegeneration (Aliev et al. 2008). Also, CRISPR technology suppresses or silences the expression of bacterial inflammatory signaling molecules and endotoxins. CRISPR-CAS9 is also used to target antibiotic resistance through phage-specific 16 s sequences of Gram-negative bacteria; these CRISPR antimicrobials can be tailored to treat patient-specific dysbiosis without changing bacterial populations, yet help in achieving effective therapeutic potential (Martella et al. 2019; Konermann et al. 2015)

Daily consumption of a fermented milk product containing probiotics such as *Bifidobacterium animalis* subspp. *sss*, *Lactobacillus bulgaricus*, *Streptococcus thermophilus*, and *Lactococcus lactis* subspp. showed significant alteration of the activity of CNS and brain regions that control the treatment of sensation and emotion (Tillisch et al. 2013), and HTLV-1-associated myelopathy/tropical spastic paraparesis (TSP/HAM) patients were treated with oral administration of *Lactobacillus casei* strain Shirota and exhibited development of motor function due to improved activity of NK cell (Finegold et al. 2002). Furthermore, probiotic administration in early age could lower the neuropsychiatric disorder risk development subsequent in infancy and the structured neurological evaluation in children administrating probiotics disclosed a substantially lower prevalence of neurological abnormalities (Pärtty et al. 2015).

10.6 Conclusion

The potential probiotics existing in the gut microbiota of the human intestinal tract are composite and live microorganisms that perform well in microvilli advancement, maintaining homeostasis, immune-potentiation, prevention of pathogen colonization in the GI tract, and gut-brain axis regulation. It could be utilized as a neuroprotective nutraceutical for treating or preventing a wide range of neurological diseases. The probiotic development emphasizes on microbial psychobiotics to potentially treat and prevent psychiatric disorders. Postbiotics which are gut microbial signaling molecules should be exploited to identify targets for therapeutic benefits. Dysbiosis in gut microbiota can cause impairment of gut barrier function and inflammation in the gut-brain axis leading to mental disorders. The gut microbiota dysbiosis and neurological dysfunction have aroused its importance in several fields, including neuroscience, immunology, bioinformatics, and microbiology, which provide potential treatment approaches for a diverse set of neurological disorders. Clinical research on the potential impact of probiotic mechanism and manipulation can positively improve the understanding of neurodevelopmental disorders and also may help evaluate the microbial effect on the gut-brain axis, this should be comprehended more in the future. It remains unclear whether the genetic modification of gut microbiota is capable of resulting in neurological diseases as there still exists only a limited amount of information concerning such protective or beneficial bacteria. The therapeutic role of psychobiotics has been studied in animals but more efficient human clinical trials are required to confirm their roles in microbiome-targeted therapies for treating gut-brain disorders. Probiotics have also shown to play a prominent role in the neuropathogenesis of neural disorders by altered gut-brain axis function.

References

Adams JB, Johansen LJ, Powell LD, Quig D, Rubin RA (2011) Gastrointestinal flora and gastrointestinal status in children with autism–comparisons to typical children and correlation with autism severity. BMC Gastroenterol 11(1):22

Adler SA, Wong-Kee-You AM (2015) Differential attentional responding in caesarean versus vaginally delivered infants. Atten Percept Psychophys 77(8):2529–2539

Aisen PS, Davis KL (1994) Inflammatory mechanisms in Alzheimer's disease: implications for therapy. Am J Psychiatry 151:1105–1113

Ait-Belgnaoui A, Colom A, Braniste V, Ramalho L, Marrot A, Cartier C, Tompkins T (2014) Probiotic gut effect prevents the chronic psychological stress-induced brain activity abnormality in mice. Neurogastroenterol Motil 26(4):510–520

Akers KG, Martinez-Canabal A, Restivo L, Yiu AP, De Cristofaro A, Hsiang HLL, Ohira K (2014) Hippocampal neurogenesis regulates forgetting during adulthood and infancy. Science 344(6184):598–602

Aliev G, Obrenovich ME, Reddy VP, Shenk JC, Moreira PI, Nunomura A, Perry G (2008) Antioxidant therapy in Alzheimer's disease: theory and practice. Mini Rev Med Chem 8(13): 1395

Argou-Cardozo I, Zeidán-Chuliá F (2018) Clostridium bacteria and autism spectrum conditions: a systematic review and hypothetical contribution of environmental glyphosate levels. Med Sci 6: 29

Arpaia N, Campbell C, Fan X, Dikiy S, van der Veeken J, Deroos P, Rudensky AY (2013) Metabolites produced by commensal bacteria promote peripheral regulatory T-cell generation. Nature 504(7480):451–455

Azpiroz F (2005) Intestinal perception: mechanisms and assessment. Br J Nutr 93(S1):S7–S12

Bachstetter AD, Van Eldik LJ, Schmitt FA, Neltner JH, Ighodaro ET, Webster SJ, Patel E, Abner EL, Kryscio RJ, Nelson PT (2015) Disease-related microglia heterogeneity in the hippocampus of Alzheimer's disease, dementia with Lewy bodies, and hippocampal sclerosis of aging. Acta Neuropathol Commun 3(1): 1–16

Bäckhed F (2011) Programming of host metabolism by the gut microbiota. Ann Nutr Metab 58 (Suppl. 2):44–52

Bajaj JS, Kassam Z, Fagan A, Gavis EA, Liu E, Cox IJ, Williams R (2017) Fecal microbiota transplant from a rational stool donor improves hepatic encephalopathy: a randomized clinical trial. Hepatology 66(6):1727–1738

Balin BJ, Hudson AP (2014) Etiology and pathogenesis of late-onset Alzheimer's disease. Curr Allergy Asthma Rep 14:013–0417

Barouei J, Moussavi M, Hodgson DM (2012) Effect of maternal probiotic intervention on HPA axis, immunity and gut microbiota in a rat model of irritable bowel syndrome. PLoS One 7(10) e46051

Barrett E, Ross RP, O'toole PW, Fitzgerald GF, Stanton C (2012) γ-Aminobutyric acid production by culturable bacteria from the human intestine. J Appl Microbiol 113(2):411–417

Beck BR, Park GS, Jeong DY, Lee YH, Im S, Song WH, Kang J (2019) Multidisciplinary and comparative investigations of potential psychobiotic effects of *Lactobacillus* strains isolated from newborns and their impact on gut microbiota and ileal transcriptome in a healthy murine model. Front Cell Infect Microbiol 9:269

Belmaker RH, Agam G (2008) Disorder MD. Major depressive disorder. N Engl J Med Overseas Ed 358:55–68

Bercik P, Park AJ, Sinclair D, Khoshdel A, Lu J, Huang X, Berger B (2011) The anxiolytic effect of Bifidobacterium longum NCC3001 involves vagal pathways for gut–brain communication. Neurogastroenterol Motil 23(12):1132–1139

Berer K, Mues M, Koutrolos M, Al Rasbi Z, Boziki M, Johner C, Krishnamoorthy G (2011) Commensal microbiota and myelin autoantigen cooperate to trigger autoimmune demyelination. Nature 479(7374):538–541

Bhattacharjee S, Lukiw WJ (2013) Alzheimer's disease and the microbiome. Front Cell Neurosci 7: 153

Boksa P, El-Khodor BF (2003) Birth insult interacts with stress at adulthood to alter dopaminergic function in animal models: possible implications for schizophrenia and other disorders. Neurosci Biobehav Rev 27(1–2):91–101

Borre YE, O'Keeffe GW, Clarke G, Stanton C, Dinan TG, Cryan JF (2014) Microbiota and neurodevelopmental windows: implications for brain disorders. Trends Mol Med 20(9): 509–518

Borthakur A, Gill RK, Tyagi S, Koutsouris A, Alrefai WA, Hecht GA, Dudeja PK (2008) The probiotic Lactobacillus acidophilus stimulates chloride/hydroxyl exchange activity in human intestinal epithelial cells. J Nutr 138(7):1355–1359

Branton WG, Lu JQ, Surette MG, Holt RA, Lind J, Laman JD, Power C (2016) Brain microbiota disruption within inflammatory demyelinating lesions in multiple sclerosis. Sci Rep 6:37344

Bravo JA, Forsythe P, Chew MV, Escaravage E, Savignac HM, Dinan TG, Cryan JF (2011) Ingestion of Lactobacillus strain regulates emotional behavior and central GABA receptor expression in a mouse via the vagus nerve. Proc Natl Acad Sci 108(38):16050–16055

Breban M, Tap J, Leboime A, Said-Nahal R, Langella P, Chiocchia G, Sokol H (2017) Faecal microbiota study reveals specific dysbiosis in spondyloarthritis. Ann Rheum Dis 76(9): 1614–1622

Cai S, Pang WW, Low YL, Sim LW, Sam SC, Bruntraeger MB, Richmond J (2015) Infant feeding effects on early neurocognitive development in Asian children. Am J Clin Nutr 101(2):326–336

Cameron J, Doucet E (2007) Getting to the bottom of feeding behaviour: who's on top? Appl Physiol Nutr Metab 32(2):177–189

Cao X, Lin P, Jiang P, Li C (2013) Characteristics of the gastrointestinal microbiome in children with autism spectrum disorder: a systematic review. Shanghai Arch Psychiatry 25(6):342

Cekanaviciute E, Yoo BB, Runia TF, Debelius JW, Singh S, Nelson CA, Crabtree-Hartman E (2017) Gut bacteria from multiple sclerosis patients modulate human T cells and exacerbate symptoms in mouse models. Proc Natl Acad Sci 114(40):10713–10718

Chen J, Chia N, Kalari KR, Yao JZ, Novotna M, Soldan MMP, Weinshenker BG (2016) Multiple sclerosis patients have a distinct gut microbiota compared to healthy controls. Sci Rep 6(1):1–10

Clarke G, Fitzgerald P, Hennessy AA, Cassidy EM, Quigley EM, Ross P, Dinan TG (2010) Marked elevations in pro-inflammatory polyunsaturated fatty acid metabolites in females with irritable bowel syndrome. J Lipid Res 51(5):1186–1192

Coccaro EF, Lee R, Groer MW, Can A, Coussons-Read M, Postolache TT (2016) Toxoplasma gondii infection: relationship with aggression in psychiatric subjects. J Clin Psychiatry 77(3): 334–341

Costedio MM, Hyman N, Mawe GM (2007) Serotonin and its role in colonic function and in gastrointestinal disorders. Dis Colon Rectum 50(3):376–388

Critchfield JW, Van Hemert S, Ash M, Mulder L, Ashwood P (2011) The potential role of probiotics in the management of childhood autism spectrum disorders. Gastroenterology research and practice

Cryan JF, Dinan TG (2012) Mind-altering microorganisms: the impact of the gut microbiota on brain and behaviour. Nat Rev Neurosci 13:701–712

Cunningham CL, Martínez-Cerdeño V, Noctor SC (2013) Microglia regulate the number of neural precursor cells in the developing cerebral cortex. J Neurosci 33(10):4216–4233

Curran EA, Khashan AS, Dalman C, Kenny LC, Cryan JF, Dinan TG, Kearney PM (2016) Obstetric mode of delivery and attention-deficit/hyperactivity disorder: a sibling-matched study. Int J Epidemiol 45(2):532–542

D'Erchia AM, Gallo A, Manzari C, Raho S, Horner DS, Chiara M, Locatelli F (2017) Massive transcriptome sequencing of human spinal cord tissues provides new insights into motor neuron degeneration in ALS. Sci Rep 7(1):1–20

Desbonnet L, Garrett L, Clarke G, Kiely B, Cryan JF, Dinan TG (2010) Effects of the probiotic Bifidobacterium infantis in the maternal separation model of depression. Neuroscience 170(4): 1179–1188

Dinan, T. G., &Cryan, J. F. (2013). Melancholic microbes: a link between gut microbiota and depression?. Neurogastroenterology & Motility, 25(9), 713–719.

Dinan TG, Stanton C, Cryan JF (2013) Psychobiotics: a novel class of psychotropic. Biol Psychiatry 74(10):720–726

D'Mello C, Ronaghan N, Zaheer R, Dicay M, Le T, MacNaughton WK, Swain MG (2015) Probiotics improve inflammation-associated sickness behavior by altering communication between the peripheral immune system and the brain. J Neurosci 35(30):10821–10830

Dominguez-Bello MG, De Jesus-Laboy KM, Shen N, Cox LM, Amir A, Gonzalez A, Mendez K (2016) Partial restoration of the microbiota of cesarean-born infants via vaginal microbial transfer. Nat Med 22(3):250

Edelmann E, Lessmann V, Brigadski T (2014) Pre- and postsynaptic twists in BDNF secretion and action in synaptic plasticity. Neuropharmacology 76(Pt C):610–627

Erny D, de Angelis ALH, Jaitin D, Wieghofer P, Staszewski O, David E, Schwierzeck V (2015) Host microbiota constantly control maturation and function of microglia in the CNS. Nat Neurosci 18(7):965

Ezendam J, De Klerk A, Gremmer ER, Van Loveren H (2008) Effects of Bifidobacterium animalis administered during lactation on allergic and autoimmune responses in rodents. Clin Exp Immunol 154(3):424–431

Fang X, Wang X, Yang S, Meng F, Wang X, Wei H, Chen T (2016) Evaluation of the microbial diversity in amyotrophic lateral sclerosis using high-throughput sequencing. Front Microbiol 7: 1479

Finegold SM, Molitoris D, Song Y, Liu C, Vaisanen ML, Bolte E, Collins MD (2002) Gastrointestinal microflora studies in late-onset autism. Clin Infect Dis 35(Supplement_1):S6–S16

Forsythe P, Sudo N, Dinan T, Taylor VH, Bienenstock J (2010) Mood and gut feelings. Brain Behav Immun 24(1):9–16

Fouhy F, Ross RP, Fitzgerald GF, Stanton C, Cotter PD (2012) Composition of the early intestinal microbiota: knowledge, knowledge gaps and the use of high-throughput sequencing to address these gaps. Gut microbes 3(3):203–220

Fung TC, Olson CA, Hsiao EY (2017) Interactions between the microbiota, immune and nervous systems in health and disease. Nat Neurosci 20(2):145

Galland L (2014) The gut microbiome and the brain. J Med Food 17(12):1261–1272

Gareau MG, Jury J, Perdue MH (2007) Neonatal maternal separation of rat pups results in abnormal cholinergic regulation of epithelial permeability. Am J Physiol Gastrointest Liver Physiol 293(1):G198–G203

Girolamo F, Coppola C, Ribatti D (2017) Immunoregulatory effect of mast cells influenced by microbes in neurodegenerative diseases. Brain Behav Immun 65:68–89

Gómez-Eguílaz M, Ramón-Trapero JL, Pérez-Martínez L, Blanco JR (2018) The beneficial effect of probiotics as a supplementary treatment in drug-resistant epilepsy: a pilot study. Benef Microbes 9(6):875–881

Grover S, Patil A, Kaur A, Garg G (2019) Probiotics: a potential immunotherapeutic approach for the treatment of schizophrenia. J Pharm Bioallied Sci 11(4):321–327

Guilliams TG, Edwards L (2010) Chronic stress and the HPA axis. The Standard 1(2):1–12

Hamer HM, Jonkers D, Venema K, Vanhoutvin S, Troost FJ, Brummer RJ (2008) Review article: the role of butyrate on colonic function. Aliment Pharmacol Ther 27(2):104–119

Hefazi, M., Patnaik, M. M., Hogan, W. J., Litzow, M. R., Pardi, D. S., & Khanna, S. (2017). Safety and efficacy of fecal microbiota transplant for recurrent Clostridium difficile infection in patients with cancer treated with cytotoxic chemotherapy: a single-institution retrospective case series. In: Mayo Clinic Proceedings (Vol. 92, No. 11, pp 1617–1624). Elsevier: Amsterdam.

Heijtz RD, Wang S, Anuar F, Qian Y, Björkholm B, Samuelsson A, Pettersson S (2011) Normal gut microbiota modulates brain development and behavior. Proc Natl Acad Sci 108(7):3047–3052

Helander HF, Fändriks L (2014) Surface area of the digestive tract–revisited. Scand J Gastroenterol 49(6):681–689
Herman JP, McKlveen JM, Ghosal S et al (2016) Regulation of the hypothalamic-pituitary-adrenocortical stress response. Compr Physiol 6(2):603–621
Ho JT, Chan GC, Li JC (2015) Systemic effects of gut microbiota and its relationship with disease and modulation. BMC Immunol 16:21
Holmqvist S, Chutna O, Bousset L, Aldrin-Kirk P, Li W, Björklund T, Li JY (2014) Direct evidence of Parkinson pathology spread from the gastrointestinal tract to the brain in rats. Acta Neuropathol 128(6):805–820
Horvath K, Papadimitriou JC, Rabsztyn A, Drachenberg C, Tildon JT (1999) Gastrointestinal abnormalities in children with autistic disorder. J Pediatr 135(5):559–563
Iannone L, Gómez-Eguílaz M, Citraro R, Russo E (2020) The potential role of interventions impacting on gut-microbiota in epilepsy. Expert Rev Clin Pharmacol 13. https://doi.org/10.1080/17512433.2020
Jiménez E, Delgado S, Maldonado A, Arroyo R, Albújar M, García N, Rodríguez JM (2008) Staphylococcus epidermidis: a differential trait of the fecal microbiota of breast-fed infants. BMC Microbiol 8(1):1–11
Johnson RL, Wilson CG (2018) A review of vagus nerve stimulation as a therapeutic intervention. J Inflamm Res 11:203–213
Joshi D, Roy S, Banerjee S (2018) Prebiotics: a functional food in health and disease. In: Mandal SC, Mandal V, Konishi T (eds) Natural Products & Drug Discovery. Elsevier, Amsterdam, pp 507–523
Jung IH, Jung MA, Kim EJ, Han MJ, Kim DH (2012) Lactobacillus pentosus var. plantarum C29 protects scopolamine-induced memory deficit in mice. J Appl Microbiol 113(6):1498–1506
Kamm K, Hoppe S, Breves G, Schröder B, Schemann M (2004) Effects of the probiotic yeast Saccharomyces boulardii on the neurochemistry of myenteric neurones in pig jejunum. Neurogastroenterol Motil 16(1):53–60
Kawashima K, Misawa H, Moriwaki Y, Fujii YX, Fujii T, Horiuchi Y, Kamekura M (2007) Ubiquitous expression of acetylcholine and its biological functions in life forms without nervous systems. Life Sci 80(24–25):2206–2209
Kleiman SC, Watson HJ, Bulik-Sullivan EC, Huh EY, Tarantino LM, Bulik CM, Carroll IM (2015) The intestinal microbiota in acute anorexia nervosa and during renourishment: relationship to depression, anxiety, and eating disorder psychopathology. Psychosom Med 77(9):969
Kohler CA, Maes M, Slyepchenko A, Berk M, Solmi M, Lanctôt KL, Carvalho AF (2016) The gut-brain axis, including the microbiome, leaky gut and bacterial translocation: mechanisms and pathophysiological role in Alzheimer's disease. Curr Pharm Des 22(40):6152–6166
Konermann S, Brigham MD, Trevino AE, Joung J, Abudayyeh OO, Barcena C, Nureki O (2015) Genome-scale transcriptional activation by an engineered CRISPR-Cas9 complex. Nature 517(7536):583–588
Krauss GL, Sperling MR (2011) Treating patients with medically resistant epilepsy. Neurol Clin Pract 1(1):14–23
Kwon OI, Woo EJ, Du YP, Hwang D (2013) A tissue-relaxation-dependent neighboring method for robust mapping of the myelin water fraction. Neuroimage 74:12–21
Landete JM, de las Rivas B, Marcobal A, Munoz R (2008) Updated molecular knowledge about histamine biosynthesis by bacteria. Crit Rev Food Sci Nutr 48:697–714
Lavasani S, Dzhambazov B, Nouri M, Fåk F, Buske S, Molin G, Weström B (2010) A novel probiotic mixture exerts a therapeutic effect on experimental autoimmune encephalomyelitis mediated by IL-10 producing regulatory T cells. PLoS One 5(2) e9009
Ledochowski M, Sperner-Unterweger B, Fuchs D (1998) Lactose malabsorption is associated with early signs of mental depression in females: a preliminary report. Dig Dis Sci 43:2513–2517
Lee DH, Lee DH, Lee JS (2007) Characterization of a new antidementia β-secretase inhibitory peptide from Saccharomyces cerevisiae. Enzyme Microb Technol 42(1):83–88

Liang S, Wang T, Hu X, Luo J, Li W, Wu X, Jin F (2015) Administration of Lactobacillus helveticus NS8 improves behavioral, cognitive, and biochemical aberrations caused by chronic restraint stress. Neuroscience 310:561–577
Lima-Ojeda JM, Rupprecht R, Baghai TC (2017) "I am I and my bacterial circumstances": linking gut microbiome, neurodevelopment, and depression. Front Psychiatry 8:153
Lin X, Chen Z, Jin L, Gao W, Qu B, Zuo Y, Yu M (2017a) Rasch analysis of the hospital anxiety and depression scale among Chinese cataract patients. PLoS One 12(9)
Lin P, Ding B, Feng C, Yin S, Zhang T, Qi X, Li Q (2017b) Prevotella and Klebsiella proportions in fecal microbial communities are potential characteristic parameters for patients with major depressive disorder. J Affect Disord 207:300–304
Logan AC, Katzman M (2005) Major depressive disorder: probiotics may be an adjuvant therapy. Med Hypotheses 64(3):533–538
Lord C, Cook EH, Leventhal BL, Amaral DG (2000) Autism spectrum disorders. Neuron 28(2): 355–363
Lu W, Mi R, Tang H, Liu S, Fan M, Wang L (1998) Over-expression of c-fos mRNA in the hippocampal neurons in Alzheimer's disease. Chin Med J (Engl) 111(1):35–37
Luczynski P, Whelan SO, O'Sullivan C, Clarke G, Shanahan F, Dinan TG, Cryan JF (2016) Adult microbiota-deficient mice have distinct dendritic morphological changes: Differential effects in the amygdala and hippocampus. Eur J Neurosci 44(9):2654–2666
Lynch SV, Pedersen O (2016) The human intestinal microbiome in health and disease. N Engl J Med 375(24):2369–2379
Lyte M, Li W, Opitz N, Gaykema RP, Goehler LE (2006) Induction of anxiety-like behavior in mice during the initial stages of infection with the agent of murine colonic hyperplasia Citrobacter rodentium. Physiol Behav 89(3):350–357
Mangalam A, Shahi SK, Luckey D, Karau M, Marietta E, Luo N, Rodriguez M (2017) Human gut-derived commensal bacteria suppress CNS inflammatory and demyelinating disease. Cell Rep 20(6):1269–1277
Martella A, Firth M, Taylor BJ, Göppert A, Cuomo EM, Roth RG, Fisher DI (2019) Systematic evaluation of CRISPRa and CRISPRi modalities enables development of a multiplexed, orthogonal gene activation and repression system. ACS Synthetic Biol 8(9):1998–2006
Mazzini L, Mogna L, De Marchi F, Amoruso A, Pane M, Aloisio I, Cantello R (2018) Potential role of gut microbiota in ALS pathogenesis and possible novel therapeutic strategies. J Clin Gastroenterol 52:S68–S70
Meyza KZ, Blanchard DC (2017) The BTBR mouse model of idiopathic autism–Current view on mechanisms. Neurosci Biobehav Rev 76:99–110
Miyake S, Kim S, Suda W, Oshima K, Nakamura M, Matsuoka T, Morita H (2015) Dysbiosis in the gut microbiota of patients with multiple sclerosis, with a striking depletion of species belonging to clostridia XIVa and IV clusters. PLoS One 10(9) e0137429
Mowry EM, Glenn JD (2018) The dynamics of the gut microbiome in multiple sclerosis in relation to disease. Neurol Clin 36(1):185–196
Murgatroyd C, Patchev AV, Wu Y, Micale V, Bockmühl Y, Fischer D, Spengler D (2009) Dynamic DNA methylation programs persistent adverse effects of early-life stress. Nat Neurosci 12(12): 1559–1566
Nagy-Szakal D, Williams BL, Mishra N, Che X, Lee B, Bateman L, Peterson DL (2017) Fecal metagenomic profiles in subgroups of patients with myalgic encephalomyelitis/chronic fatigue syndrome. Microbiome 5(1):1–17
Neufeld KM, Kang N, Bienenstock J, Foster JA (2011) Reduced anxiety-like behavior and central neurochemical change in germ-free mice. Neurogastroenterol Motil 23(3):255. e119
Noseworthy JH, Lucchinetti C, Rodriguez M, Weinshenker BG (2000) Multiple sclerosis. N Engl J Med 343:938–952
O'Mahony SM, Hyland NP, Dinan TG, Cryan JF (2011) Maternal separation as a model of brain–gut axis dysfunction. Psychopharmacology (Berl) 214(1):71–88

Ogbonnaya ES, Clarke G, Shanahan F, Dinan TG, Cryan JF, O'Leary OF (2015) Adult hippocampal neurogenesis is regulated by the microbiome. Biol Psychiatry 78(4):e7–e9
Okun E, Barak B, Saada-Madar R, Rothman SM, Griffioen KJ, Roberts N, Arumugam TV (2012) Evidence for a developmental role for TLR4 in learning and memory. PLoS One 7(10)
Ota K, Matsui M, Milford EL, Mackin GA, Weiner HL, Hafler DA (1990) T-cell recognition of an immuno-dominant myelin basic protein epitope in multiple sclerosis. Nature 346(6280): 183–187
Pärtty A, Kalliomäki M, Wacklin P, Salminen S, Isolauri E (2015) A possible link between early probiotic intervention and the risk of neuropsychiatric disorders later in childhood: a randomized trial. Pediatr Res 77(6):823–828
Perner C, Perner F, Stubendorff B, Förster M, Witte OW, Heidel FH, Grosskreutz J (2018) Dysregulation of chemokine receptor expression and function in leukocytes from ALS patients. J Neuroinflammation 15(1):99
Petimar J, O'Reilly É, Adami HO, van den Brandt PA, Buring J, English DR, Larsson SC (2019) Coffee, tea, and caffeine intake and amyotrophic lateral sclerosis mortality in a pooled analysis of eight prospective cohort studies. Eur J Neurol 26(3):468–475
Petrov VA, Saltykova IV, Zhukova IA, Alifirova VM, Zhukova NG, Dorofeeva YB, Mironova YS (2017) Analysis of gut microbiota in patients with Parkinson's disease. Bull Exp Biol Med 162(6):734–737
Petschow B, Dore J, Hibberd P, Dinan T, Reid G, Blaser M et al (2013) Probiotics, prebiotics, and the host microbiome: the science of translation. Ann N Y Acad Sci 1306:1–17
Qian Y, Yang X, Xu S, Wu C, Song Y, Qin N, Xiao Q (2018) Alteration of the fecal microbiota in Chinese patients with Parkinson's disease. Brain Behav Immun 70:194–202
Quigley EM (2013) Gut bacteria in health and disease. Gastroenterol Hepatol 9(9):560
Quigley MA, Hockley C, Carson C, Kelly Y, Renfrew MJ, Sacker A (2012) Breastfeeding is associated with improved child cognitive development: a population-based cohort study. J Pediatr 160(1):25–32
Ranuh R, Athiyyah AF, Darma A et al (2019) Effect of the probiotic *Lactobacillus plantarum* IS-10506 on BDNF and 5HT stimulation: role of intestinal microbiota on the gut-brain axis. Iran J Microbiol 11(2):145–150
Rao JS, Ertley RN, Lee H-J et al (2007) n-3 polyunsaturated fatty acid deprivation in rats decreases frontal cortex BDNF *via* a p38 MAPK-dependent mechanism. Mol Psychiatry 12(1):36–46
Reichenberg A, Yirmiya R, Schuld A, Kraus T, Haack M, Morag A, Pollmächer T (2001) Cytokine-associated emotional and cognitive disturbances in humans. Arch Gen Psychiatry 58(5): 445–452
Rolls A, Shechter R, London A, Ziv Y, Ronen A, Levy R, Schwartz M (2007) Toll-like receptors modulate adult hippocampal neurogenesis. Nat Cell Biol 9(9):1081–1088
Rooks MG, Garrett WS (2016) Gut microbiota, metabolites and host immunity. Nat Rev Immunol 16:341–352
Rumah KR, Linden J, Fischetti VA, Vartanian T (2013) Isolation of Clostridium perfringens type B in an individual at first clinical presentation of multiple sclerosis provides clues for environmental triggers of the disease. PLoS One 8(10) e76359
Sánchez B, Delgado S, Blanco-Míguez A, Lourenço A, Gueimonde M, Margolles A (2017) Probiotics, gut microbiota, and their influence on host health and disease. Mol Nutr Food Res 61(1):1600240
Sandman CA, Davis EP, Buss C, Glynn LM (2012) Exposure to prenatal psychobiological stress exerts programming influences on the mother and her fetus. Neuroendocrinology 95(1):8–21
Saresella M, Mendozzi L, Rossi V, Mazzali F, Piancone F, LaRosa F, Clerici M (2017) Immunological and clinical effect of diet modulation of the gut microbiome in multiple sclerosis patients: a pilot study. Front Immunol 8:1391
Savignac HM, Tramullas M, Kiely B, Dinan TG, Cryan JF (2015) Bifidobacteria modulate cognitive processes in an anxious mouse strain. Behav Brain Res 287:59–72

Sayar GH, Cetin M (2016) Psychobiotics: the potential therapeutic promise of microbes in psychiatry. Klinik Psikofarmakol Bülteni 26(2):93–102

Schwarcz R, Bruno JP, Muchowski PJ, Wu HQ (2012) Kynurenines in the mammalian brain: when physiology meets pathology. Nat Rev Neurosci 13(7):465–477

Shahi SK, Freedman SN, Mangalam AK (2017) Gut microbiome in multiple sclerosis: the players involved and the roles they play. Gut microbes 8(6):607–615

Sharma V, Kaur S (2020) The Effect of Probiotic Intervention in Ameliorating the Altered Central Nervous System Functions in Neurological Disorders: A Review. Open Microbiol J 14(1)

Shishov VA, Kirovskaya TA, Kudrin VS, Oleskin AV (2009) Amine neuromediators, their precursors, and oxidation products in the culture of Escherichia coli K-12. Appl Biochem Microbiol 45(5):494–497

Skaper SD, Facci L, Zusso M, Giusti P (2018) An inflammation-centric view of neurological disease: beyond the neuron. Front Cell Neurosci 12:72

Smith SM, Vale WW (2006) The role of the hypothalamic-pituitary-adrenal axis in neuroendocrine responses to stress. Dialogues Clin Neurosci 8(4):383

Smith CJ, Emge JR, Berzins K et al (2014) Probiotics normalize the gutbrain- microbiota axis in immunodeficient mice. Am J Physiol Gastrointest Liver Physiol 307(8):G793–G802

Steyn FJ, Ioannides ZA, van Eijk RP, Heggie S, Thorpe KA, Ceslis A, Henderson RD (2018) Hypermetabolism in ALS is associated with greater functional decline and shorter survival. J Neurol Neurosurg Psychiatry 89(10):1016–1023

Sudo N, Chida Y, Aiba Y, Sonoda J, Oyama N, Yu XN, Koga Y (2004) Postnatal microbial colonization programs the hypothalamic–pituitary–adrenal system for stress response in mice. J Physiol 558(1):263–275

Takata K, Kinoshita M, Okuno T, Moriya M, Kohda T, Honorat JA, Sakoda S (2011) The lactic acid bacterium Pediococcusacidilactici suppresses autoimmune encephalomyelitis by inducing IL-10-producing regulatory T cells. PLoS One 6(11) e27644

Tan AH, Chong CW, Song SL, Teh CSJ, Yap IKS, Loke MF, Lim SY (2018) Altered gut microbiome and metabolome in patients with multiple system atrophy. Mov Disord Clin Pract Soc 33(1):174

Tap J, Derrien M, Törnblom H, Brazeilles R, Cools-Portier S, Doré J, Simrén M (2017) Identification of an intestinal microbiota signature associated with severity of irritable bowel syndrome. Gastroenterology 152(1):111–123

Thomas CA, Hong T, van Pijkeren JP, Hemarajata P, Trinh DV, Hu W, Britton RA, Kalkum M, Versalovic J (2012) Histamine derived from probiotic Lactobacillus reuteri suppress TNF via modulation of PKA and ERK signalling. PLoS One 7(2) e31951

Thursby E, Juge N (2017) Introduction to the human gut microbiota. Biochem J 474(11): 1823–1836

Tillisch K, Labus J, Kilpatrick L, Jiang Z, Stains J, Ebrat B, Mayer EA (2013) Consumption of fermented milk product with probiotic modulates brain activity. Gastroenterology 144(7): 1394–1401

Toepfer CF, Klauser A, Riepl RL, Muller-Felber W, Pongratz D, M. (2000) Gastrointestinal dysfunction in amyotrophic lateral sclerosis. Amyotroph Lateral Scler Other Motor Neuron Disord 1(1):15–19

Tomova A, Husarova V, Lakatosova S, Bakos J, Vlkova B, Babinska K, Ostatnikova D (2015) Gastrointestinal microbiota in children with autism in Slovakia. Physiol Behav 138:179–187

Tripathi AK, Ray AK, Mishra SK (2022) Molecular and pharmacological aspects of piperine as a potential molecule for disease prevention and management: evidence from clinical trials. Beni-Suef Univ J Basic Appl Sci 11(1):1–24

Tsavkelova EA, Botvinko IV, Kudrin VS, Oleskin AV (2000) Detection of neurotransmitter amines in microorganisms with the use of high-performance liquid chromatography. **Dokl Biochem 372(1–6):115**

Turner JR (2009) Intestinal mucosal barrier function in health and disease. Nat Rev Immunol 9(11): 799–809

Varghese FP, Brown ES (2001) The hypothalamic-pituitary-adrenal axis in major depressive disorder: a brief primer for primary care physicians. Prim Care Companion J Clin Psychiatry 3(4):151

Wall R, Ross RP, Shanahan F, O'Mahony L, Kiely B, Quigley E, Stanton C (2010) Impact of administered bifidobacterium on murine host fatty acid composition. Lipids 45(5):429–436

Wall R, Marques TM, O'Sullivan O, Ross RP, Shanahan F, Quigley EM, Fouhy F (2012) Contrasting effects of Bifidobacterium breve NCIMB 702258 and Bifidobacterium breve DPC 6330 on the composition of murine brain fatty acids and gut microbiota. Am J Clin Nutr 95(5):1278–1287

Wallace CJ, Milev R (2017) The effects of probiotics on depressive symptoms in humans: a systematic review. Ann Gen Psychiatry 16(1):14

Wang L, Christophersen CT, Sorich MJ, Gerber JP, Angley MT, Conlon MA (2011) Low relative abundances of the mucolytic bacterium *Akkermansiamuciniphila* and *Bifidobacterium* spp. in feces of children with autism. Appl Environ Microbiol 77(18):6718–6721

Wu S, Yi J, Zhang YG, Zhou J, Sun J (2015) Leaky intestine and impaired microbiome in an amyotrophic lateral sclerosis mouse model. Physiol Rep 3(4)

Yan F, Polk DB (2011) Probiotics and immune health. Curr Opin Gastroenterol

Yeon SW, You YS, Kwon HS, Yang EH, Ryu JS, Kang BH, Kang JH (2010) Fermented milk of Lactobacillus helveticus IDCC3801 reduces beta-amyloid and attenuates memory deficit. J Funct Foods 2(2):143–152

Zareie M, Johnson-Henry K, Jury J, Yang PC, Ngan BY, McKay DM, Sherman PM (2006) Probiotics prevent bacterial translocation and improve intestinal barrier function in rats following chronic psychological stress. Gut 55(11):1553–1560

Zhang YG, Wu S, Yi J, Xia Y, Jin D, Zhou J, Sun J (2017) Target intestinal microbiota to alleviate disease progression in amyotrophic lateral sclerosis. Clin Ther 39(2):322–336

Zhang M, Ma W, Zhang J, He Y, Wang J (2018) Analysis of gut microbiota profiles and microbe-disease associations in children with autism spectrum disorders in China. Sci Rep 8(1):1–9

Zhao C, Deng W, Gage FH (2008) Mechanisms and Functional Implications of Adult Neurogenesis. Cell 132

Zhou T, Ahmad TK, Gozda K, Truong J, Kong J, Namaka M (2017) Implications of white matter damage in amyotrophic lateral sclerosis. Mol Med Rep 16(4):4379–4392

Zhuang ZQ, Shen LL, Li WW, Fu X, Zeng F, Gui L, Zheng P (2018) Gut microbiota is altered in patients with Alzheimer's disease. J Alzheimers Dis 63(4):1337–1346

Chapter 11
Reversal of Metabolic Disorder Through the Restoration of Gut Microbiota

Prabhat Upadhyay, Diya Kalra, Sarika Gupta, and Sunil Kumar Mishra

Abstract Dysbiosis is a condition in which the makeup of the gut microbiota varies as a result of food habits and lifestyle changes caused by external or internal influences. This metabolic disorder is associated with a variety of etiologies, depending on age, immune system, and other associated triggers. Phytochemicals can modulate gut microbiota and reduce short-chain inflammation by inducing microbial fermentative metabolites, thereby preventing inflammation-related degenerative and chronic diseases. The function of phytochemicals in the modulation of the gut microbiota, which has implications for the prognosis of a variety of metabolic aliments, including obesity, Alzheimer's disease, and diabetes, is discussed in this chapter.

Keywords Phytochemical compound · Short-chain fatty acid · Metabolites

11.1 Introduction

With widespread westernization of dietary patterns and lifestyles, changes in intestinal microbial composition caused by extrinsic or intrinsic factors can lead to dysbiosis in the host and promote the development of metabolic disorders. There is, in particular, mild inflammation stimulated by metabolic disorders occurred after dysbiosis. The microbial organelle is two orders of magnitude larger than the

P. Upadhyay · S. Gupta
Molecular Science Lab, National Institute of Immunology, New Delhi, India
e-mail: upadhyayprabhat@nii.ac.in; sarika@nii.ac.in

D. Kalra
Department of Biotechnology, Jamia Hamdard University, New Delhi, India

S. K. Mishra (✉)
Department of Pharmaceutical Eng and Technology, Indian Institute of Technology, Banaras Hindu University, Varanasi, India
e-mail: skmishra.phe@itbhu.ac.in

A. K. Tripathi, M. Kotak (eds.), *Gut Microbiome in Neurological Health and Disorders*, Nutritional Neurosciences,
https://doi.org/10.1007/978-981-19-4530-4_11

microbiota in the host's nuclear genome, and it controls the host's key pharmacological functions (Ribeiro et al. 2020).

Long-term symbiotic host associations of good bacteria are involved in gene regulation, intestinal barrier endothelial function, nutrition, metabolism, neurotransmitter release, and regulation of the host immune system. Obesity, Alzheimer's disease, and diabetes are all diseases that start with dysbiosis. Probiotics, prebiotics, fecal microflora transplantation (FMT), metabolic surgery, modern medicine, and botanicals are now considered current approaches to gut microbiota healing. Alternatively, traditional plant-based formulations represent an effective form of intervention to ameliorate metabolic disorders (Illiano et al. 2020). Herbal medicines (HM) are mostly classified according to their pharmacological activity. Large amounts of phytochemicals are biosynthesized in plants that function to be present in natural laboratories (Upadhyay et al. 2018a).

Phytochemicals or auxiliary metabolites showing natural properties as a dynamic fixing, e.g., alkaloids, glycosides, tannins, triterpenes, flavonoids, sterols, against creepy crawlies, microbes, organisms and their metabolites are detailed for different restorative exercises such as antioxidant, anti-inflammatory, neurotransmitter control, homeostasis of metabolic activity.(Raju et al. 2013; Upadhyay et al. 2018b). As of late, inquire about is coordinated towards the distinguishing proof of unused dynamic substances of common beginning, which have tremendous restorative signs for a number of irresistible and systemic illness in people and creatures. Genetically, mutations can be observed today after the long-term use of synthetic substances in terms of side effects. Therefore, many researchers point to the need to redirect the preparation of natural products as substitutes or adjuvants to the discriminated use of antibiotics and their antimicrobial potential (Upadhyay et al. 2019). At present, however, it remains unclear how phytochemicals exert multifunctional activity in the human body when multiple metabolites are present. The part of the intestine microbiota is to preserve the astuteness and work of the gut. Dysbiosis causes severe systemic and local lesions. In healthy individuals, bacterial strains Bacteroides and Actinomycetes predominate. *Clostridium leptum* and *Clostridium coccoides* seeds are clearly predominant in the life of a one-year-old child. There are several secondary metabolites called phytochemicals that can bolster the development of microscopic organism within the digestive tract beneath solid food materials or have unhealthy conditions (Yin et al. 2019).

The most reason of this chapter is to deliver short-chain fatty acids and increment colonies of solid microbes to reestablish the intestinal vegetation of obsessive conditions such as Alzheimer's illness, weight, and diabetes through metabolites or herbs or nature. It is to characterize the part of the item. Endotoxin levels in the digestive system have decreased.

11.2 Role of Phytochemicals in the Gut Restoration

In adults, the microbiota, already present in the gut, has different physiological roles such as supplement retention through food fermentation and the production of several metabolites that help to regulate the immune system, cytokines at different axes in the host against pathogens. The very stable adult gut microbiota is *Lactobacillus*, which predominates in the *Bacteroidetes* and *Firmicutes* groups, with the coccoides *Clostridium leptum* and *Clostridium* being the most common. *E coli* is considered a predominant species in infants and a predominant population in adults, but the GIT is estimated to contain 5,001,000 different species of microorganisms, with the most elevated concentrations within the intestine (up to 1 gram of stool per gram). *Bifidobacterium* spp. maintain the stumbling block activities of the intestinal epithelium and health. Lipopolysaccharide (LPS) (endotoxin) produced by population reduction of the bacterial genus causes persistent low-grade aggravation. Furthermore, in the elderly, a host health problem arises since a critical diminish within the composition of *Bacteroidetes* and *Bifidobacterium* leads to a diminish in *Lactobacilli.* The composition of mutable anaerobic bacteria shows variations in the elderly and in adults, highlighted by several authors (Yin et al. 2019).

Dysbiosis caused by species of bacteria that are the associated flora of the gut, and by other highly pathogenic species such as *Enterobacter* spp., *Shigella dysenteriae*, *Salmonella Enteritidis*, *Salmonella heidelberg*, *Salmonella enterica*, *Clostridium difficile*, *Helicobacter pylori*, *Yersinia pestis*, *Vibrio* spp., *Proteus* spp., *Bacillus cereus*, *Campylobacter coli*, *Campylobacter jejuni*, and enterotoxigenic *Escherichia coli*. Various pathogens that cause GI disorders are activated by these microorganisms. The toll-like receptor (TLR) is a critical immunoreceptor for lipopolysaccharide (LPS), which is an important part of Gram-negative bacteria's cell wall. These mechanisms are involved in causing harm to the intestinal barrier and the immune system of this organ (Figs. 11.1, 11.3, 11.4, and 11.5) (Ley et al. 2006).

Many diseases are associated with multifactorial pathogens and require their treatment with a multi-targeted approach. Phytochemicals have the ability to act on multiple targets in a single dose. Most diseases have common factors such as inflammation and oxidative stress. However, most plants have flavonoids that act directly against them (Mukherjee et al. 2019) (Table 11.1).

Several researchers have published detailed reports showing the effects of phytochemicals against microbial pathogens in terms of bactericidal and microbial inhibition. Metabolites produced by the microbiota as a by-product have also shown pharmacological effects for the treatment of diseases. These phytochemicals (Fig. 11.2) serve to protect genetic material by preventing harmful germs from growing and regulating signal translation or gene expression in cell membranes (Singh et al. 2019).

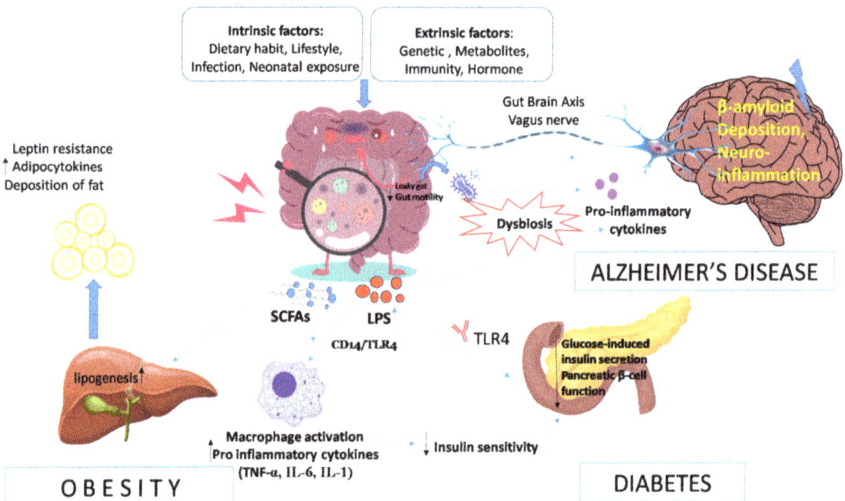

Fig. 11.1 Dysregulation of gut microbiota in Alzheimer's disease (AD), diabetes, and obesity

11.3 Restoration of Gut Microbiota in AD Via Phytomolecules

The gut-brain axis is decoded by the bidirectional interaction between the central nervous system (CNS) and the enteric nervous system (ENS) in the gastrointestinal tract, which results in the creation of active peptides with biological properties as well as various hormones. The ENS, also known as the "second brain," has roughly the same number of nerves as the spinal cord and is involved in the restoration of gut functions such as intestinal permeability, peristalsis, endocrine signaling and immune activity. The gut via mucosal activity share brain biogenic amines and signaling molecules to the brain (Semar et al. 2013). The brain communication with the GI through the sympathetic and parasympathetic divisions of the hypothalamic-pituitary-adrenal (HPA) axis, the autonomic nervous system (ANS), and the para-sympathetic adrenocortical axis (Ranjana and Da 2019).

Communication from the gut to the brain is most not unusual place via vagus muscle fibers and transfer for 80–90% of all vagal fibers. Vaginal afferent fibers also play an essential function in talking to the HPA axis for stress response control and anti-inflammatory activities. Immune system activation, such as inflammatory marker signaling triggered by microbial LPS or peptidoglycan, is also an example of brain communication. The translocation of these metabolites to the periphery can lead to changes in intestinal barrier integrity, leading to downstream bacterial cell activation as well as neuroinflammation (Carabotti et al. 2015).

Hormonal secretion by enteroendocrine cells (EECs) is influenced by synergistic metabolites of the intestinal microbiota and influences processes. The interplay among the EEC and intestinal glial cells, which is important for protective, possibly

Table 11.1 List of phytochemicals altered the gut bacteria and modulates the Pharmacological functions through metabolites

Phytochemicals	Bacterial species	Metabolites (molecules)	Function	References
Stilbenoids, Flavonols, Anthocyanins, Phenolic acids, Dihydroflavonols	↑*Enterococcus* spp., *Prevotella* spp., *Bifidobacterium* spp., *B. uniformis*, *E. lento*, *B. coccoides*, *E. rectale*	Desaminotyrosine (General Flavonoids)	Modulation of type I interferon	Descamps et al. (2019)
		Equol Daidzein (Soy isoflavones)	Regulate the platelet function in prevention of thrombotic events	Sánchez-Calvo et al. (2013)
		Daidzein (Puerarin)	Estrogenic and antioxidant activity	Mayo et al. (2019)
		Norathyriol (Mangiferin)	Suppresses the skin cancers. Restore obesity-induced insulin resistance	Sanugul et al. (2005)
		Hesperetin (Hesperidin)	Anti-inflammatory and antioxidation effect	Estruel-Amades et al. (2019)
		Baicalein oroxylin A (Baicalein)	Anti-pruritic, Anti-inflammatory	Noh et al. (2016)
Chlorogenic acid-polyphenols	↑*Bacteroides*, *Bifidobacterium*, *Lactobacillus*, *Ruminococcaceae* ↓*Desulfovibrionaceae*, *Lachnospiraceae*, *Erysipelotrichaceae*	Caffeic acid, Quinic acid, Hippuric acid, Hydroxyhippuric acid	Anti-oxidant, anticarcinogenic, and suppresses the adherence of pathogenic bacteria such as *H. pylori*	Sova and Saso (2020)
Phenolic compounds	↓*S. epidermidis*, *K. pneumoniae*		Activation of SCFA excretion and improve the immune function	Singh et al. (2017)
Tannins	↓*Clostridium leptum*, ↑*Bacteroides*	Gallotannin and ellagitannin	Produced the short-chain fatty acids by intestinal bacteria in the colon.	Kawabata et al. (2019)
Anthocyanidins	↓*H. pylori*, *S. epidermis*, *K. pneumoniae*, *S. epidermis*,	Cyanidin	Anti-inflammatory	

(continued)

Table 11.1 (continued)

Phytochemicals	Bacterial species	Metabolites (molecules)	Function	References
	K. pneumoniae, *E. coli*, *Salmonella* spp., ↑*Bifidobacterium* spp., *Lactobacillus* spp			Huang et al. (2020)
Gallic acids	↓*E. coli*, *P. aeruginosa*	Esters and catechin derivatives Methyl gallate epicatechin	Modulate the immune response, Anti-inflammatory, antimicrobial	Yang et al. (2020)
Coumarins	↓*Staphylococcus aureus* ↓*E. coli*, ↓*Salmonella* spp.↓*E. faecalis*, *B. subtilis*, *P. aeruginosa*↑*Bifidobacterium*, *Enterococcus*, *Eggerthella lenta*	Scopoletin	Anti-inflammatory, antiproliferative, inhibition of inducible NO synthase	Stassen et al. (2020)
Triterpenes	↓*H. pylori*, ↓*B. subtilis*, *S. aureus*, *Candida albicans*↓*M. fortuitum*, *S. pyogenes*, *E. faecalis*, *S. Typhi*		Reduce the number of sulfate containing bacteria, results anti-inflammatory in gut	Chang et al. (2018)
Alkaloids	↓*E. faecalis*, *S. aureus*, *E. coli*, *S.* Typhimurium, *Salmonella* spp. ↓*E. coli*, *P. aeruginosa*	Aconitine	Anti-inflammatory, analgesic	Peng et al. (2019)
Glycosides	*Bacteroides* sp.	Rg3, Rh2, and compound K (Ginsenoside)	Cardiovascular protection, Suppress the tumor growth	Ratan et al. (2020)
		Glycyrrhetic acid monoglucuronide	Tumor suppression, protection from inflammation, Antiviral	

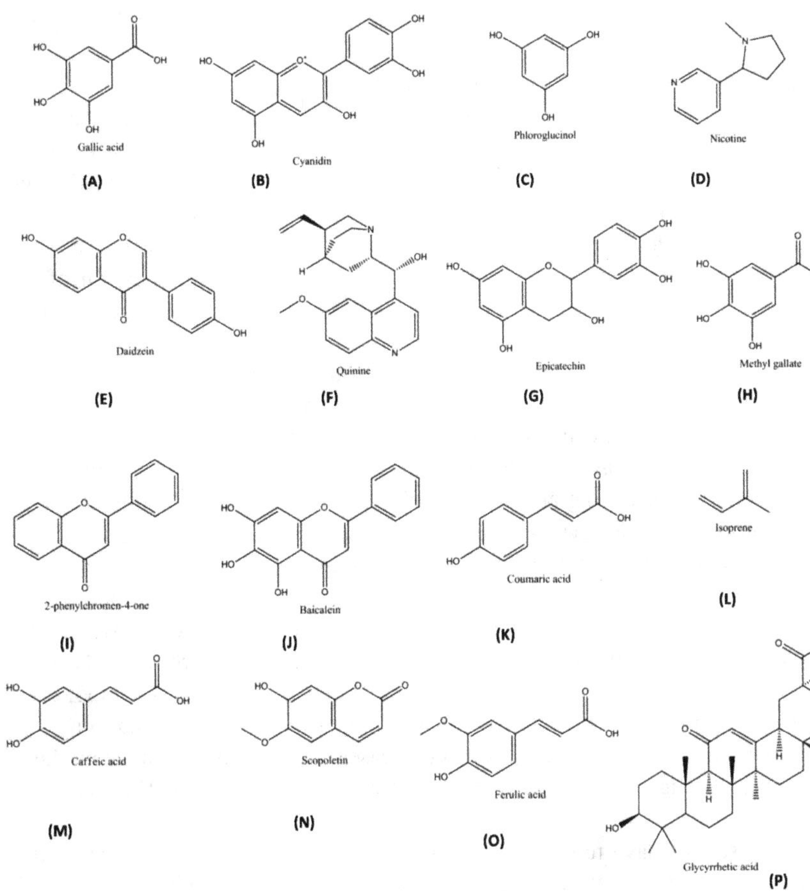

Fig. 11.2 (**a–p**) Phytochemicals/metabolites basic structures and their names

regulates hormone secretion and enteric nervous system (ENS) signaling (Yu et al. 2020).

In a recent study involving the pathogenesis of *E. coli*, *Staphylococcus aureus* and *Candida albicans* were shown to actuate amyloidosis, the metabolites and discharge of several cytokines in human immunodeficiency consciousness compared with the control group. The *Salmonella* brain infection model used during the development of AD in mice shows the formation of Aβ plaques in the hippocampus and temporal cortex after infection, which were permeabile BBB and *Salmonella* were recouped from these plates (Endres 2020; Upadhyay et al. 2020).

The role of the intestinal microbiota in regulating complex brain functions, such as those related to behavior and effects, has been extensively reported in the literature. In view of this, probiotic supplementation to combat dysbacteriosis and other restoration in the gut microbiota has recently been the subject of clinical studies and is now considered an effective and alternative approach. Potential to prevent combat emotional or cognitive disturbances in both healthy individuals and in AD subjects (Angelucci et al. 2019).

There are a number of phytochemicals found in medicinal herbs that have significant prebiotic effects on the gut microbial community. Phenols, triterpene flavonoids, sterols, etc. have therapeutic potential for the nervous system. The range of phytochemicals listed have tremendous therapeutic effects in modulating gut microbial communities by inducing metabolites. These metabolites produce neurotrophins such as epinephrine, serotonin, etc. and reduced the LPS level. Phytochemicals are thought to counteract the beginning and progression of AD through different mechanisms, such as their propensity to scavenge free radicals, promote cellular stress reactions, and enhance learning ability and memory (Prakash Reddy et al. 2020). Reduced bioavailability after intake and extremely low levels in tissues, including the central nervous system, are important limitations that have hampered the therapeutic use of these herbal medicines in AD patients. Coffee supplementation showcasing neuroprotective properties is mainly due to two mechanisms; polyphenols having antioxidant activity as well as the imapctofcoffee bean fibers contributing to decreasing the ratio of *Firmicutes* to *Bacteroidetes*, associated with inflammation was reduced. Even nutraceuticals, notably polyphenols found in fruits, vegetables, and some beverages, have been shown to be neuroprotective. Another innovative study of polyphenols in people with moderate cognitive decline, often known as the prodromic stage of vascular dementia or early AD, showed a sliver of strong evidence that they can positively affect gut bacteria. It's important to be aware of potential interactions between botanical medicines and drugs for Alzheimer's disease or other concomitant diseases. Flavonoids can easily modulate the activity of drug-metabolizing enzymes (Serra et al. 2020) (Figs. 11.3, 11.4, and 11.5).

Several studies have reported that ashwagandha and bacopa have been used to restore the gut microbiota by increasing *Bifidobacterium* colonies and similarly supplementing *Bacteroides spp*. For example, *B. vulgatus* and *B. Uniformis*. Kapikacchu uniquely chose *Ruminococcus bromii*, an organism having resistant starch degradation potential. Bacopa also chose a few dominant advantageous species, including *Bacteroides xylanolyticus*, *B. uniformis*, and *Butyrivibrio crossotus*. These three HM species followed the SCFA pathways, which may affect the production of luminal butyrate followed by an increase in the quantity of neurotransmitters and diminishing in neuroinflammation in AD. In the analysis of psychoactive herbs, the ability to strongly modulate the microbiome and simultaneously reconstruct metabolic preferences and products was observed. Signaling through the ENS may be attenuated by gut microbiota modulation by these psychoactive herbs. However, further studies are needed to confirm these expectations and their effects on subjects' health (Peterson 2020).

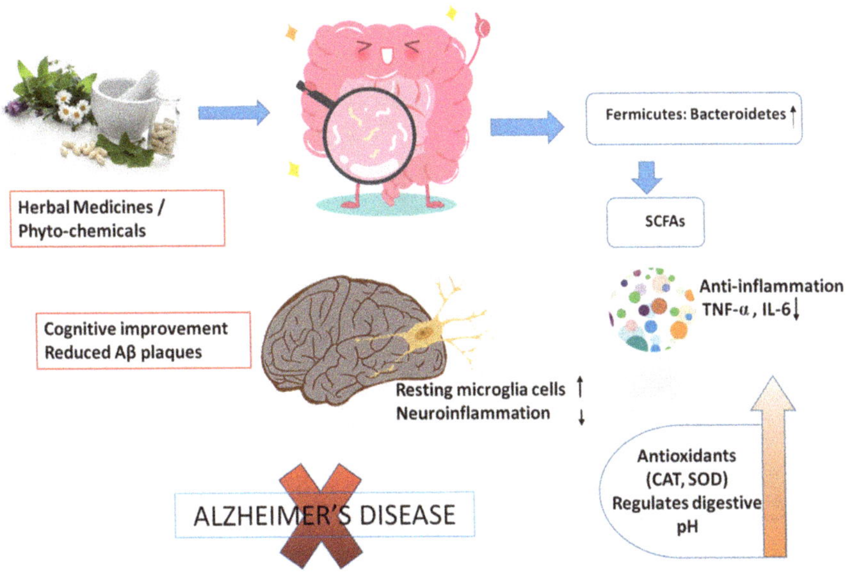

Fig. 11.3 Plants/plant-based diets make contribution in the prevention of Alzheimer's disease through the restoration of the gut microbiota: mechanistic pathways. *CAT* catalase, *SOD* superoxide dismutase, *SCFA* short-chain fatty acids, *Aβ* beta-amyloid

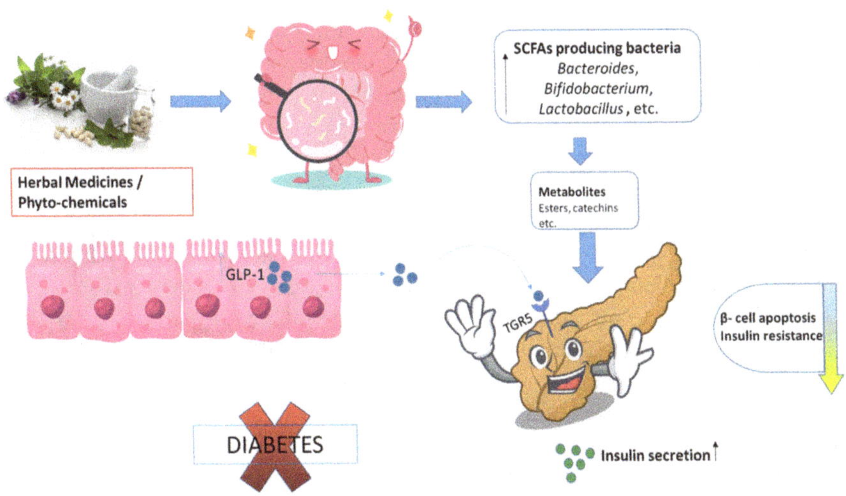

Fig. 11.4 Plants/plant-based diets contribute to diabetes prevention through restoration of gut microbiota: mechanistic pathways

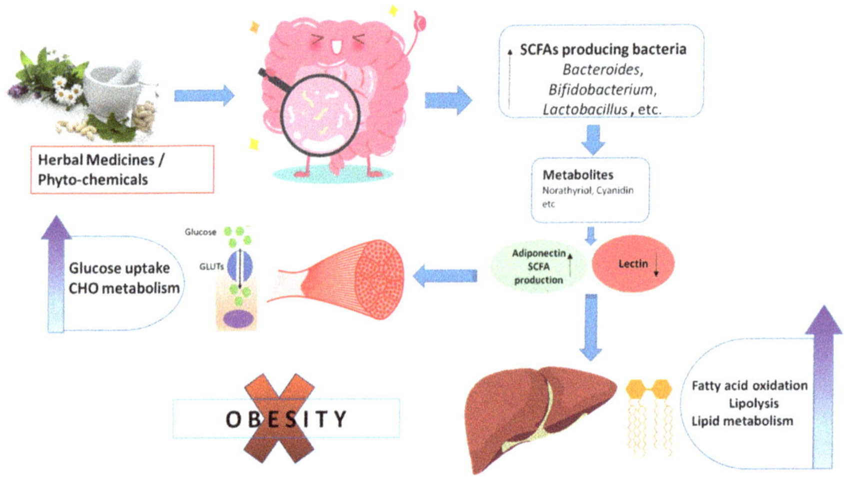

Fig. 11.5 Plants/plant-based diets contribute to obesity prevention through gut microbiota restoration: mechanistic pathways. *CHO* cholesterol, *SCFAs* Short-chain fatty acids

11.4 Restoration of Gut Microbiota in Diabetes Via Phytomolecules

Diabetes, one of the twenty-first century's global health issues, continues to grow very rapidly. In 2019, diabetes affects 463 million people globally, with the figure estimated to increase 0.7 billion by 2045. Insulin is considered an important hormone responsible for regulating blood sugar levels and in a healthy state. The pancreas secretes this chemical under normal circumstances, which lowers glucose synthesis by the liver and enhances glucose absorption by skeletal muscle. This condition is known as hyperglycemia—a coorelation between insulin absorption and insulin secretion. However, elevated systemic glucose levels are caused by pancreatic-cell failure, hyperglycemia, skeletal muscle, or insulin resistance in liver (Sirdah and Reading 2020).

The number and composition of the gut microbiota differs between people with diabetes and non-diabetic, according to the research, and blood glucose levels are substantially associated with diabetes prevalence. *Bacteroidetes: Firmicutes* as well as the ratio *Bacteroides: Prevotella: Clostridium: coccoides* group of rectal bacteria.

Furthermore, the percentage of the phylum *Firmicutes* and the *Clostridia* group showed a significant decrease in the diabetic group. In type II diabetics with the metabolic syndrome chronic kidney disease, another study found an improvement in the amount of *Clostridium clostridioforme* opportunistic bacteria, a drop in butyrate amount, and a further improvement in insulin sensitivity. The significant improvement in metabolic syndrome-related markers, including insulin sensitivity in a model induced by a high-fat diet, was credited with the polyphenol-rich extract and positive effects attributed to the relative growth in the proportion of *Akkermansia* spp. gut microbiota, by reducing inflammation and oxidative stress. Baicalein as a metabolite, the major flavone glycoside, shows significant effects on diabetes by restoring the gut microbiota. Blackberries have higher antidiabetic activity by generating metabolites during microbial fermentation. Several phytochemicals and herbal medicines have been reported (Fig. 11.2) to induce anti-inflammatory markers from metabolites, modulate glucose uptake mechanisms, and reduce insulin resistance by restoring the microbiota intestinal flora (Gurung et al. 2020).

11.5 Restoration of Gut Microbiota in Obesity Via Phytomolecules

Overweight or obesity condition is characterized as a body mass index (BMI) of thirty kilogram per meter square or higher, which has gained popularity, especially in developed countries, due to the adoption of the Western lifestyle, in which people engage in less physical activity and consume fat-rich foods. The inequality between the energy obtained, and the energy expended in the body is profoundly caused by the consumption of high-energy foods and a restrictive lifestyle, so that excessive storage of body fat occurs due to excess energy. Obesity is thought to be the result of an intersection of environmental, genetic, and physiological factors that disrupt energy acquisition and consumption. Central obesity, a debilitating medical condition, is linked to a few metabolic issues, including hypertension, hyperglycemia, and dyslipidemia. Obesity is caused by a complicated interaction between heredity, hormones, and the environment, which is known as a multifactorial etiology. Although many candidate genes are involved in the pathogenesis of obesity, these results are unreliable. Various hormones, including intestinal hormones, adipokines, and others, are related in the regulation and etiology of obesity (Blüher 2019). Ghrelin, the only identified peripheral activity-directing hormone, is a cyclic peptide hormone produced in the stomach. It aids in the stimulation of hunger. PYY (peptide YY) is released by L-cells in the distal small intestine and colon, and it is found in increasing amounts throughout the gut, with the highest concentrations in the colon and rectum. PYY is produced after a meal and sends messages to the brain, delaying stomach emptying and decreasing gastric output. Reduced feed consumption due to the usage of PYY before meals. In reaction to dietary fats, the gallbladder, pancreas, and stomach create cholecystokinin (CCK), which is concentrated in the small intestine. Gallbladder contraction, stomach emptying, exocrine pancreatic secretion, and intestinal motility are all controlled by it. CCK regulates satiety signaling through CCKA accessory receptors on brain afferent vagal fibers, and works centrally by enhancing fullness and decreasing hunger (Kaila and Raman 2008). Similar observations are also observed for glucagon-like peptides, i.e., the 29 amino acid fragments of glucagon. Intravenous management in people complements satiety and decreases meal intake. Adipocytes produce a variety of hormones known collectively as adipokines. Important secretory molecules include interleukin-6 (IL-6), adiponectin, tumor necrosis factor-alpha (TNF α-), and leptin. TNF-α is linked to insulin resistance in obesity by releasing free fatty acids, decreasing adiponectin production, and decreasing insulin signaling. On activation of nuclear factor-kappa B via TNF-α, a collection of inflammatory alterations in vascular tissue occurs. The higher the visceral fact, the higher the levels of IL-6, TNF-α, and C-reactive protein, and the lower the levels of adiponectin and interleukin-10, leading to a proinflammatory environment that causes endothelial dysfunction and insulin resistance, culminating in metabolic syndrome, diabetes, and atherosclerosis. Visceral fat calms these essential inflammation-controlling organs and has anti-inflammatory action equivalent to or greater than macrophages. Drugs and neuro-endocrine illnesses (hypothalamic, pituitary, thyroid, and adrenal) are thought to be secondary causes of weight issues (Oussaada et al. 2019).

The gut microbiota will be one of the environmental elements that contribute to obesity progression. The gut microbiota has been linked to the formation of fat mass and energy homeostasis. Firmicutes are on the rise: The Bacteroidetes ratio was shown to be lower in obese people compared to lean people after they lost weight on both types of calorie-restricted diets (Bouter et al. 2017).

Several research groups have used phenolic compounds to target the gut microbiota in obesity to have better understanding of interrelationships among phenolic chemicals and the microbiome. Dietary polyphenols and their metabolites improve health by modulation of the gut bacteria. The consumption of green and black tea polyphenols changed the composition of the gut microbiota, with an increase in Bifidobacterium spp., Lactobacillus spp., and Enterococcus spp. and a decrease in phylogenetics of Prevotella, Bacteroides, and Clostridium histolyticum. The phenolic extract in plums helps to regulate the growth of particular bacteria in the gut microbiome, which helps to reduce weight gain. Quercetin, one of the flavonoids, has been shown to reduce body weight increase and lower Firmicutes: Bacteroidetes ratio by reducing the growth of related bacteria such as *Eubacterium*, *Tenericutes*, *Bacillus*, and *Erysipelotrichaceae*. Obesity caused by diet. Thus, colon polyphenol fermentation may decrease obesity-related inflammatory markers while simultaneously promoting Bifidobacterium spp. growth (Castro-Barquero et al. 2018).

11.6 Conclusions

The significance of phytochemicals in gut microbial community restoration is the subject of this chapter. Microbial fermentation produces metabolites from phyto-chemicals. These metabolites protect against bacteria, insects, and fungi, and their metabolites have been linked to several medicinal benefits, such as antioxidant, anti-inflammatory, and regulatory properties. Metabolic activity, transporters, neuro-transmitters, in blood balance by increasing a manifestation of *Bifidobacterium* spp. and *Lactobacillus* spp., which helps alleviate dysbacteriosis, bacterial translocation, intestinal barrier damage, and gastrointestinal problems. Various pathogenic bacteria are involved in the disruption of epithelial membrane, leading to several diseases pathogenesis. Through numerous biochemical and physiological routes, beneficial chemicals meet the requirement to prevent the growth of harmful bacteria present in the GI.

Acknowledgement Prabhat Upadhyay has received financial grant from the Department of Health Research, Ministry of Health and Family Welfare, Government of India Scholarship for Young Scientists under the Human Resource Development Scheme.

References

Angelucci F, Cechova K, Amlerova J, Hort J (2019) Antibiotics, gut microbiota, and Alzheimer's disease. J Neuroinflammation 16:108. https://doi.org/10.1186/s12974-019-1494-4

Blüher M (2019) Obesity: global epidemiology and pathogenesis. Nat Rev Endocrinol 15:288–298. https://doi.org/10.1038/s41574-019-0176-8

Bouter KE, van Raalte DH, Groen AK, Nieuwdorp M (2017) Role of the gut microbiome in the pathogenesis of obesity and obesity-related metabolic dysfunction. Gastroenterology. https://doi.org/10.1053/j.gastro.2016.12.048

Carabotti M, Scirocco A, Maselli MA, Severi C (2015) The gut-brain axis: interactions between enteric microbiota, central and enteric nervous systems. Ann Gastroenterol 28:203–209

Castro-Barquero S, Lamuela-Raventós RM, Doménech M, Estruch R (2018) Relationship between mediterranean dietary polyphenol intake and obesity. Nutrients 10:1523. https://doi.org/10.3390/nu10101523

Chang TS, Chiang CM, Wang TY et al (2018) New triterpenoid from novel triterpenoid 15-O-glycosylation on ganoderic acid a by intestinal bacteria of zebrafish. Molecules 23:2345. https://doi.org/10.3390/molecules23092345

Descamps HC, Herrmann B, Wiredu D, Thaiss CA (2019) The path toward using microbial metabolites as therapies. EBio Medicine 44:747–757. https://doi.org/10.1016/j.ebiom.2019.05.063

Endres K (2020) Amyloidogenic peptides in human neuro-degenerative diseases and in microorganisms: a sorrow shared is a sorrow halved? Molecules 25:925. https://doi.org/10.3390/molecules25040925

Estruel-Amades S, Massot-Cladera M, Pérez-Cano FJ et al (2019) Hesperidin effects on gut microbiota and gut-associated lymphoid tissue in healthy rats. Nutrients 11:324. https://doi.org/10.3390/nu11020324

Gurung M, Li Z, You H et al (2020) Role of gut microbiota in type 2 diabetes pathophysiology. EBioMedicine 51:1–9. https://doi.org/10.1016/j.ebiom.2019.11.051

Huang F, Zhao R, Xia M, Shen GX (2020) Impact of cyanidin-3-glucoside on gut microbiota and relationship with metabolism and inflammation in high fat-high sucrose diet-induced insulin resistant mice. Microorganisms 8:1238. https://doi.org/10.3390/microorganisms8081238

Illiano P, Brambilla R, Parolini C (2020) The mutual interplay of gut microbiota, diet and human disease. FEBS J 287:833–855

Kaila B, Raman M (2008) Obesity: A review of pathogenesis and management strategies. Can J Gastroenterol 22:61–68. https://doi.org/10.1155/2008/609039

Kawabata K, Yoshioka Y, Terao J (2019) Role of intestinal microbiota in the bioavailability and physiological functions of dietary polyphenols. Molecules 24:370. https://doi.org/10.3390/molecules24020370

Ley RE, Peterson DA, Gordon JI (2006) Ecological and evolutionary forces shaping microbial diversity in the human intestine. Cell 124:837–848

Mayo B, Vázquez L, Flórez AB (2019) Equol: a bacterial metabolite from the Daidzein isoflavone and its presumed beneficial health effects. Nutrients 11:2231–2251. https://doi.org/10.3390/nu11092231

Mukherjee S, Tripathi AK, Kumar G et al (2019) Neuroprotective potential of small molecule phytochemicals in stroke therapy. In: Ranjana P, Ashish TAKD (eds) Advancement in the pathophysiology of cerebral stroke. Springer, p 155

Noh K, Kang Y, Nepal MR et al (2016) Role of intestinal microbiota in baicalin-induced drug interaction and its pharmacokinetics. Molecules 21:337. https://doi.org/10.3390/molecules21030337

Oussaada SM, van Galen KA, Cooiman MI et al (2019) The pathogenesis of obesity. Metabolism 92:26–36. https://doi.org/10.1016/j.metabol.2018.12.012

Peng J, Zheng TT, Li X et al (2019) Plant-derived alkaloids: the promising disease-modifying agents for inflammatory bowel disease. Front Pharmacol 10:351. https://doi.org/10.3389/fphar.2019.00351

Peterson CT (2020) Dysfunction of the microbiota-gut-brain axis in neurodegenerative disease: the promise of therapeutic modulation with prebiotics, medicinal herbs, probiotics, and synbiotics. J Evidence-Based Integr Med 25:1–19. https://doi.org/10.1177/2515690X20957225

Prakash Reddy V, Aryal P, Robinson S et al (2020) Polyphenols in alzheimer's disease and in the gut–brain axis. Microorganisms 8:199. https://doi.org/10.3390/microorganisms8020199

Raju GS, Moghal MR, Dewan SMR et al (2013) Characterization of phytoconstituents and evaluation of total phenolic content, anthelmintic, and antimicrobial activities of Solanum violaceum Ortega. Avicenna J phytomedicine 3:313–320

Ranjana P, Da TAK (2019) Advancement in the pathophysiology of cerebral stroke. Springer, Cham

Ratan ZA, Haidere MF, Hong YH et al (2020) Pharmacological potential of ginseng and its major component ginsenosides. J Ginseng Res 1–12. https://doi.org/10.1016/j.jgr.2020.02.004

Ribeiro CFA, Silveira GGDOS, Cândido EDS et al (2020) Effects of antibiotic treatment on gut microbiota and how to overcome its negative impacts on human health. ACS Infect Dis 6:2544–2559

Sánchez-Calvo JM, Rodríguez-Iglesias MA, Molinillo JMG, Macías FA (2013) Soy isoflavones and their relationship with microflora: Beneficial effects on human health in equol producers. Phytochem Rev 12:979–1000

Sanugul K, Akao T, Li Y et al (2005) Isolation of a human intestinal bacterium that transforms mangiferin to norathyriol and inducibility of the enzyme that cleaves a C-glucosyl bond. Biol Pharm Bull 28:1672–1678. https://doi.org/10.1248/bpb.28.1672

Semar S, Klotz M, Letiembre M et al (2013) Changes of the enteric nervous system in amyloid-β protein precursor transgenic mice correlate with disease progression. J Alzheimers Dis 36:7–20. https://doi.org/10.3233/JAD-120511

Serra D, Almeida LM, Dinis TCP (2020) Polyphenols in the management of brain disorders: Modulation of the microbiota-gut-brain axis. Adv Food Nutr Res 91:27. https://doi.org/10.1016/bs.afnr.2019.08.001

Singh RK, Chang HW, Yan D et al (2017) Influence of diet on the gut microbiome and implications for human health. J Transl Med 15:1–17

Singh AK, Cabral C, Kumar R et al (2019) Beneficial effects of dietary polyphenols on gut microbiota and strategies to improve delivery efficiency. Nutrients 11:2216–2237. https://doi.org/10.3390/nu11092216

Sirdah MM, Reading NS (2020) Genetic predisposition in type 2 diabetes: a promising approach toward a personalized management of diabetes. Clin Genet 98:525–547. https://doi.org/10.1111/cge.13772

Sova M, Saso L (2020) Natural sources, pharmacokinetics, biological activities and health benefits of hydroxycinnamic acids and their metabolites. Nutrients 12:2190. https://doi.org/10.3390/nu12082190

Stassen MJJ, Hsu SH, Pieterse CMJ, Stringlis IA (2020) Coumarin communication along the microbiome–root–shoot axis. Trends Plant Sci S1360-1385:30286–30287. https://doi.org/10.1016/j.tplants.2020.09.008

Upadhyay P, Shukla R, Mishra SK, Dubey GP, Verma R, Purohit S (2018a) Medicinal roles of phytomolecules in the treatment and management of cancer. In: Phytochemistry. Volume 2, pharmacognosy, nanomedicine, and contemporary issues. Chapter 15. CRC Press. ISBN no 9781771887601, pp 333–366

Upadhyay P, Purohit S, Mishra SK et al (2018b) Pharmacognostic standardization of Asian folk medicinal plant Reinwardtia Indica Dumort. Int J Green Pharm 12:S380–S385. https://doi.org/10.22377/IJGP.V12I02.1790

Upadhyay P, Mishra SK, Mishra AK et al (2019) Evaluation of antioxidant and antimicrobial potential of a novel Himalayan plant Reinwardtia indica dumort: Scientifically unexplored. Microb Pathog 127:326–334. https://doi.org/10.1016/j.micpath.2018.12.024

Upadhyay P, Shukla R, Tiwari KN et al (2020) Neuroprotective effect of Reinwardtia indica against scopolamine induced memory-impairment in rat by attenuating oxidative stress. Metab Brain Dis. https://doi.org/10.1007/s11011-019-00479-0

Yang K, Zhang L, Liao P et al (2020) Impact of gallic acid on gut health: focus on the gut microbiome, immune response, and mechanisms of action. Front Immunol 11:580208. https://doi.org/10.3389/fimmu.2020.580208

Yin R, Kuo HC, Hudlikar R et al (2019) Gut microbiota, dietary phytochemicals, and benefits to human health. Curr Pharmacol Reports 5:332–344

Yu Y, Yang W, Li Y, Cong Y (2020) Enteroendocrine cells: sensing gut microbiota and regulating inflammatory bowel diseases. Inflamm Bowel Dis 26:11–20. https://doi.org/10.1093/ibd/izz217

Chapter 12
Gut Microbiome and Diet: Promising Approach for Treatment of Cognitive Impairment

Awakash Soni, Priya Gupta, and Ankit Verma

Abstract Human gastrointestinal microbiota is a key regulator of the brain and behavior in both healthy and disordered conditions. Neurotransmitters, change in the intestinal barrier and enteric sensors, bacteria-derived short-chain fatty acids, and immune regulators support in bidirectional cross-talk between brain and gut. Gut microbiota and their derived metabolites elicit host immune response by inducing cytokines and chemokine production, which further cause inflammation in the central nervous system and associated with the pathogenesis of brain disorders for instance pain, anxiety, depression, and age-associated neurodegenerative disorders with cognitive impairment. An increase in proinflammatory microbes and a decrease in anti-inflammatory microbes in the microbiome cause gut dysbiosis and strongly reflected in numerous disease conditions including neurological diseases with cognitive impairment. The treatment availability of these diseases is limited. Here, in this chapter, we are discussing different ways to improve gut microbiome comprising high-fiber diet, probiotics, genetically modified probiotic bacteria (GMP), fecal microbiota transplantation (FMT), and physical workout and exploring their therapeutic potential to cure neurological diseases including cognitive impairment.

Keywords Microbiota · Cognitive impairment · Genetically modified probiotics · Fecal microbiota transplantation · High-fiber diet

A. Soni (✉)
Department of Microbiology and Molecular Genetics, The Kuvin Center for the Study of Infectious and Tropical Diseases, Hebrew University Hadassah Medical School, Jerusalem, Israel

P. Gupta
Department of Developmental Biology and Cancer Research, Hebrew University-Hadassah Medical School, Jerusalem, Israel

A. Verma
Department of Life Sciences and the National Institute for Biotechnology in the Negev, Ben Gurion University of Negev, Beer Sheva, Israel

A. K. Tripathi, M. Kotak (eds.), *Gut Microbiome in Neurological Health and Disorders*, Nutritional Neurosciences,
https://doi.org/10.1007/978-981-19-4530-4_12

12.1 Introduction

A diverse array of microbes is the inhabitant of our gastrointestinal tract, together called gut microbiota. The effects of these microbes on gastrointestinal physiology and different biological processes are well established. Many studies have proven that gut microbiota contributes to the proper brain functions and its development. The disturbances in the microbiome composition increase the host susceptibility to different diseases including neurological diseases (Bienenstock et al. 2015; Collins and Bercik 2009; Collins et al. 2012; Cryan and Dinan 2012; De Palma et al. 2015; Sampson and Mazmanian 2015). Even though gut microbiota and brain communication are still poorly understood, host metabolic, immune, endocrine, and neuronal host functions are believed to interact with the gut microbiome and, in turn, affect brain functions (Collins et al. 2012; Cryan and Dinan 2012; Sharon et al. 2014). Trouble in learning new skills, recalling, concentrating, and making decisions is known as cognitive impairment. Risk of age-related diseases, including cognitive impairment, also emerged with a higher percentage of elderly people in the population around the globe. Accumulating evidence suggests that cognitive impairment suffering people have an additional occurrence of functional disability and poor quality of life (Schmidt et al. 2018; Panza et al. 2018). It was also reported that cognitive impairment has been experienced in clinical settings in higher proportion (~11–40%) compared to the general community setting (1.0–12%) (Schmidt et al. 2018). It has been predicted that ~5.1 million elderly people in the United States may be suffering from cognitive impairment (Alzheimer's disease) and assumed number may rise ~13 million by 2050 (Hebert et al. 2003). The increasing population of elderly people with cognitive impairment will demand huge medical expenses and harshly affect the economy of several countries around the world. To prevent social and economic damage, the development of new therapeutic interventions to treat cognitive impairment is essential.

The gut microbes are vital for many physiological processes in the human body including energy generation from undigested food (Canfora et al. 2015), synthesis of vitamins and neurotransmitters, regulation (Yano et al. 2015), metabolism of bile salts (Devlin and Fischbach 2015), the host immune response (Fulde and Hornef 2014), the swaying of host-cell proliferation and vascularization (Kamada et al. 2013), regulation of endocrine functions,, and neurologic signaling (Neuman et al. 2015). As listed, gut flora is crucial to maintain functional homeostasis in the body. Mental health, cardiovascular, metabolic and inflammatory diseases are found to be associated with dysbiosis (loss of balance of healthy gut microbiome) and more interestingly they have also been connected to cognitive impairment (Panza et al. 2018; Sender et al. 2016). The gut microbiome and brain interactions are known as the gut-brain axis. This consists of the communication among the central nervous system, the enteric nervous system, the autonomic nervous system, the endocrine system, and the immune system. Cognitive centers of the brain and peripheral intestinal function happen in tightly regulated loop (Foster and McVey Neufeld 2013; El Aidy et al. 2015) and this bidirectional cross-talk between brain and gut is assisted by neurotransmitters, enteric sensors, alteration in the intestinal barrier, bacterial short-chain fatty acids, and immune regulators (Forsythe et al. 2014; Calvani et al. 2018; Kennedy et al. 2017). The majority of neurological and metabolic diseases/disorders (i.e., Parkinson's disease, Alzheimer's disease, cardiovascular diseases, diabetes, cancer, arthritis, and others) are characterized by a chronic low level of inflammation. This chronic inflammation can serve as a connecting link between gut flora and cognitive impairment. Taken together, as proven in studies, the gut microbiome fairly affects nearly all biological processes in the human body and is associated with many disease conditions. In this regard, given valuable connections between gut and brain, the hidden potential of healthy gut microbes could be exploited to develop new therapeutic interventions for many conditions, including cognitive impairment. The conceivable details about the same are discussed below.

12.2 Potential of Modified Diet for Treatment of Cognitive Dysfunction

Accumulating evidence suggested a strong link between one's food habit (diet) and cognitive functions both in short term and long term (Spencer et al. 2017; Davidson et al. 2019; Dominguez and Barbagallo 2018; Beilharz et al. 2016). The beneficial effect for diets has already been explored in different disease conditions including neurological disorders (Lichtwark et al. 2014; Reddel et al. 2019). High-fiber, low-fermentable and gluten-free (GFD) diets help to maintain healthy gut flora, thus improving cognitive functions. Mediterranean diet (MeDi) which includes fish, vegetables, fruits, nuts, olive oil, etc. has proven role to diminish cognitive impairment (Kim and Yun 2018). The experimental evidence of use of probiotic microbes (*B. longum*, *B. breve*, *B. infantis*, *L. casei*, and others) have been found to improve cognitive functions and also useful in improving CNS function as well as ameliorating CNS disorders' symptoms (anxiety, depression, stress) (Tsai et al. 2019; Wang et al. 2016). TLR receptors mediate pathways in neurological disorders and the microbiota-gut-brain axis is widely explored (Ma et al. 2019). Many studies indicate the beneficial role of probiotics in preventing the growth of pathogenic microorganisms by helping the restoration of the disrupted intestinal barrier and the stimulation of the human mucosal immune system. Furthermore, probiotics-derived TLR ligands seem to induce the production of anti-inflammatory cytokines (Thomas et al. 2017; Caputi and Giron 2018).

12.2.1 High-Fiber Diet

The portion of plant-based food which can't be digested by human enzyme but by gut microbes is known as dietary fiber. The addition of dietary fiber in the diet is the most extensively used diet modification to improve gut microbiota (Mailing et al. 2019a). The lumen microbes process dietary fiber into short-chain fatty acids (SCFAs) which have several health benefits. Children taking a high-fiber diet displayed better performance on a cognitive task, indicating the importance of diet in cognitive function (Khan et al. 2015). Mice on a high-fiber diet have shown attenuated inflammation in microglia which contributes to 15% of total cells in the brain. High-fiber diets and their fermentation product SCFAs reduce the risk of many inflammatory diseases such as colon cancer, obesity, and cardiovascular disease, thus emphasizing the capability of gut microbes in maintaining homeostasis and anti-inflammatory effects (Mailing et al. 2019a). It has been reported that a high-fiber diet alters the gut microbiota by enhancing the population of acetate-producing bacteria (Marques et al. 2017). And separately in another study that both fiber and acetate or colonization with butyrate-production bacteria decreased gut dysbiosis and restored blood–brain barrier permeability (Marques et al. 2017; Braniste et al. 2014). GABA supplementation also enhances the concentrations of acetate, propionate, and butyrate and total SCFAs in colonic, cecal contents and thus lowers colonic pH. Acetate, propionate, and butyrate individually have also been linked to gut health. Moreover, butyrate promotes several neurotransmitters (gamma-aminobutyric acid and serotonin) synthesis, confers neuroprotective activity, and highlights the anti-inflammatory effects of SCFAs in the brain (Xie et al. 2017). Hence, these pieces of evidence signify the importance of a high-fiber diet and further can be exploited for managing cognitive impairment.

12.2.2 Potential of Probiotics for Cognitive Impairment Therapy

Probiotics are live microorganisms, possessing health benefits through improving gut microbiota after consumption (Reid 2016). Nobel laureate Élie Metchnikoff was the first to notice the positive impact of certain bacteria on health in the year 1907. Gut microbes benefit the human body in several ways and play an indispensable role to maintain homeostasis. Therapeutic efficacy of probiotic bacteria has been tested to treat many gastrointestinal diseases including antibiotic-associated diarrhea, infectious diarrhea, *Clostridium difficile* colitis, ulcerative colitis, Crohn's disease, irritable bowel syndrome, and others. The mechanistic details regarding communications between the microbiome and the host are still under investigation. It is proposed that probiotics exert their beneficial effect by preventing gut dysbiosis (Ng et al. 2009). The neurological and metabolic diseases have a key feature of low chronic inflammation and probiotic bacteria support to reduce the inflammation (Hakansson and

Molin 2011), thus probiotics seem to have great potential for the treatment of cognitive impairment. Tryptophan, an essential amino acid, catabolizes into kynurenine metabolites (kynurenine pathway). These metabolites regulate several biological processes including immune response, inflammation, host-microbiome signaling, and neurotransmission (Cervenka and Agudelo 2017). With this, the expression of kynurenine pathway enzymes is regulated through nutritional and inflammatory signals and found to be associated with depression and schizophrenia. A study showed that gut microbes produce tryptophan metabolites which equate mucosal reactivity. It was found that when mice switched from sugar to tryptophan for energy source, increased lactobacilli content in gut microbiome which produces an aryl hydrocarbon receptor (AhR) ligand, indole-3-aldehyde which in turn leads AhR-dependent IL-22 transcription. Subsequent IL-22-dependent balanced mucosal response permits the existence of mixed microbial societies and protects the mucosa from inflammation (Zelante et al. 2013).

The beneficial effect of probiotics on cognitive impairment is still under investigation. But there are many shreds of evidence where beneficial effects of probiotics on different age-related diseases have been established which include AD (Yang et al. 2020; Nimgampalle and Yellamma 2017; Bonfili et al. 2017), neuroinflammation (Musa et al. 2017), diabetes (Davari et al. 2013), and vascular dementia (Liu et al. 2015). Probiotic bacteria *Bifidobacterium breve* strain A1 has shown to prevent cognitive dysfunction in the AD mice model. In addition, bacterial metabolite acetate also partially improved cognitive decline. The repressed hippocampal expressions of inflammation and immune-reactive genes after consumption of *B. breve* A1 have proven its beneficial effect (Kobayashi et al. 2017). It has also been established that probiotic *L. plantarum* MTCC1325 displayed anti-Alzheimer properties. *L. plantarum* MTCC1325 treatment not only ameliorates cognitive deficits in diseased rats but also restored the level of acetylcholine and histopathological features (Nimgampalle and Yellamma 2017). Another investigation with probiotic agents *Lactobacillus acidophilus*, *L. fermentum*, *Bifidobacterium lactis*, and *B. longum* in rat found to improve memory and inhibit the AD pathology (Azm et al. 2018). Thus, probiotic bacteria *B. breve*, *L. plantarum*, *L. acidophilus*, *L. fermentum*, *B. lactis*, and many others appear to have therapeutic potential for cognitive impairment.

Many animal studies strongly support the beneficial use of probiotic bacteria in different disease conditions but the limited data on humans is available to support the hypothesis. A human trial with probiotic species such as *Lactobacillus acidophilus*, *Lactobacillus casei*, *Bifidobacterium bifidum*, and *Lactobacillus fermentum* showed a positive impact on cognitive function and metabolic homeostasis in AD patients (Akbari et al. 2016). In another human study, the supplementation of multispecies probiotic to AD patients leads to an increase in *Faecalibacterium prausnitzii* in gut flora and influenced tryptophan metabolism in serum (Leblhuber et al. 2018). Thus, studies showed the potential of probiotic bacteria as an additional treatment for cognitive impairment but still needs to go long ways to establish its effect on humans.

12.2.3 Potential of Genetically Modified Probiotics (GMP) for Cognitive Impairment Therapy

By using genetic engineering tools, probiotics can be modified into Genetically Modified Probiotics (GMP) to deliver drugs, therapeutic proteins, and even genes with better site-specificity. GMP with anti-inflammatory and other specific properties could be exploited to benefit human health. Thus, bioengineered probiotics may become a therapeutic strategy to cure/manage neurological diseases. The engineered cytokine IL-10 secreting probiotic bacteria *Lactococcus lactis* (*L. lactis*) has shown to improve colitis in mice. The bacteria not only reduced the therapeutic dose of IL-10, but also caused a 50% reduction of colitis in mice (Steidler et al. 2000). In another study, *Bifidobacterium longum*, engineered to secrete human IL-10 (BL-hIL-10), was administered into dextran sulfate sodium-induced ulcerative colitis mice model, and resulted in alleviation of symptoms of colitis syndrome (Yao et al. 2011). These supportive effects of GMP on animal diseases are encouraging for investigating deeper GMP safety for human administration. In a study with human subjects, human IL-10 secreting genetically altered *Lactococcus lactis* (LL-Thy12) was used to treat patients of Crohn's disease and found innocuous and effective. Further, the effect of GMPs on other age-related diseases has also been determined in some studies. The binding of angiotensin II to its receptor angiotensin II receptor type 1 (AT1) has proposed to control the blood pressure through the renin-angiotensin system (RAS). Lately, the alternate RAS axis has obtained more attention, where angiotensin-converting enzyme 2 (ACE2) converts angiotensin I to angiotensin (1–7). With this, anti-inflammatory effect with other favorable physiological effects of ACE2/Ang (1–7)/MAS axis has been reported (Simon et al. 2015). Furthermore, *Lactobacillus paracasei* (LP) was genetically modified to synthesize Ang (1–7) (LP-A) as a fusion protein with a transepithelial carrier and given orally to diabetic mice. Improved glucose tolerance by enhancing insulin level, reduced retina and kidney damage were observed in LP-A-treated diabetic mice (Li et al. 2018).

Recently, the involvement of RAS in aging has been established (Carter et al. 2020). It has also been found that the treatment of altered *L. paracasei* (LP-A) which can secrete Ang (1–7) in mice model of aging increases RAS metabolites with elevated levels of Ang (1–7) in circulation. This study gives clues on how GMPs can be further explored to treat age-related neurological diseases (Carter et al. 2020). There is a strong association between hypertension and cardiovascular diseases (Weber 1994; Escobar 2002). In two independent studies, antihypertensive effects of *Lactobacillus plantarum* (Lb. plantarum) and *Escherichia coli* expressing angiotensin-converting enzyme inhibitory peptides (ACEIPs) have been determined on the spontaneously hypertensive rats (SHRs) model. Oral administration of recombinant *Lb. plantarum* and *E. coli* showed promising antihypertensive effects (Yang et al. 2015; Huang et al. 2012). Given all the previous points, there is supporting evidence for GMPs to hold the promising therapeutic potential to treat neurological and metabolic diseases. Still, more research is needed to unfold its full competence.

12.3 Fecal Microbiota Transplantation (FMT) as a Cognitive Impairment Therapy

Because of the strong involvement of gut dysbiosis in different diseases, improvement in the content of gut microbiota may reverse the disease phenotype. Infusion of stool in the colon from a healthy individual to a diseased individual who has an unhealthy gut microbiome (gut dysbiosis) is known as fecal microbiota transplantation (FMT) (Brandt and Aroniadis 2013; Bakken et al. 2011). FMT was first described in 1958, but it is receiving more attention and rapid approval in modern science (Eiseman et al. 1958; Chin et al. 2017). Endoscopy, nasogastric and nasoenteric tubes, and capsules are few routes for delivering FMTs. Administering a new healthy gut microbial community helps with achieving normal gut functions and restoring its colonization resistance (Gerding 2005; Wenisch et al. 1996; Sadowsky and Khoruts 2016). *Clostridium difficile* infection (CDI) has been successfully treated by repopulating the gut with a healthy microbiota through FMT for six decades. Because of proven safety, efficacy, and high success rates (92%) of FMT in treatment of recurrent CDI (rCDI) in several randomized clinical trials, now FMT has been recommended as a second-line treatment (Cammarota et al. 2014; Kelly et al. 2016; Van Nood et al. 2013; Youngster et al. 2014; Trubiano et al. 2016; McDonald et al. 2018). Given the increasing success of FTMs, it has been also recommended for the treatment of other diseases like inflammatory bowel disease (IBD), allergic diseases, metabolic disorders, and autoimmune disorders (Choi and Cho 2016). Inflammatory bowel diseases (IBDs) and ulcerative colitis have peculiar feature of chronic inflammation of the lumen and have a prevalence of 0.5–1.0% among European adults with increasing rates around the globe (Molodecky et al. 2012). Decreased anti-inflammatory phyla *Bacteroides*, *Firmicute* and increase in proinflammatory *Proteobacteria* in gut dysbiosis is a leading cause for IBDs (Konturek et al. 2015). Similar alterations in gut microbiota are associated with other disease conditions including neurodegenerative diseases (Calvani et al. 2018), obesity, metabolic diseases (Konturek et al. 2015; Vindigni and Surawicz 2017), chronic kidney disease and hypertension (Chin et al. 2017; Richards et al. 2017).

Clinical trial in Australia has suggested the increased sustained microbial diversity after fecal microbiota transplantation. A healthy clinical outcome of this study was associated with different bacterial taxa including *Fusobacterium spp.* (Paramsothy et al. 2017). Some other studies also concluded the positive effect of FMT, but not concrete needs more extensive studies (Sokol et al. 2020; Imdad et al. 2018). A recent study has demonstrated that enhanced neuroinflammation and motor dysfunction were observed in germ-free mice who received fecal microbiota from PD patients (Sampson et al. 2016), supporting the link of dysbiosis and disease conditions. Another study also showed increased insulin sensitivity along with elevated levels of butyrate-producing intestinal microbiota when intestinal microbiota from lean donors was transferred to metabolic syndrome patients (Vrieze et al. 2012). Similarly, FMT from hypertensive individuals to germ-free mice elevated blood pressure in mice and substantiated the role of gut microbiota in hypertension (Li et al. 2017). Alzheimer's disease is characterized by cognitive dysfunction. The observations of cognitive function in SAMP8 (senescence-accelerated mouse prone 8) mice model have told that cognitive function in SAMP8 mice was severely decreased as compared with senescence-accelerated mouse resistant 1 (SAMR1) mice. Furthermore, fecal microbiota transplantation from SAMP8 or SAMR1 mice to pseudo-germ-free mice led to a significant decrease in cognitive function. However, pseudo-germ-free mice receiving fecal bacteria from SAMR1 mice showed behavioral improvements (Zhan et al. 2018). Together, these studies intensely support the involvement of abnormal gut microbiota (gut dysbiosis) in neurological (including cognitive impairment) and metabolic diseases. Identification of key microbes and their benefits on different pathological conditions will offer a great tool in the form of FMTs to cure multiple illnesses. The availability of highly screened, effective fecal bacteria catalog may also support the treatment. Furthermore, extensive and systematic studies will be compulsory before approval of the use of FMTs on humans.

12.4 Potential of Physical Training/Exercise for Cognitive Impairment Therapy

A regular physical exercise practice reduces the risk of acute and chronic inflammatory diseases because of its confirmed anti-inflammatory effects. Recent studies established that exercise has an independent capability to modify the gut microbiota composition and its positive functional ability (Kang et al. 2014; Kern et al. 2020; Mailing et al. 2019b). When gut microbiota transfer from exercise-trained mice into germ-free mice, it leads to a reduction in colitis risk which is evidenced by enhanced expression of cytokines involved in the resolution of inflammation, tissue regeneration and attenuated mucus depletion, colon shortening (Allen et al. 2018). In another study, exercise preconditioning-induced modification in gut microbiota composition showed a beneficial effect on septic shock by regulating host response to sepsis in mice, which is defined by the improved balance between pro- and anti-inflammatory responses, less organ damage, and increased survival (Kim and Kang 2019). Consistently, mice on wheel running exercise for 12 weeks showed reduced intestinal inflammation, increased antioxidant enzymes in intestinal lymphocytes, and improved epithelial membrane integrity by increasing microbial diversity in mice (Campbell et al. 2016; Allen et al. 2015). *Bifidobacterium* spp. is one of the important gut microbes highly affected by physical exercise in normal mice (Lambert et al. 2015). In fact, increase in *Bifidobacterium* in the fecal content was observed in mice went through workout on treadmill (Lambert et al. 2015). Many inflammatory conditions of the gut have been associated with decreased level of neurotrophic factor derived from brain in the hippocampal area and *Bifidobacterium* supplementary diet has shown to increase brain-derived neurotrophic factor expression. This study strongly suggests that physical activity affects brain function through hippocampus neurogenesis by modulating gut microbiota (Cassilhas et al. 2016). Similarly, in human volunteers (professional athletes), improved microbiome diversity has been observed which was associated physical workout (Clarke et al. 2014). Aerobic exercise training in older adults efficiently contraries age-related loss in hippocampal volume and improved learning and memory by enhancing serum levels of BDNF, which is a crucial neurogenesis mediator in dentate gyrus (Erickson et al. 2011). Studies have proven that *Akkermansia* and *Proteobacteria* are the most exercises responsive taxa. 5–6 weeks endurance exercise training in older people and overweight woman has shown to increase *Akkermansia*, *Oscillospira* and decrease *Proteobacteria* and *C. difficile* which are detrimental to health (Taniguchi et al. 2018; Munukka et al. 2018). Together, these studies strengthen physical training/exercise as significantly improving the fraction of anti-inflammatory microbes in the gut microbiota and having benefits on neurological diseases. Preliminary research on the health effects of physical exercise on mice and humans is promising and creating a foundation to explore its therapeutic importance for cognitive impairment.

12.5 Conclusion

Gut microbiota and human health are complementary to one another. The past two decades of studies on microbiome revolutionized our opinion about gut microbes and established its involvement with immunological, neurological, and metabolic diseases/disorders. Gut microbiota influences the bidirectional communication between the central and the enteric nervous system (gut-brain axis). It is a well-recognized fact now that gut dysbiosis is allied with the majority of illnesses including neurological disease and cognitive impairment. Thus, healthy gut microbes appear to have the potential to manage/treat cognitive impairment. Probiotic *B. breve, L. plantarum, L. acidophilus, L. fermentum*, and others have proven benefits on neurological and metabolic diseases. Also, these healthy probiotics can be modified into GMPs for delivering certain therapeutic agents specific to disease conditions. Above all, gastrointestinal infections and gut dysbiosis are already treated with FMT. FMT potential to restore a healthy gut microbiome can also be exploited in age-related diseases including cognitive impairment. The beneficial role of physical exercise on overall health by influencing gut microbes has already been confirmed and could be further explored for neurological disorders. The gut microbiota also secretes a variety of metabolites that have positive effects on host physiology and in turn are also affirmatively influenced by diet, especially high-fiber diets. Thus, managing cognitive impairment through administering a high-fiber diet with other health supplements seems to have great potential to be explored. Probiotics, GMPs, FMT, exercise, and a specific diet plan appear to have the potential to treat neurological diseases and cognitive impairment (Fig. 12.1) but further systematic and extensive animal studies and human trials on these factors (taken each separately or in combination) will be required before confirming them as a therapeutic agent for cognitive impairment.

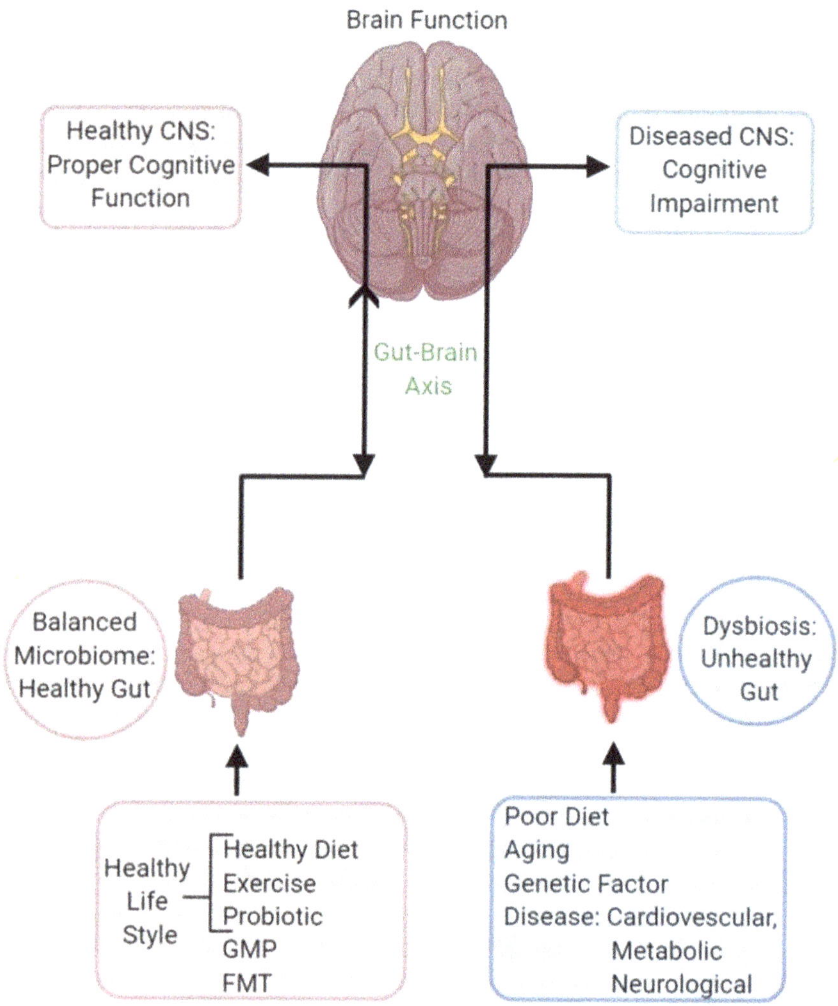

Fig. 12.1 Schematic representation of different gut microbiota influencing factors in both healthy, dysbiosis conditions and their impacts on cognitive functions through gut-brain axis

Acknowledgments Critical reading and suggestions to improve the chapter by Dr. Pradeep Kumar, Ms. Chiara Mazzoni, and Ms. Avital Cher are highly appreciated.

References

Akbari E, Asemi Z, Kakhaki RD et al (2016) Effect of probiotic supplementation on cognitive function and metabolic status in Alzheimer's disease: a randomized, double-blind and controlled trial. Front Aging Neurosci 8:256

Allen JM, Miller MEB, Pence BD et al (2015) Voluntary and forced exercise differentially alters the gut microbiome in C57BL/6J Mice. J Appl Physiol 118:1059–1066
Allen JM, Mailing LJ, Cohrs J et al (2018) Exercise training-induced modification of the gut microbiota persists after microbiota colonization and attenuates the response to chemically-induced colitis in gnotobiotic mice. Gut Microbes 9:115–130
Azm SAN, Djazayeri A, Safa M et al (2018) Lactobacilli and bifidobacteria ameliorate memory and learning deficits and oxidative stress in β-amyloid (1–42) injected rats. Appl Physiol Nutr Metab 43:718–726
Bakken JS, Borody T, Brandt LJ et al (2011) Treating clostridium difficile infection with fecal microbiota transplantation. Clin Gastroenterol Hepatol 9:1044–1049
Beilharz JE, Kaakoush NO, Maniam J et al (2016) The effect of short-term exposure to energy-matched diets enriched in fat or sugar on memory, gut microbiota and markers of brain inflammation and plasticity. Brain Behav Immun 57:304–313
Bienenstock J, Kunze W, Forsythe P (2015) Microbiota and the gut-brain axis. Nutr Rev 73:28–31
Bonfili L, Cecarini V, Berardi S et al (2017) Microbiota modulation counteracts Alzheimer's disease progression influencing neuronal proteolysis and gut hormones plasma levels. Sci Rep 7:2426
Brandt LJ, Aroniadis OC (2013) An overview of fecal microbiota transplantation: techniques, indications, and outcomes. Gastrointest Endosc 78:240–249
Braniste V, Al-Asmakh M, Kowal C et al (2014) The gut microbiota influences blood-brain barrier permeability in mice. Sci Transl Med 6:26ra158
Calvani R, Picca A, Lo Monaco MR et al (2018) Of microbes and minds: a narrative review on the second brain aging. Front Med 5:53
Cammarota G, Ianiro G, Gasbarrini A (2014) Fecal microbiota transplantation for the treatment of clostridium difficile infection: a systematic review. J Clin Gastroenterol 48:693–702
Campbell SC, Wisniewski PJ, Noji M et al (2016) The effect of diet and exercise on intestinal integrity and microbial diversity in mice. PLoS One 11. e0150502
Canfora EE, Jocken JW, Blaak EE (2015) Short-chain fatty acids in control of body weight and insulin sensitivity. Nat Rev Endocrinol 11:577–591
Caputi V, Giron MC (2018) Microbiome-gut-brain axis and toll-like receptors in parkinson's disease. Int J Mol Sci 19:1689
Carter CS, Morgan D, Verma A et al (2020) Therapeutic delivery of ang(1-7) via genetically modified probiotic: a dosing study. J Gerontol A Biol Sci Med Sci 75:1299–1303
Cassilhas RC, Tufik S, De Mello MT (2016) Physical exercise, neuroplasticity, spatial learning and memory. Cell Mol Life Sci 73:975–983
Cervenka I, Agudelo LZ, Ruas JL. Kynurenines: Tryptophan's metabolites in exercise, inflammation, and mental health. Science 2017;357:eaaf9794
Chin SM, Sauk J, Mahabamunuge J et al (2017) Fecal microbiota transplantation for recurrent clostridium difficile infection in patients with inflammatory bowel disease: a single-center experience. Clin Gastroenterol Hepatol 15:597–599
Choi HH, Cho YS (2016) Fecal microbiota transplantation: current applications, effectiveness, and future perspectives. Clinical Endoscopy 49:257–265
Clarke SF, Murphy EF, O'Sullivan O et al (2014) Exercise and associated dietary extremes impact on gut microbial diversity. Gut 63:1913–1920
Collins SM, Bercik P (2009) The relationship between intestinal microbiota and the central nervous system in normal gastrointestinal function and disease. Gastroenterology 136:2003–2014
Collins SM, Surette M, Bercik P (2012) The interplay between the intestinal microbiota and the brain. Nat Rev Microbiol 10:735–742
Cryan JF, Dinan TG (2012) Mind-altering microorganisms: the impact of the gut microbiota on brain and behaviour. Nat Rev Neurosci 13:701–712
Davari S, Talaei SA, Alaei H et al (2013) Probiotics treatment improves diabetes-induced impairment of synaptic activity and cognitive function: behavioral and electrophysiological proofs for microbiome-gut-brain axis. Neuroscience 240:287–296

Davidson TL, Jones S, Roy M et al (2019) The cognitive control of eating and body weight: it's more than what you "think.". Front Psychol 10:62
De Palma G, Blennerhassett P, Lu J et al (2015) Microbiota and host determinants of behavioural phenotype in maternally separated mice. Nat Commun 6:7735
Devlin AS, Fischbach MA (2015) A biosynthetic pathway for a prominent class of microbiota-derived bile acids. Nat Chem Biol 11:685–690
Dominguez LJ, Barbagallo M (2018) Nutritional prevention of cognitive decline and dementia. Acta Biomed 89:276–290
Eiseman B, Silen W, Bascom GS et al (1958) Fecal enema as an adjunct in the treatment of pseudomembranous. Surgery 44:854–859
El Aidy S, Dinan TG, Cryan JF (2015) Gut microbiota: the conductor in the orchestra of immune-neuroendocrine communication. Clin Ther 37:954–967
Erickson KI, Voss MW, Prakash RS et al (2011) Exercise training increases size of hippocampus and improves memory. Proc Natl Acad Sci U S A 108:3017–3022
Escobar E (2002) Hypertension and coronary heart disease. J Hum Hypertens 16:S61–S63
Forsythe P, Bienenstock J, Kunze WA (2014) Vagal pathways for microbiome-brain-gut axis communication. Adv Exp Med Biol 817:115–133
Foster JA, McVey Neufeld KA (2013) Gut-brain axis: how the microbiome influences anxiety and depression. Trends Neurosci 36:305–312
Fulde M, Hornef MW (2014) Maturation of the enteric mucosal innate immune system during the postnatal period. Immunol Rev 260:21–34
Gerding DN (2005) Metronidazole for Clostridium difficile-associated disease: is it okay for mom? Clin Infect Dis 40:1598–1600
Hakansson A, Molin G (2011) Gut microbiota and inflammation. Nutrients 3:637–687
Hebert LE, Scherr PA, Bienias JL et al (2003) Alzheimer disease in the US population: prevalence estimates using the 2000 census. Arch Neurol 60:1119–1122
Huang L, Ma H, Li Y et al (2012) Antihypertensive activity of recombinant peptide IYPR expressed in Escherichia coli as inclusion bodies. Protein Expr Purif 83:15–20
Imdad A, Nicholson MR, Tanner-Smith EE et al (2018) Fecal transplantation for treatment of inflammatory bowel disease. Cochrane Database Syst Rev 11:CD012774
Kamada N, Chen GY, Inohara N et al (2013) Control of pathogens and pathobionts by the gut microbiota. Nat Immunol 14:685–690
Kang SS, Jeraldo PR, Kurti A et al (2014) Diet and exercise orthogonally alter the gut microbiome and reveal independent associations with anxiety and cognition. Mol Neurodegener 9:36
Kelly CR, Khoruts A, Staley C et al (2016) Effect of fecal microbiota transplantation on recurrence in multiply recurrent clostridium difficile infection a randomized trial. Ann Intern Med 165: 609–616
Kennedy PJ, Cryan JF, Dinan TG et al (2017) Kynurenine pathway metabolism and the microbiota-gut-brain axis. Neuropharmacology 112:399–412
Kern T, Blond MB, Hansen TH et al (2020) Structured exercise alters the gut microbiota in humans with overweight and obesity—A randomized controlled trial. Int J Obes 44:125–135
Khan NA, Raine LB, Drollette ES et al (2015) Dietary fiber is positively associated with cognitive control among prepubertal children. J Nutr 145:143–149
Kim D, Kang H (2019) Exercise training modifies gut microbiota with attenuated host responses to sepsis in wild-type mice. FASEB J 33:5772–5781
Kim KY, Yun JM (2018) Association between diets and mild cognitive impairment in adults aged 50 years or older. Nutr Res Pract 12:415–425
Kobayashi Y, Sugahara H, Shimada K et al (2017) Therapeutic potential of Bifidobacterium breve strain A1 for preventing cognitive impairment in Alzheimer's disease. Sci Rep 7:13510
Konturek PC, Haziri D, Brzozowski T et al (2015) Emerging role of fecal microbiota therapy in the treatment of gastrointestinal and extra-gastrointestinal diseases. J Physiol Pharmacol 66:483–491

Lambert JE, Myslicki JP, Bomhof MR et al (2015) Exercise training modifies gut microbiota in normal and diabetic mice. Appl Physiol Nutr Metab 40:749–752
Leblhuber F, Steiner K, Schuetz B et al (2018) Probiotic supplementation in patients with Alzheimer's dementia - an explorative intervention study. Curr Alzheimer Res 15:1106–1113
Li J, Zhao F, Wang Y et al (2017) Gut microbiota dysbiosis contributes to the development of hypertension. Microbiome 5:14
Li Q, Xu K, Du T et al (2018) Recombinant probiotics expressing angiotensin-(1-7) improves glucose metabolism and diabetes-induced renal and retinal injury. Diabetes, 67. 33-LB
Lichtwark IT, Newnham ED, Robinson SR et al (2014) Cognitive impairment in coeliac disease improves on a gluten-free diet and correlates with histological and serological indices of disease severity. Aliment Pharmacol Ther 40:160–170
Liu J, Sun J, Wang F et al (2015) Neuroprotective effects of clostridium butyricum against vascular dementia in mice via metabolic butyrate. Biomed Res Int 2015:412946
Ma Q, Xing C, Long W et al (2019) Impact of microbiota on central nervous system and neurological diseases: the gut-brain axis. J Neuroinflammation 16:53
Mailing LJ, Allen JM, Pence BD et al (2019a) Behavioral response to fiber feedingis cohort-dependent and associated with gut microbiota composition in mice. Behav Brain Res 359:731–736
Mailing LJ, Allen JM, Buford TW et al (2019b) Exercise and the gut microbiome: a review of the evidence, potential mechanisms, and implications for human health. Exerc Sport Sci Rev 47:75–85
Marques FZ, Nelson E, Chu PY et al (2017) High-fiber diet and acetate supplementation change the gut microbiota and prevent the development of hypertension and heart failure in hypertensive mice. Circulation 135:964–977
McDonald LC, Gerding DN, Johnson S et al (2018) Clinical practice guidelines for clostridium difficile infection in adults and children: 2017 update by the infectious diseases society of America (IDSA) and society for healthcare epidemiology of America (SHEA). Clin Infect Di 66:987–994
Molodecky NA, Soon IS, Rabi DM et al (2012) Increasing incidence and prevalence of the inflammatory bowel diseases with time, based on systematic review. Gastroenterology 142. https://doi.org/10.1053/j.gastro.2011.10.001
Munukka E, Ahtiainen JP, Puigbó P et al (2018) Six-week endurance exercise alters gut metagenome that is not reflected in systemic metabolism in over-weight women. Front Microbiol 9:2323
Musa NH, Mani V, Lim SM et al (2017) Lactobacilli-fermented cow's milk attenuated lipopolysaccharide-induced neuroinflammation and memory impairment in vitro and in vivo. J Dairy Res 84:488–495
Neuman H, Debelius JW, Knight R et al (2015) Microbial endocrinology: the interplay between the microbiota and the endocrine system. FEMS Microbiol Rev 39:509–521
Ng SC, Hart AL, Kamm MA et al (2009) Mechanisms of action of probiotics: Recent advances. Inflamm Bowel Dis 15:300–310
Nimgampalle M, Yellamma K (2017) Anti-Alzheimer properties of probiotic, Lactobacillus plantarum MTCC 1325 in Alzheimer's disease induced albino rats. J Clin Diagnostic Res 11: KC01–KC05
Panza F, Lozupone M, Solfrizzi V et al (2018) Different cognitive frailty models and hHealth-and cognitive-related outcomes in older age: from epidemiology to prevention. J Alzheimers Dis 62: 993–1012
Paramsothy S, Kamm MA, Kaakoush NO et al (2017) Multidonor intensive faecal microbiota transplantation for active ulcerative colitis: a randomised placebo-controlled trial. Lancet 389: 1218–1228
Reddel S, Putignani L, Del Chierico F (2019) The impact of low-FODMAPs, Gluten-free, and ketogenic diets on gut microbiota modulation in pathological conditions. Nutrients 11:373

Reid G (2016) Probiotics: definition, scope and mechanisms of action. Best Practice and Research: Clinical Gastroenterology 30:17–25
Richards EM, Pepine CJ, Raizada MK et al (2017) The gut, its microbiome, and hypertension. Curr Hypertens Rep 19:36
Sadowsky MJ, Khoruts A (2016) Faecal microbiota transplantation is promising but not a panacea. Nat Microbiol 1:16015
Sampson TR, Mazmanian SK (2015) Control of brain development, function, and behavior by the microbiome. Cell Host and Microbe 17:565–576
Sampson TR, Debelius JW, Thron T et al (2016) Gut microbiota regulate motor deficits and neuroinflammation in a model of Parkinson's disease. Cell 167:1469–1480.e12
Schmidt TSB, Raes J, Bork P (2018) The human gut microbiome: from association to modulation. Cell 172:1198–1215
Sender R, Fuchs S, Milo R (2016) Revised estimates for the number of human and bacteria cells in the body. PLoS Biol 14. e1002533
Sharon G, Garg N, Debelius J et al (2014) Specialized metabolites from the microbiome in health and disease. Cell Metab 20:719–730
Simon CB, Lee-McMullen B, Phelan D et al (2015) The renin–angiotensin system and prevention of age-related functional decline: where are we now? Age (Omaha) 37:1–11
Sokol H, Landman C, Seksik P et al (2020) Fecal microbiota transplantation to maintain remission in Crohn's disease: a pilot randomized controlled study. Microbiome 8:12
Spencer SJ, Korosi A, Layé S et al (2017) Food for thought: how nutrition impacts cognition and emotion. npj Sci Food 1:7
Steidler L, Hans W, Schotte L et al (2000) Treatment of murine colitis by Lactococcus lactis secreting interleukin-10. Science 289:1352–1355
Taniguchi H, Tanisawa K, Sun X et al (2018) Effects of short-term endurance exercise on gut microbiota in elderly men. Physiol Rep 6:e13935
Thomas S, Izard J, Walsh E et al (2017) The host microbiome regulates and maintains human health: a primer and perspective for non-microbiologists. Cancer Res 77:1783–1812
Trubiano JA, Cheng AC, Korman TM et al (2016) Australasian society of infectious diseases updated guidelines for the management of Clostridium difficile infection in adults and children in Australia and New Zealand. Intern Med J 46:479–493
Tsai YL, Lin TL, Chang CJ et al (2019) Probiotics, prebiotics and amelioration of diseases. J Biomed Sci 26:3
Van Nood E, Vrieze A, Nieuwdorp M et al (2013) Duodenal infusion of donor feces for recurrent clostridium difficile. N Engl J Med 368:407–415
Vindigni SM, Surawicz CM (2017) Fecal microbiota transplantation. Gastroenterol Clin N Am 46: 171–185
Vrieze A, Van Nood E, Holleman F et al (2012) Transfer of intestinal microbiota from lean donors increases insulin sensitivity in individuals with metabolic syndrome. Gastroenterology 143: 913–6.e7
Wang H, Lee IS, Braun C et al (2016) Effect of probiotics on central nervous system functions in animals and humans: a systematic review. Journal of Neurogastroenterology and Motility 22: 589–605
Weber MA (1994) Coronary heart disease and hypertension. Am J Hypertens 7:146S–153S
Wenisch C, Parschalk B, Hasenhündl M et al (1996) Comparison of vancomycin, teicoplanin, metronidazole, and fusidic acid for the treatment of Clostridium difficile-associated diarrhea. Clin Infect Dis 22:813–818
Xie M, Chen HH, Nie SP et al (2017) Gamma-aminobutyric acid increases the production of short-chain fatty acids and decreases pH values in mouse colon. Molecules 22:653
Yang G, Jiang Y, Yang W et al (2015) Effective treatment of hypertension by recombinant Lactobacillus plantarum expressing angiotensin converting enzyme inhibitory peptide. Microb Cell Factories 14:202

Yang X, Yu D, Xue L et al (2020) Probiotics modulate the microbiota–gut–brain axis and improve memory deficits in aged SAMP8 mice. Acta Pharm Sin B 10:475–487

Yano JM, Yu K, Donaldson GP et al (2015) Indigenous bacteria from the gut microbiota regulate host serotonin biosynthesis. Cell 161:264–276

Yao J, Wang JY, Lai MG et al (2011) Treatment of mice with dextran sulfate sodium-induced colitis with human interleukin 10 secreted by transformed Bifidobacterium longum. Mol Pharm 8:488–497

Youngster I, Sauk J, Pindar C et al (2014) Fecal microbiota transplant for relapsing clostridium difficile infection using a frozen inoculum from unrelated donors: a randomized, open-label, controlled pilot study. Clin Infect Dis 58:1515–1522

Zelante T, Iannitti RG, Cunha C et al (2013) Tryptophan catabolites from microbiota engage aryl hydrocarbon receptor and balance mucosal reactivity via interleukin-22. Immunity 39:372–385

Zhan G, Yang N, Li S et al (2018) Abnormal gut microbiota composition contributes to cognitive dysfunction in SAMP8 mice. Aging (Albany NY) 10:1257–1267

Chapter 13
Nanoplastics, Gut Microbiota, and Neurodegeneration

Ananya Rai

Abstract Plastics are high molecular weight organic polymers and their production has increased drastically, i.e., 150 times from 1950 to 2015. Plastics in the environment are persistent form of a pollutant. Especially, Microplastics (MPs) (Diameter < 0.001 mm) and Nanoplastics (diameters <0.1–1 mm) are one of the major environmental concerns. The present chapter describes the important sources, routes of exposure, and the impact of MPs/NPs as well as additives on mammalian gut microbiota that ultimately hampers neurological functioning. Microbes, light, heat, various chemical-induced biodegradation, ultraviolet B radiation of sunlight, and hydrolytic properties of water play a major role in the generation of MPs as well as in conversion of MPs into NPs. In addition to these minuscule plastic particles, other components such as polycarbonate (PC), polyethylene terephthalate (PET), and polyethylene (PE) have been detected in several human samples. Many research corroborated the effect of MPs/NPs on infertility, cancer, inflammation, gastrointestinal abnormalities, etc. This chapter will provide information on toxicological mechanisms such as oxidative stress, inflammation, metabolic impairment, and alteration in the secondary and tertiary structure of proteins, due to MPs/NPs. Pathophysiological consequences of acute and chronic exposure of MPs/NPs on the human system are yet unclear. This chapter includes a detailed understanding of how MPs/NPs induce harmful effects via reactive oxygen species (ROS) mediated signaling of MAPK-HIF-1/NF-kB. It also provides recent advancements related to polystyrene NP surface modification. Although, even being high-risk environmental factors these plastic particles are being overlooked. There is a need to explore and scrutinize the cellular and molecular mechanisms induced by MPs/NPs, leading to several maladies.

Keywords Nanoplastics · Oxidative stress · Neurodegenerative diseases

A. Rai (✉)
Molecular Biology Unit, Institute of Medical Sciences, Banaras Hindu University, Varanasi, India

A. K. Tripathi, M. Kotak (eds.), *Gut Microbiome in Neurological Health and Disorders*, Nutritional Neurosciences,
https://doi.org/10.1007/978-981-19-4530-4_13

Abbreviations

AD	Alzheimer's disease
ALS	Amyotrophic lateral sclerosis
BBB	Blood–brain barrier
BFRs	Brominated flame retardants
BPA	Bisphenol A
CAT	Catalase
CNS	Central nervous System
DEHP	Di (2-ehtylhexyl) phthalate
ENS	Enteric nervous System
GBA	Gut-brain axis
GI	Gastrointestinal tract
GPx	Glutathione peroxidase
GSH	Glutathione
HD	Huntington disease
HPA	Hypothalamic pituitary adrenal
IBD	Inflammatory bowel disease
JNK	c-Jun N-terminal kinase
MAPK	Mitogen-activated protein kinase
MPs	Microplastics
MS	Multiple sclerosis
NADP	Nicotinamide adenine dinucleotide phosphate
NADPH	Nicotinamide adenine dinucleotide phosphate hydrogen
NPs	Nanoplastics
Nrf2	Nuclear factor erythroid 2-related factor 2
PBDEs	Polybrominated diphenyl ethers
PCB	Polychlorinated biphenyls
PD	Parkinson's disease
PSNPs	polystyrene nanoparticles
PVC	Polyvinyl chloride
RNS	Reactive nitrogen species
ROS	Reactive oxygen species
SCFAs	Short-chain fatty acids
SOD	Superoxide dismutase
VNS	Vagus Nerve Stimulation

13.1 Introduction

Plastic products are the most common human creation and are visible in approximately every field such as health, construction, agriculture, textiles as well as in houses. Global production of plastic has increased enormously from

1.7 million tons in the 1950s to over 322 million tons in 2016 (Revel et al. 2018). In 2019, the global plastic market size was valued at USD 568.9 billion and is expected to grow at a compound annual growth rate (CAGR) of 3.2% from 2020 to 2027 (Amobonye et al. 2020). This drastic increase has left a huge amount of waste in our environment. In 2015, 5000 million metric tons of waste was produced and the experts estimate that this number will increase up to 25,000 million metric tons in 2050 (Geyer et al. 2017). Moreover, the percentage of waste undergoing recycling is quite negligible and it often leaks into the environment.

India is one of the leading producers of plastic products in the global plastic-production market. These products lead to approximately 26,000 tons of waste per day. Most of these plastic pollutants end up in the nearby water sources such as rivers and then ultimately to oceans and seas.

In the environment, these plastics interact with various degrading agents that break the larger plastic pieces into smaller fragments and result in the generation of MPs/NPS. These chemically inert MPs/NPs are recalcitrant, potential eco-toxic. MPs/NPs negatively affect the growth of plants and the animals that consume them, which could have adverse effects at subsequent higher trophic levels (Shen et al. 2019; Chae and An 2020). Defect in metabolism, reproductive system, larvae structure, and changes in brain appearance are the most common effects seen in aquatic organisms (Yong et al. 2020). The liver, lungs, kidney, brain, etc. are commonly affected organs (Bouwmeester et al. 2015). In humans, these particles are responsible for disrupting the gut barriers and cell receptors, and also altering the gene expression in the nucleus. It can also compromise the endocytic pathway function as well as our immune response (Rubio et al. 2020). Inflammation, genotoxicity, and oxidative stress responses have also been reported (Cortés et al. 2020). NPs are compelling because of their small size, specific physicochemical properties, their protein corona characteristics that permit the immune system to detect them, and in the case of metal oxide-based NPs: their magnetic characteristics as well (Barbero et al. 2017; Katas et al. 2018; Teleanu et al. 2018, 2019). Since the gut microbiota is ultra-sensitive to the changes, they become the major toxicological target for MPs/NPs. Oxidative stress caused due to the influence of MPs/NPs can alter the gut microbiota and affect dietary metabolism.

13.2 Plastic, Microplastic, and Nanoplastic: Origin and Its Chemical Composition

High molecular weight organic polymers reunite together to give rise to plastic products and are malleable. Predominately plastics are synthetic, generally composed of petrochemicals although some are cellulose-based. Plastics are categorized broadly into two groups: macroplastics and microplastics (Zalasiewicz et al. 2016). *Macroplastics* are greater than 5 mm in size, hence they are easily visible with naked eyes. When these macroplastics get exposed to the UVB radiation of the sunlight,

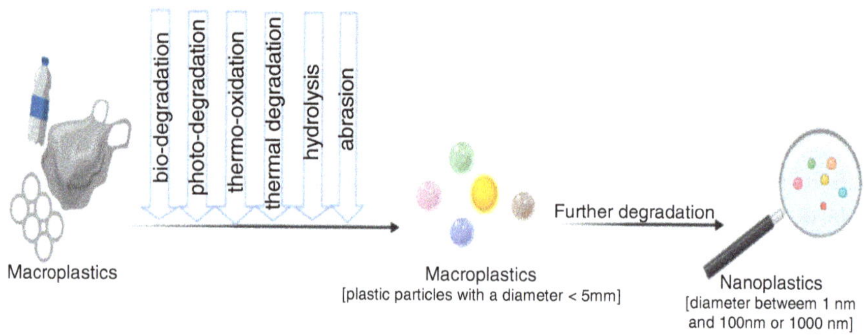

Fig. 13.1 Diagrammatic representation of conversion of macroplastics into nanoplastics. Macroplastics when exposed to specific microbes, light, UV, heat, and other chemical agents in the environment, undergoes the fragmentation and results in production of microplastics; these microplastics further degrades into nanoplastics

they degrade into smaller pieces in the presence of seawater (hydrolysis) and environmental oxygen (oxidation) and are known as microplastics (Moore 2008). *Microplastics (MPs)* (<5 mm) are typically imperceptible to the naked eye. These are too small to filter out either by any machinery or a simple sewage plant. Because of their small size microplastics can effortlessly peregrinate far with the assistance of river and sea current, and can get settled in sediment layers (Browne et al. 2011). Recently, a new plastic pollutant has been detected whose dimensions are smaller than MPs as well, these are named *Nanoplastics (NPs)*, little attentiveness has been paid to these minuscule particles (>100 nm) (Gigault et al. 2018). Clear definition of NPs has not been proposed yet.

Primarily NPs are produced from larger plastics or MPs as the result of their fragmentation (Eerkes-Medrano et al. 2015) (Fig. 13.1). Studies of NPs have shown that they have a great capacity to absorb organic compounds because of their huge surface-to-volume ratio, which potentially allow them to penetrate through cell wall (Gregory 1996; Fendall and Sewell 2009). Hence, NPs are being manufactured purposefully as well, for drug delivery, detergents, or cosmetic use. Nanoplastics are chemically inert and present significant ecological and health concerns because of their persistence in the environment and their ability to function as vectors for chemical pollutants as well as pathogens (Rakesh et al. 2020).

13.2.1 Sources of Nanoplastics

Various sources of microplastics/nanoplastics have been suggested in Table 13.1. Textiles, tires and city dust alone accounts for 80% of microplastics in the environment.

Table 13.1 Tabular representation of sources of microplastics/nanoplastics with their corresponding usage

S. no.	Source of polymeric microplastics or nanoplastics	Usages	References
1	Domestic wastewater	Cosmetics, cleaning products discharged.	Gregory (1996), Fendall and Sewell (2009), Chang (2015)
2	Industrial emergence	Feedstock in plastic manufacturing.	Sadri and Thompson (2014)
3	Plastic resin powders or pellets	Used for air blasting.	Claessens et al. (2011), Zbyszewski et al. (2014)
4	Breakdown of larger plastic items before entering the environment	Fragmentation of synthetic fibers during the washing of clothes.	Browne et al. (2011)
5	Depletion of the plastic items	Via degradation, When the expanded polystyrene undergoes mechanical abrasion, Persistent wearing by sand particles on the beaches.	Corcoran et al. (2009), Tosin et al. (2012)
6	UV radiation and microbiological activity		Lambert et al. (2014)
7	Sewage sludge that are contaminated and are used as fertilizer, agricultural polyethylene (PE) foil's degradation, clothes drying		Liebezeit and Liebezeit (2014)
8	Thermal cutting of polystyrene foam in the range of ~20–220 nm		Zhang et al. (2012)
9	3D printing (11.5 nm–116 nm)	Quick prototyping and small-scale manufacturing	Stephens et al. (2013)
10	Use of polymeric nanoparticles and nano capsules	Biomedical applications, such as drug delivery	Pohlmann et al. (2013)

13.2.2 Routes of Exposure

Drinking water: MPs occurrence has been seen in the soil and freshwater ecosystem and their amount has also been measured, their presence has also been seen at the drinking water sources as these particles are way too small, hence often pass through the filtration system (Eriksen et al. 2013; Carr et al. 2016). Range (daily discharge)—50,000 up to 15 million (Mason et al. 1987).

Food chain: If any particular aquatic environment has somehow been contaminated with MPs/NPs, then the consumption of water or any contaminated organisms may lead to exposure. In addition, storing the fish in plastic containers for storage or transportation may also add up to this (Revel et al. 2018). In crabs and mussels, translocation of MPs has been demonstrated in the laboratory across the gastrointestinal tract (Browne et al. 2008; Watts et al. 2016). In fishes, gastrointestinal tracts were found contaminated with MPs/NPs but beyond that MPs/NPs have not been evaluated yet (Bouwmeester et al. 2015).

Dermal: Water contaminated with MPs/NPs can come in contact with humans during various usual routine work such as washing or through facial/body scrubs harboring MPs/NPs (Sykes et al. 2014). Since MPs have a dimension of >5 mm approximately, they cannot penetrate through the stratum corneum as this layer allows the uptake of fine particles smaller than 100 nm, hence absorption via the skin is unlikely to occur for MPs. However, NPs could gradually diffuse through human skin (Sykes et al. 2014).

Inhalation: (Air) MPs/NPs can become airborne, may be because of wave action in the aquatic environment or as a result of wastewater treatment and these can be inhaled by humans as well as other organisms (Revel et al. 2018).

Other sources: Ingestion of various products or particles may introduce MPs/NPs into human system, for example, honey (Liebezeit and Liebezeit 2013), beer (Liebezeit and Liebezeit 2014), salt (Karami et al. 2017), ingested indirectly through personal care products as well (toothpaste, scrubs) (Revel et al. 2018).

13.2.3 Additives

Plastics are mainly composed of monomers, which are the by-products of the petroleum industry (Chen and Patel 2012). To improve performance and achieve better stability or durability, plastic monomers are blended with some additives such as plasticizers, light and heat stabilizers, lubricants, and pigments. Various other agents, as well as fillers such as kaolin, clay, calcium and carbonate are used to enhance the strength or to alter the texture (Hahladakis et al. 2018). During the degradation, these additives leach out in the surroundings with MPs/NPs.

13.2.4 Impact of Nanoplastic on Gut Microbiota and Its Molecular Mechanism

Gut Microbiota: Microorganisms are an inseparable part of our life, not only in the surroundings but also reside inside our bodies. The vast variety of microorganisms (e.g., bacteria, archaea, viruses, unicellular eukaryotic) in the number of trillions exists in close relationship with our gut in the form of complex microbial communities and are termed as "gut microbiota" (Bäckhed et al. 2005; Hsiao et al. 2008). The number of bacterial cells inhabiting the GI tract has been estimated to be more than 10^{14}, which means 10 times more than the number of human cells, and over 100 times genomic content (microbiome) as the human genome (Gill et al. 2006; Thursby and Juge 2017). *Firmicutes* and *Bacteroidetes* were mainly present in human gut microbiota, whereas the *Actinobacteria, Proteobacteria, Fusobacteria, Verrucomicrobia*, and *Cyanobacteria* account for a small proportion (Eckburg et al. 2005). Similarly, in the gut microbiota of mice *Firmicutes* and *Bacteroidetes* were mainly observed, while in the gut microbiota of zebrafish *Proteobacteria* and *Fusobacteria* were in the majority (Rawls et al. 2006).

Considering microorganisms always a foe would not be right as these micro unicellular organisms play a very crucial role in maintaining the ecological as well as personal wellbeing. The abundance of gut microbiota harbors in the intestinal area and shows complex trophic relationships varying from symbiosis to parasitism (Ventura et al. 2009). These microbes make the inaccessible nutrients available by the breakdown of food particles and further helps in absorption (Milani et al. 2017), promote host cell differentiation (Sommer and Bäckhed 2013), prevent the colonization of pathogens (Buffie and Pamer 2013; Kamada et al. 2013; Pickard et al. 2017), and stimulate the immune system as well (Cerf-Bensussan and Gaboriau-Routhiau 2010; Martin et al. 2010; Maranduba et al. 2015; Pickard et al. 2017; Lazar et al. 2018).

Various environmental factors such as temperature (Fontaine et al. 2018), pH (Duncan et al. 2009; Hansen et al. 2018), nutrient availability (Hansen et al. 2015), oxygen level/redox state (Qiao et al. 2013; Jones and Neish 2017; Ma et al. 2019) influence the gastrointestinal microbiota. Foreign particulates inhaled from the environment also hampers the gut microbiota. The drastic increase in plastic production has made our life easy, but at the same time, its abuse has left a huge amount of waste in our environment. Gut microbiota is one of the major toxicological targets of MPs/NPs due to its sensitivity. MPs/NPs were found to alter the composition of beta diversity (but no significant changes were induced by treatment groups in the alpha diversity), functioning of gut microbiota that leads to physiological dysfunction of the host and causes various diseases such as inflammatory bowel disease (IBD), malnutrition, obesity, diabetes, etc. (Lu et al. 2019).

Since the mice are common mammalian model. Hence, to understand the effect of MPs/NPs on the mammalian system it is the favored one. Various studies showed that exposure to MPs/NPs reduces body weight, liver and lipid weight, gut inflammation (Li et al. 2020a), and a decrease in the relative abundance of *Firmicutes* and *α-Proteobacteria* in the feces of mice (Lu et al. 2018). A study on soil oligochaetes (Zhu et al. 2018) and soil springtails (Ju et al. 2019) showed gut dysbiosis, and altered bacterial diversity, which was further accompanied by negative health effects, such as stunted growth and reproduction, and changes in gut metabolic profiles linked to oxidative stress and inflammation in zebrafish (Jin et al. 2018). MPs/NPs ingestion altered the gut microbiota of mussels (*Mytilus edulis*), higher concentrations were shown to have other ill effects (Li et al. 2020b).

The major query is how these MPs/NPs influence the gut microbiota.

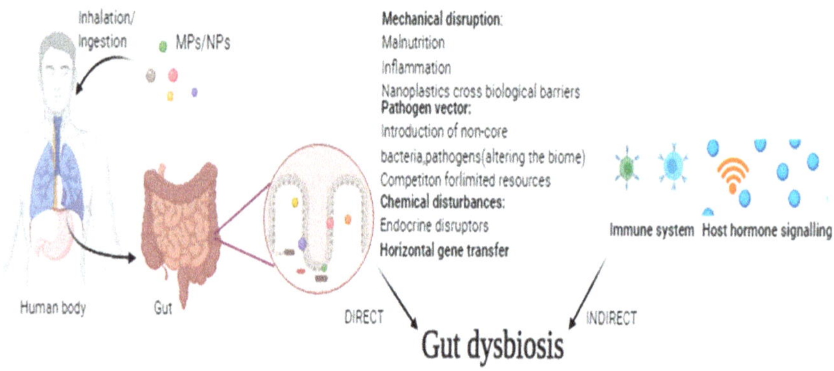

Fig. 13.2 Diagrammatic representation of intake of MPs/NPs, and their direct and indirect impact on gut microbiota. MPs/NPs can (1) cause mechanical disruption, leading to malnutrition, inflammation of the GI tract, nanoplastics can cross biological barriers as well; (2) act as a vector for pathogens and foreign, non-core bacteria, that will eventually lead to competition for limited resources with resident bacteria; (3) carry chemicals known to disrupt the endocrine system, altering host hormone signaling; act as a carrier for various environmental pollutants as well. They may activate the immune system; the dysbiosis may trigger chronic diseases, diminishing host's health, and alteration in the gene capacity and expression of gut microbiota

Various factors such as malnutrition, inflammation, pathogens, endocrine disruptors, and environmental chemicals, influence the gut microbiota accompanied by ingestion of MPs/NPs (Fig. 13.2). An increase in potential pathogenic bacteria, a reduction in beneficial bacteria, and the rise of negative interactions between essential gut bacteria, and potential pathogens are the major factors that influence the structure of the gut microbiome (Fackelmann and Sommer 2019). It's known that healthy microbiomes are alike in composition, and the microbiomes in dysbiosis exhibit no uniform pattern of dissimilarity (Zaneveld et al. 2017).

Prediction is that organisms ingesting the MPs/NPs also take in microbes (e.g., bacteria) that are able to break down plastics or partially detoxify the chemicals associated with them. In a study, MPs/NPs associated with other chemicals or polymers such as PCB were found to be colonized by PCB transforming microbes rapidly (by reductive de-chlorination) (Harrison et al. 2014; Michels et al. 2018), hydrogen degrading genera were also found to be colonized (Harrison et al. 2014). Hence, when the ingestion of MPs/NPs increases, it alters the core microbiome simultaneously. Furthermore, not only the gut microbiome but also its hologenome must be studied propely. As, the microbes can very easily access the plasmids on the surface of MPs/NPs via horizontally gene transfer (transfer of genetic material between organisms, other than the transfer from parent to offspring) (Soucy et al. 2015) and can harness various antibiotic resistance genes (Arias-Andres et al. 2018). Hence, not only the composition, but also the gene structure as well as their expression pattern should be a major parameter to evaluate.

13.2.5 Impact of Additives on Gut Microbiome

Since additives (except for some flame retardants) are not bound to the polymer matrix, they may leach out of polymer effortlessly into the surrounding, especially if they are of low molecular weight (von Moos et al. 2012; Liebezeit and Liebezeit 2014; Wright and Kelly 2017; Campanale et al. 2020). The most harmful chemicals are recalcitrant, carcinogenic, act as a mutagen for DNA, have toxic reproductive effects, are capable of bioaccumulation, and have other harmful properties, such as disrupting hormones (Locke n.d.; Schubert 1972). The major pathway of exposure to these chemicals is in association with MPs/NPs. These additives can be extremely hazardous for gut microbiota (Table 13.2).

13.3 Molecular Mechanism

The reactive oxygen species or ROS are the radicals derived from oxygen in the biological system. Unrestricted production of ROS can escalate membrane permeability and cellular stress, eventually resulting in the amplification of facultative anaerobic bacteria (previously dominated by obligate anaerobic bacteria), gut barrier dysfunction, and inflammation (Sung et al. 2011; Mao et al. 2019). MPs/NPs can result in the generation of free radicals (Liu et al. 2020); these free radicals have a high tendency to be involved in different reactions, including oxidative chemical reactions.

13.3.1 Initiation Events (IE)

Studies have shown that MPs/NPs can induce the generation of ROS at different sizes, doses, surface characteristics, and exposure times (Jeong et al. 2016, 2017; Paul-Pont et al. 2016). ROS generated as the result of interaction with MPs/NPs fall under two categories (Fig. 13.3): extracellular and intracellular (Hu and Palić 2020a). Extracellularly, the encounter of the plastic polymer to various weathering events such as photo, thermal degradation, or UV radiation results in several chemical modifications (Celina 2013; Jahnke et al. 2017). It eventually leads to the generation of free radicals on the surfaces of MPs/NPs, via various pathways such as removal of the hydrogen atom from a macromolecular chain, or addition to an unsaturated carbon chain group (crosslinking reaction) (Yousif and Haddad 2013). These free radicals once produced can react with atmospheric oxygen and produce polymer peroxy radicals with the further generation of secondary polymer alkyl radicals (Allen and Bevington 1996).

Enormous intracellular generation of ROS by virgin MPs/NPs was observed using a broad range of model systems, from mammalian cell lines to marine

Table 13.2 Tabular representation of the types of additives, their uses and potential ill effect on gut microbiota

Additives	Use	Impact	References
Plasticizers	Make plastics more flexible. Mostly used in PVC and cellulose-based polymers.	Phthalates are endocrine disruptors and possess estrogenic activities; di (2-Ethylhexyl) phthalate have toxic impacts on reproductive organs, the heart, liver, kidneys, and lungs, even at low exposures. • Hormones secreted by the host, affect the gut bacteria due to the presence of necessary hormone receptors on it, ultimately disturbing the gut microbiome (Hughes and Sperandio 2008; Neuman et al. 2015). • Binding of plasticizers to microbial hormone receptors, or interfering with host hormone signaling, both can alter gut microbiota.	Diamanti-Kandarakis et al. (2009), Caldwell (2012), De Toni et al. (2017), Godwin (2017)
Flame retardants	Used to reduce the chance of ignition and the spread of fire when plastics are used in close association with critical conditions like construction, electrical, transport.	BFRs can induce endocrine, reproductive, and behavioral effects. PBDEs affect the microbial homeostasis of human gut microbiota. It leads to lower bacterial densities, high doses showed the shift at phylum and family levels.	Lyche et al. (2015), Cruz et al. (2020)
Antioxidants, ultraviolet antioxidants, stabilizers	Phenolics generally added at lower amounts. Oxidation can cause discoloration, loss of impact strength, etc. Hence, antioxidants are used to prevent the thermal oxidation (high temperatures) as well as light oxidation (when plastics are exposed to UV light).	BPA intake can reduce species diversity, as it favors the growth of specific bacteria and inhibits others, for example enhancing *proteobacteria*, *Helicobacteraceae*, etc., but reduces *clostridia* in the gut microbiome. BPA can induce colon and liver inflammation via dysbiosis and altering metabolite profiles. It can also induce changes in	Javurek et al. (2016), Lai et al. (2016), Reddivari et al. (2017), Galloway et al. (2019)

(continued)

Table 13.2 (continued)

Additives	Use	Impact	References
		the central nervous system permanently.	
Antimicrobial	Microbial attack can cause staining, discoloration, odor, loss of electrical insulation, hygiene, and mechanical properties of the product. Antimicrobials prevent such attacks.	These antimicrobials can effortlessly result in dysbiosis, loss of beneficial bacteria, pathobionts expansion; disrupt gut barrier, metabolism, increases inflammation response, biofilm development/Quorum sensing, antibiotic resistance.	D'Arcy (2001), Ribeiro et al. (2020)
Heat stabilizers	Processing of the polymer requires the treatment of high temperature well above 180 °C. Hence, heat stabilizers are used to prevent the decomposition of the polymer during processing.	Organophosphites are commonly used heat stabilizers in PVC polymer. OPs exposure significantly alters the gut microbiome, which further contributes to the neurological disorder. The altered gut microbiome affects the physiological properties of the gut permeability and many biological processes such as the production of important metabolites (vitamins and SCFAs.)	Polymer Solutions News Team (2015), Roman et al. (2019)

invertebrate and living fish models (Cole and Galloway 2015; Schirinzi et al. 2017; Qiao et al. 2019). MPs/NPs can be engulfed via the process of endocytosis or pinocytosis (Geys et al. 2006; von Moos et al. 2012), once inside these MPs/NPs are treated as an outsider, hence triggering the innate immune defense mechanism (Greven et al. 2016; Wen et al. 2018). During attempts of the immune cell to neutralize potentially infectious foreign particles, ROS (superoxide and hydrogen peroxide) are generated in a high amount as products of NADPH oxidase or other enzymatic reactions (Yang et al. 2013). Both O_2 and H_2O_2 are produced to serve as a key mediator driving oxidative stress series and play a significant role in signal transduction (Yang et al. 2013; Qiao et al. 2019). The fluctuation in surface charge potential and mass/surface ratio in NPs permits easier engagement of free radicals as well as allows easier shifting through membranes. Demonstrating a strong interrelationship between particle size and ROS generation potential, the smaller the particle, the higher ROS generation potential (Jeong et al. 2016; Jeong et al. 2017; Lei et al. 2018; Yu et al. 2019).

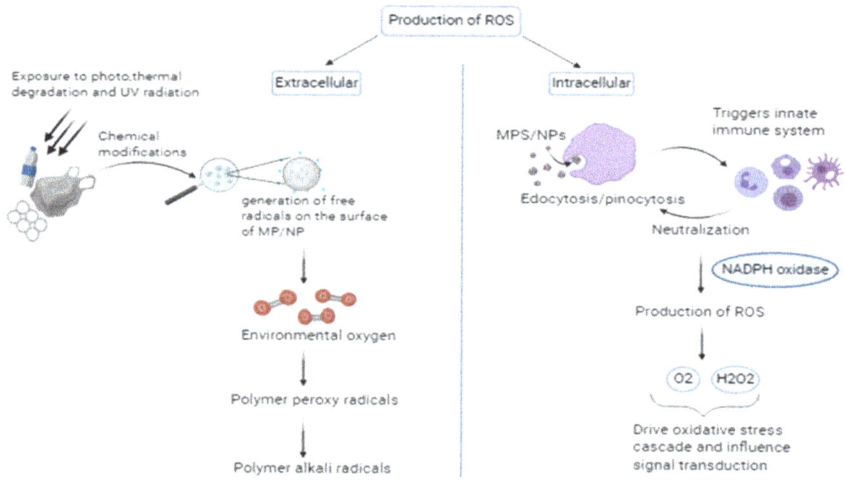

Fig. 13.3 Diagrammatic representation for the production of ROS under the influence of MPs/NPs. Generation of these radicals occur extracellularly as a result of environmental agents and aging processes; intracellularly engulfed MPs/NPs are treated as foreign substances, so it triggers innate immune system; and during the process of neutralization huge amount of ROS are generated as the product of NADPH oxidase. Created with BioRender.com

13.3.2 Key Event: Oxidative Stress

Oxidative stress is the result of disproportion between the generation of ROS and antioxidants-based detoxification. Here, we will discuss the oxidative stress pathway concerning MPs/NPs.

Various animal models including rotifers (B. koreanus) (Jeong et al. 2016), water fleas (Daphnia magna) (Liu et al. 2020), fishes (D. rerio) (Lu et al. 2016; Chen et al. 2017), and mammals (M. musculus) (Deng et al. 2017) were observed and the involvement of antioxidants system following the exposure of MPs/NPs was found, indicating the occurrence of oxidative stress, supporting the fact that MPs and NPs can induce oxidative stress as the fundamental mechanism. Antioxidant systems are complex and include various compounds such as vitamins (e.g., C, E, and D3); together with multi-enzyme pathways, that produce antioxidants and eliminate ROS and RNS. SOD/CAT antioxidation mechanisms play a major role as a tool against the adverse effect of ROS and respond to these signals (Hu and Palić 2020a).

13.3.3 Activation of Oxidative Stress Pathway

One major stress responsible for the activation of MAPK (mitogen-activated protein kinase) pathways in oxidative stress is caused due to increased ROS (McCubrey

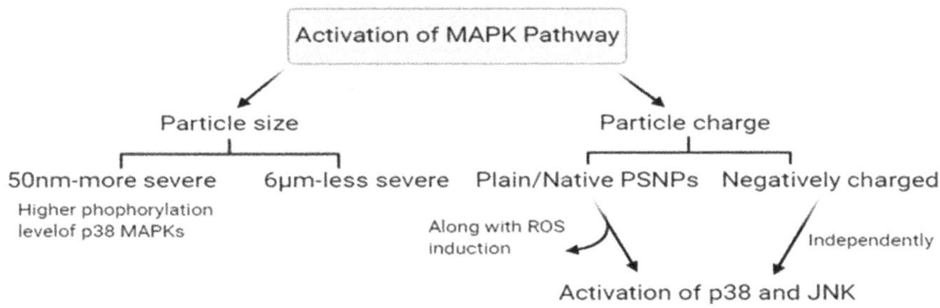

Fig. 13.4 Diagrammatic representation showing the dependency of MAPK pathway activation on the particle size and charge; study on *P. nana* and *B. koreanus* showed that particles of size 50 nm generate more severe oxidative stress as compared to smaller particles (size-6 μm), causing higher phosphorylation of p38 MAPKs; surface charge came out to be condemnatory for MAPK cascades as well, plain PSNPs (Poly Styrene Nano Plastics) were found to activate the p38 and JNK along with ROS induction while negative charge holder molecules activated independently

et al. 2006). Concurrent elevation in Nrf2 activity can downregulate the MAPK activation by increased ROS (Shi and Zhou 2010), still, the exact mechanism remains unclear. Although due to prolonged exposure to ROS, MAPK can remain activated for longer, which will eventually have several negative consequences such as autophagy, inhibition of ERK, and activation of p38MAPK components. Activation of MAPK pathway is influenced by particle size and charge as well (Fig. 13.4).

In recent studies, brominated flame retardants are found to activate the MAPK pathway and induce oxidative stress (Park et al. 2017). Exposure of MPs/NPs in marine copepod (*P. nana*) and Mitten crab (*E. sinensis*) predicted the induction of MAPK downstream pathways and enhanced phosphorylation level of ERK and p38 kinase showed a positive interrelationship with the production of ROS (Jeong et al. 2017). Furthermore, exposure to MPs/NPs showed an increase in transcription factor Nrf-2 (nuclear factor erythroid 2-related factor 2) (Jeong et al. 2017).

ROS generated is responsible for several other molecular key events, such as lipid peroxidation, DNA damage, acetylcholine inhibition causing inflammation, genotoxicity, and neurotoxicity, which in turn may add to oxidative stress. In addition, mitochondrial dysfunction and liposomes disruption are two major cellular key events triggered due to MPs/NPs (Hu and Palić 2020b).

13.3.4 Impact of MPs/NPs Induced Oxidative Stress on Gut Microbiota

Gut microbiota contributes to managing the host's antioxidant response via the modulation of reduced glutathione (GSH) synthesis (Mardinoglu et al. 2015). However, the gut microbiota can be affected by oxidative stress. A study on non-pathogenic gut bacterium *Enterococcus durans* (MTCC 3031) showed that oxidative stress caused due to cancer therapy leads to gut microbiota alteration

(Jose et al. 2018). Similarly, the oxidative stress, caused due to the influence of MPs/NPs, may alter the intracellular redox status of the gut microbiota, which is determined by the NADPH/NADP ratio. This intracellular redox status is related to various significant functions varying from growth, reductive biosynthesis of several metabolites to apoptosis (Ivarsson et al. 2005). Fluctuations in the redox status can result in the alteration of gut microbiota.

With decreased redox ratio, the reductive power of the cell also decreases, which will make gut microbiota unable to perform their usual role in various functioning for the host. Decreased bacterial growth would lead to a significant reduction in the bacterial population under such circumstances, altering the gut microbiome. It is already known that maintenance of the cellular redox status is crucial for the proper functioning of the gut microbiome.

13.3.5 Gut Microbiome and Neurodegenerative Disorder

Gut microbiota is in close association with the central nervous system (CNS) through the Gut-Brain Axis (GBA). Various components of this axis are in coordination with each other through a communicational pathway linked from brain to gut and vice versa (Bercik et al. 2012). It usually communicates by passing on the signals via several routes which consist of Vegas Nervous System (VNS), Central Nervous System (CNS), Enteric Nervous System (ENS), metabolites, immune system, Hypothalamic Pituitary Adrenal (HPA) (ScienceDirect Topics, n.d.). Aggregated evidence showed the interrelationship of gut microbiota dysbiosis and several neurodegenerative disorders such as Parkinson's disease (PD), Alzheimer's disease (AD), Amyotrophic Lateral Sclerosis (ALS), Multiple Sclerosis (MS), and Huntington's disease (HD) (Kelly et al. 2017; Follmer 2020; Raval et al. 2020; Tilocca et al. 2020a, 2020b).

13.3.6 Impact of Altered Gut Microbiota Due to Ingested MPs/NPs on Neurodegenerative Diseases

Ingested MPs/NPs are responsible for causing neurodegenerative disorders by two major pathways, primarily via creating oxidative stress, which is itself firmly associated with mitochondrial dysfunctions and neurodegenerations, or secondarily, via altering the gut microbiota. This altered gut microbiota hence will not be able to perform its usual functions in GBA and may result in neurodegenerative disorders. Dysbiosis is one of the leading factors for the increase in inflammatory cytokines and bacterial metabolites, which ultimately alter the gut and BBB permeability, and cause neuro-inflammation aiding various neurodegenerative disorders (Fig. 13.5).

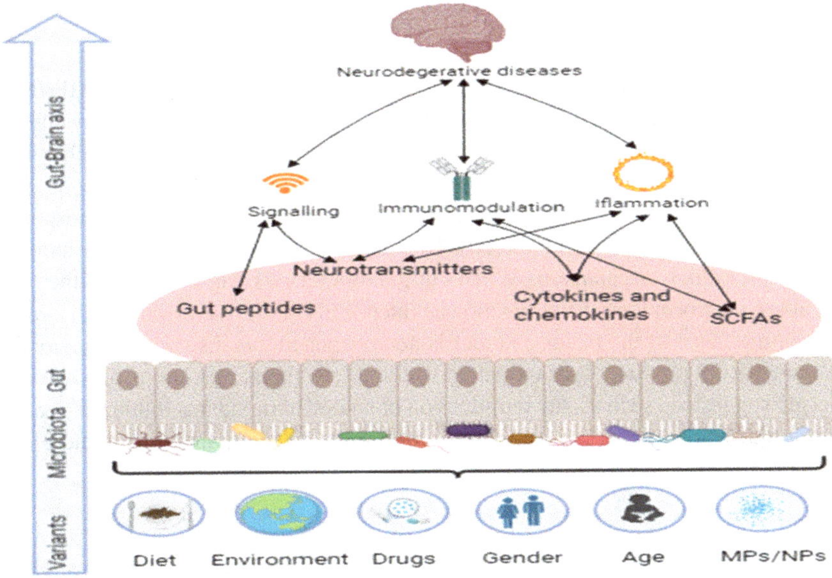

Fig. 13.5 Diagrammatic representation of various endogenous factors such as diet, drugs, gender, and age as well as environmental factors and MPs/NPs altering the gut microbiota. Ultimately, this dysbiosis leads to neurodegenerative diseases via various mediators such as neuroactive molecules (e.g., short-chain fatty acid (SCFAs), neurotransmitters); Abbreviation: *Ach* acetylcholine, *His*, histidine, *DA* dopamine, *5-HT* serotonin, *NpY* neuropeptide-Y, *CcK* cholecystokinin, *ILs* interleukins, *TNF* tumor necrosis factor, *CRP* C-reactive protein

Dysbiosis can enhance the levels of reactive oxygen species (ROS) that add on the OS scenario and neuronal inflammation.

13.3.7 Role of Antioxidants

Antioxidant plays a vital role in preventing damage from the free radicals. Hence, intake of natural antioxidants such as vitamins (vitamin A, C, and E), carotenoids (β-carotene, lycopene, and astaxanthin), polyphenols (tea polyphenols and red wine polyphenols), and flavonoids (flavonoids, isoflavone, xanthones, and anthocyanins) in our diet is the most appreciated way to prevent various critical disorders including neurodegeneration (Li et al. 2014). Lycopene is a non-provitamin carotenoid. The major dietary sources of lycopene for humans are tomatoes and tomato products (Sies and Stahl 1998). Studies have shown the protective effect of lycopene against the oxidative stress caused by various additives such as BPA, DEHP, etc. (Abdel-Rahman et al. 2018; Zhao et al. 2020). Another organic compound, named Piperidine, obtained from black pepper with curcumin showed approximately 69% increment in gut microbial species (Tripathi et al. 2022).

13.4 Conclusion

MPs/NPs are the results of the fragmentation of larger plastic pieces. MPs/NPs ingested via several routes are responsible for hampering the normal physiology and also serve as a vector for various unwanted guests (such as pathogens, chemical additives, environmental toxins, etc.). The ingested/inhaled MPs/NPs can interact and hinder the gut residential microbes indirectly (by employing immune response and hormone signaling) and directly by causing oxidative stress via activating several molecular mechanisms. When gut microbiota gets hampered, then they are not able to perform their usual functions, which leads to the rise in various disorders in the hosts. One of the major disorder is neurodegenerative disorder. As a cure for increased ROS production or oxidative stress, a few alterations in our lifestyle and diet are suggested, such as, the introduction of natural antioxidants found very easily in affordable food items for e.g.: fruits, vegetables, whole grains, green and black tea, coffee, herbs and spices.

References

Abdel-Rahman HG, Abdelrazek HMA, Zeidan DW, Mohamed RM, Abdelazim AM (2018) Lycopene: hepatoprotective and antioxidant effects toward bisphenol a-induced toxicity in female wistar rats. Oxidative Med Cell Longev 2018:e5167524. https://doi.org/10.1155/2018/5167524

G. Allen and J.C. Bevington (1996) Comprehensive polymer science and supplements: comprehensive polymer science https://books.google.co.in/books/about/Comprehensive_Polymer_Science_and_Supple.html?id=p1ZBngAACAAJ&redir_esc=y. Accessed 22 Dec 2020

Amobonye A, Bhagwat P, Singh S, Pillai S (2020) Plastic biodegradation: frontline microbes and their enzymes. Sci Total Environ 143536. https://doi.org/10.1016/j.scitotenv.2020.143536

Arias-Andres M, Klümper U, Rojas-Jimenez K, Grossart H-P (2018) Microplastic pollution increases gene exchange in aquatic ecosystems. Environ Pollut 237:253–261. https://doi.org/10.1016/j.envpol.2018.02.058

Bäckhed F, Ley RE, Sonnenburg JL, Peterson DA, Gordon JI (2005) Host-bacterial mutualism in the human intestine. Science 307:1915–1920. https://doi.org/10.1126/science.1104816

Barbero F, Russo L, Vitali M, Piella J, Salvo I, Borrajo ML et al (2017) Formation of the protein corona: the interface between nanoparticles and the immune system. Semin Immunol 34:52–60. https://doi.org/10.1016/j.smim.2017.10.001

P. Bercik, S. M. Collins, E. F. Verdu. 2012. Microbes and the gut-brain axis. Neurogastroenterol Motility. https://onlinelibrary.wiley.com/doi/full/10.1111/j.1365-2982.2012.01906.x. Accessed 30 Dec 2020

Bouwmeester H, Hollman PCH, Peters RJB (2015) Potential health impact of environmentally released micro- and nanoplastics in the human food production chain: experiences from nanotoxicology. Environ Sci Technol 49:8932–8947. https://doi.org/10.1021/acs.est.5b01090

Browne MA, Dissanayake A, Galloway TS, Lowe DM, Thompson RC (2008) Ingested microscopic plastic translocates to the circulatory system of the mussel, Mytilus edulis (L.). Environ Sci Technol 42:5026–5031

Browne MA, Crump P, Niven SJ, Teuten E, Tonkin A, Galloway T et al (2011) Accumulation of microplastic on shorelines woldwide: sources and sinks. Environ Sci Technol 45:9175–9179. https://doi.org/10.1021/es201811s

CG Buffie, EG Pamer. (2013). Microbiota-mediated colonization resistance against intestinal pathogens. Nat Rev Immunol. https://www.nature.com/articles/nri3535?page=10. Accessed 11 Dec 2020

Caldwell, JC. (2012). DEHP: genotoxicity and potential carcinogenic mechanisms—a review. Mutation Res/Rev Mutation Res https://www.sciencedirect.com/science/article/pii/S1383574212000178. Accessed 16 Dec 2020

Campanale C, Massarelli C, Savino I, Locaputo V, Uricchio VF (2020) A detailed review study on potential effects of microplastics and additives of concern on human health. Int J Environ Res Public Health 17:1212. https://doi.org/10.3390/ijerph17041212

Carr SA, Liu J, Tesoro AG (2016) Transport and fate of microplastic particles in wastewater treatment plants. Water Res 91:174–182. https://doi.org/10.1016/j.watres.2016.01.002

Celina MC (2013) Review of polymer oxidation and its relationship with materials performance and lifetime prediction. Polym Degrad Stab 98:2419–2429. https://doi.org/10.1016/j.polymdegradstab.2013.06.024

N Cerf-Bensussan, V Gaboriau-Routhiau (2010). The immune system and the gut microbiota: friends or foes?. Nat Rev Immunol. https://www.nature.com/articles/nri2850. Accessed 11 Dec 2020

Chae Y, An Y-J (2020) Nanoplastic ingestion induces behavioral disorders in terrestrial snails: trophic transfer effects via vascular plants. Environ Sci Nano 7:975–983. https://doi.org/10.1039/C9EN01335K

Chang M (2015) Reducing microplastics from facial exfoliating cleansers in wastewater through treatment versus consumer product decisions. Mar Pollut Bull 101:330–333. https://doi.org/10.1016/j.marpolbul.2015.10.074

Guo-Qiang Chen, Martin K. Patel (2012). Plastics derived from biological sources: present and future: a technical and environmental review. Chem Rev. https://pubs.acs.org/doi/full/10.1021/cr200162d. Accessed 16 Dec 2020

Chen Q, Gundlach M, Yang S, Jiang J, Velki M, Yin D et al (2017) Quantitative investigation of the mechanisms of microplastics and nanoplastics toward zebrafish larvae locomotor activity. Sci Total Environ 584–585:1022–1031. https://doi.org/10.1016/j.scitotenv.2017.01.156

Claessens M, Meester SD, Landuyt LV, Clerck KD, Janssen CR (2011) Occurrence and distribution of microplastics in marine sediments along the Belgian coast. Mar Pollut Bull 62:2199–2204. https://doi.org/10.1016/j.marpolbul.2011.06.030

Cole M, Galloway TS (2015) Ingestion of nanoplastics and microplastics by Pacific oyster larvae. Environ Sci Technol 49:14625–14632. https://doi.org/10.1021/acs.est.5b04099

Corcoran PL, Biesinger MC, Grifi M (2009) Plastics and beaches: a degrading relationship. Mar Pollut Bull 58:80–84. https://doi.org/10.1016/j.marpolbul.2008.08.022

Cortés C et al (2020) Nanoplastics as a potential environmental health factor: effects of polystyrene nanoparticles on human intestinal epithelial Caco-2 cells. Environ Sci Nano. https://doi.org/10.1039/C9EN00523D

Cruz R, Palmeira JD, Martins ZE, Faria MA, Ferreira H, Marques A et al (2020) Multidisciplinary approach to determine the effect of polybrominated diphenyl ethers on gut microbiota. Environ Pollut 260:113920. https://doi.org/10.1016/j.envpol.2020.113920

D'Arcy N (2001) Antimicrobials in plastics: a global review. Plast Addit Compd 3:12–15. https://doi.org/10.1016/S1464-391X(01)80328-7

De Toni L, Tisato F, Seraglia R, Roverso M, Gandin V, Marzano C et al (2017) Phthalates and heavy metals as endocrine disruptors in food: a study on pre-packed coffee products. Toxicol Rep 4:234–239. https://doi.org/10.1016/j.toxrep.2017.05.004

Deng Y, Zhang Y, Lemos B, Ren H (2017) Tissue accumulation of microplastics in mice and biomarker responses suggest widespread health risks of exposure. Sci Rep 7:46687. https://doi.org/10.1038/srep46687

Diamanti-Kandarakis E, Bourguignon J-P, Giudice LC, Hauser R, Prins GS, Soto AM et al (2009) Endocrine-disrupting chemicals: an endocrine society scientific statement. Endocr Rev 30:293–342. https://doi.org/10.1210/er.2009-0002

Duncan SH, P Louis, JM Thomson, HJ Flint (2009). The role of pH in determining the species composition of the human colonic microbiota. Environ Microbiol. https://sfamjournals.onlinelibrary.wiley.com/doi/full/10.1111/j.1462-2920.2009.01931.x. Accessed 11 Dec 2020

Eckburg PB, Bik EM, Bernstein CN, Purdom E, Dethlefsen L, Sargent M et al (2005) Diversity of the human intestinal microbial flora. Science 308:1635–1638. https://doi.org/10.1126/science.1110591

Eerkes-Medrano D, Thompson RC, Aldridge DC (2015) Microplastics in freshwater systems: a review of the emerging threats, identification of knowledge gaps and prioritisation of research needs. Water Res 75:63–82. https://doi.org/10.1016/j.watres.2015.02.012

Eriksen M, Mason S, Wilson S, Box C, Zellers A, Edwards W et al (2013) Microplastic pollution in the surface waters of the Laurentian Great Lakes. Mar Pollut Bull 77:177–182. https://doi.org/10.1016/j.marpolbul.2013.10.007

Gloria Fackelmann, Simone Sommer. (2019). Microplastics and the gut microbiome: how chronically exposed species may suffer from gut dysbiosis. Mar Pollut Bull. https://www.sciencedirect.com/science/article/pii/S0025326X19302917?casa_token=mSBwIFTAzs8AAAAA:98wEmCA3u0I93bDy5GVlPZTScvtPneU_v2GyDgL-c_x6HBj5b-hkhiuf1x1P-LZhaTZjBNDnrHQ. Accessed 12 Dec 2020

Fendall LS, Sewell MA (2009) Contributing to marine pollution by washing your face: microplastics in facial cleansers. Mar Pollut Bull 58:1225–1228. https://doi.org/10.1016/j.marpolbul.2009.04.025

Follmer C (2020) Viral infection-induced gut dysbiosis, neuroinflammation, and α-synuclein aggregation: updates and perspectives on covid-19 and neurodegenerative disorders. ACS Chem Neurosci 11:4012–4016. https://doi.org/10.1021/acschemneuro.0c00671

SS Fontaine, AJ Novarro, KD Kohl. (2018). Environmental temperature alters the digestive performance and gut microbiota of a terrestrial amphibian. J Exp Biol. https://jeb.biologists.org/content/221/20/jeb187559.short. Accessed 11 Dec 2020

T. S. Galloway, et al. 2019. Plastics additives and human health: a case study of bisphenol A (BPA). https://www.researchgate.net/publication/330075159_Plastics_Additives_and_Human_Health_A_Case_Study_of_Bisphenol_A_BPA. Accessed 16 Dec 2020

Geyer R, Jambeck JR, Law KL (2017) Production, use, and fate of all plastics ever made. Sci Adv 3:e1700782. https://doi.org/10.1126/sciadv.1700782

Geys J, Coenegrachts L, Vercammen J, Engelborghs Y, Nemmar A, Nemery B et al (2006) In vitro study of the pulmonary translocation of nanoparticles: a preliminary study. Toxicol Lett 160:218–226. https://doi.org/10.1016/j.toxlet.2005.07.005

Gigault J, ter Halle A, Baudrimont M, Pascal P-Y, Phi T-L, El Hadri H et al (2018) Current opinion: what is a nanoplastic? Environ Pollut 235:1030–1034. https://doi.org/10.1016/j.envpol.2018.01.024

Gill SR, Pop M, DeBoy RT, Eckburg PB, Turnbaugh PJ, Samuel BS et al (2006) Metagenomic analysis of the human distal gut microbiome. Science 312:1355–1359. https://doi.org/10.1126/science.1124234

Godwin AD (2017) 24 - plasticizers. In: Kutz M (ed) Appl. Plast. Eng. Handb, 2nd edn. William Andrew Publishing, pp 533–553. https://doi.org/10.1016/B978-0-323-39040-8.00025-0

Gregory MR (1996) Plastic 'scrubbers' in hand cleansers: a further (and minor) source for marine pollution identified. Mar Pollut Bull 32:867–871. https://doi.org/10.1016/S0025-326X(96)00047-1

Greven A-C, Merk T, Karagöz F, Mohr K, Klapper M, Jovanović B et al (2016) Polycarbonate and polystyrene nanoplastic particles act as stressors to the innate immune system of fathead minnow (Pimephales promelas). Environ Toxicol Chem 35:3093–3100. https://doi.org/10.1002/etc.3501

Hahladakis JN, Velis CA, Weber R, Iacovidou E, Purnell P (2018) An overview of chemical additives present in plastics: migration, release, fate and environmental impact during their use, disposal and recycling. J Hazard Mater 344:179–199. https://doi.org/10.1016/j.jhazmat.2017.10.014

Hansen, NCK, et al. (2015). High nutrient availability reduces the diversity and stability of the equine caecal microbiota. Microb Ecol Health Dis. https://www.tandfonline.com/doi/full/10.3402/mehd.v26.27216. Accessed 11 Dec 2020

Hansen, TH, et al. (2018). The effect of drinking water pH on the human gut microbiota and glucose regulation: results of a randomized controlled cross-over intervention. Sci Rep. 8. https://www.nature.com/articles/s41598–018–34761-5. Accessed 11 Dec 2020

JP Harrison, M Schratzberger, M Sapp, AM Osborn (2014). Rapid bacterial colonization of low-density polyethylene microplastics in coastal sediment microcosms. BMC Microbiol. https://bmcmicrobiol.biomedcentral.com/articles/10.1186/s12866-014-0232-4. Accessed 12 Dec 2020

WWL Hsiao, et al. (2008). The microbes of the intestine: an introduction to their metabolic and signaling capabilities. Endocrinol Metab Clin N Am. https://www.ncbi.nlm.nih.gov/pmc/articles/PMC4411945/. Accessed 11 Dec 2020

Hu M., Palić D. (2020a). Micro- and nano-plastics activation of oxidative and inflammatory adverse outcome pathways. Redox Biol. doi:10.1016/j.redox.2020.101620

Hu M, Palić D (2020b) Micro- and nano-plastics activation of oxidative and inflammatory adverse outcome pathways. Redox Biol 37:101620. https://doi.org/10.1016/j.redox.2020.101620

Hughes DT, Sperandio V (2008) Inter-kingdom signalling: communication between bacteria and their hosts. Nat Rev Microbiol 6:111–120. https://doi.org/10.1038/nrmicro1836

Ivarsson R, Quintens R, Dejonghe S, Tsukamoto K, in 't Veld P, Renström E et al (2005) Redox control of exocytosis: regulatory role of NADPH, thioredoxin, and glutaredoxin. Diabetes 54:2132–2142. https://doi.org/10.2337/diabetes.54.7.2132

Jahnke, A., et al. (2017). Reducing uncertainty and confronting ignorance about the possible impacts of weathering plastic in the marine environment. Environ Sci Technol Lett. https://pubs.acs.org/doi/10.1021/acs.estlett.7b00008. Accessed 22 Dec 2020

Javurek AB, Spollen WG, Johnson SA, Bivens NJ, Bromert KH, Givan SA et al (2016) Effects of exposure to bisphenol A and ethinyl estradiol on the gut microbiota of parents and their offspring in a rodent model. Gut Microbes 7:471–485. https://doi.org/10.1080/19490976.2016.1234657

Jeong C-B, Won E-J, Kang H-M, Lee M-C, Hwang D-S, Hwang U-K et al (2016) Microplastic size-dependent toxicity, oxidative stress induction, and p-JNK and p-p38 activation in the monogonont rotifer (Brachionus koreanus). Environ Sci Technol 50:8849–8857. https://doi.org/10.1021/acs.est.6b01441

Jeong C-B, et al. (2017). Adverse effects of microplastics and oxidative stress-induced MAPK/Nrf2 pathway-mediated defense mechanisms in the marine copepod Paracyclopina nana. Sci Rep. https://www.nature.com/articles/srep41323. Accessed 22 Dec 2020

Jin Y, Xia J, Pan Z, Yang J, Wang W, Fu Z (2018) Polystyrene microplastics induce microbiota dysbiosis and inflammation in the gut of adult zebrafish. Environ Pollut 235:322–329. https://doi.org/10.1016/j.envpol.2017.12.088

RM Jones, AS Neish. (2017). Redox signaling mediated by the gut microbiota. Free Radic Biol Med. https://www.sciencedirect.com/science/article/pii/S0891584916309777. Accessed 11 Dec 2020

Jose S, Bhalla P, Suraishkumar GK (2018) Oxidative stress decreases the redox ratio and folate content in the gut microbe, Enterococcus durans (MTCC 3031). Sci Rep 8:12138. https://doi.org/10.1038/s41598-018-30691-4

H Ju, D Zhu, M Qiao. (2019). Effects of polyethylene microplastics on the gut microbial community, reproduction and avoidance behaviors of the soil springtail, Folsomia Candida. Environ Pollut. https://www.sciencedirect.com/science/article/abs/pii/S0269749118353193. Accessed 12 Dec 2020

N Kamada, GY Chen, N Inohara, G Núñez. (2013). Control of pathogens and pathobionts by the gut microbiota. Nat Immunol. https://www.nature.com/articles/ni.2608?WT?message-global=remove&WT. Accessed 11 Dec 2020

Karami A, Golieskardi A, Keong Choo C, Larat V, Galloway TS, Salamatinia B (2017) The presence of microplastics in commercial salts from different countries. Sci Rep 7:46173. https://doi.org/10.1038/srep46173

Katas H, Moden NZ, Lim CS, Celesistinus T, Chan JY, Ganasan P et al (2018) Biosynthesis and potential applications of silver and gold nanoparticles and their chitosan-based nanocomposites in nanomedicine. J Nanotechnol 2018:e4290705. https://doi.org/10.1155/2018/4290705

Kelly JR, Minuto C, Cryan JF, Clarke G, Dinan TG (2017) Cross talk: the microbiota and neurodevelopmental disorders. Front Neurosci 11:490. https://doi.org/10.3389/fnins.2017.00490

Lai K-P, Chung Y-T, Li R, Wan H-T, Wong CK-C (2016) Bisphenol A alters gut microbiome: Comparative metagenomics analysis. Environ Pollut 218:923–930. https://doi.org/10.1016/j.envpol.2016.08.039

Lambert S, Sinclair C, Boxall A (2014) Occurrence, degradation, and effect of polymer-based materials in the environment. In: Whitacre DM (ed) Rev. Environ. Contam. Toxicol. Vol. 227. Springer International Publishing, Cham, pp 1–53. https://doi.org/10.1007/978-3-319-01327-51

Lazar V, Ditu L-M, Pircalabioru GG, Gheorghe I, Curutiu C, Holban AM et al (2018) Aspects of gut microbiota and immune system interactions in infectious diseases, immunopathology, and cancer. Front Immunol 9. https://doi.org/10.3389/fimmu.2018.01830

Lei L, Liu M, Song Y, Lu S, Hu J, Cao C et al (2018) Polystyrene (nano)microplastics cause size-dependent neurotoxicity, oxidative damage and other adverse effects in Caenorhabditis elegans. Environ Sci Nano 5:2009–2020. https://doi.org/10.1039/C8EN00412A

Li S, Chen G, Zhang C, Wu M, Wu S, Liu Q (2014) Research progress of natural antioxidants in foods for the treatment of diseases. Food Sci Hum Wellness 3:110–116. https://doi.org/10.1016/j.fshw.2014.11.002

B. Li et al. (2020a). Polyethylene microplastics affect the distribution of gut microbiota and inflammation development in mice. Chemosphere. https://www.sciencedirect.com/science/article/abs/pii/S0045653519327328. Accessed 12 Dec 2020

Li L-L, Amara R, Souissi S, Dehaut A, Duflos G, Monchy S (2020b) Impacts of microplastics exposure on mussel (Mytilus edulis) gut microbiota. Sci Total Environ 745:141018. https://doi.org/10.1016/j.scitotenv.2020.141018

Liebezeit G, Liebezeit E (2013) Non-pollen particulates in honey and sugar. Food Addit Contam Part A 30:2136–2140. https://doi.org/10.1080/19440049.2013.843025

Liebezeit G, Liebezeit E (2014) Synthetic particles as contaminants in German beers. Food Addit Contam Part A 31:1574–1578. https://doi.org/10.1080/19440049.2014.945099

Liu Z, Huang Y, Jiao Y, Chen Q, Wu D, Yu P et al (2020) Polystyrene nanoplastic induces ROS production and affects the MAPK-HIF-1/NFkB-mediated antioxidant system in Daphnia pulex. Aquat Toxicol 220:105420. https://doi.org/10.1016/j.aquatox.2020.105420

Locke D. (n.d.). Understanding toxic substances. 17

Lu Y, Zhang Y, Deng Y, Jiang W, Zhao Y, Geng J et al (2016) Uptake and accumulation of polystyrene microplastics in Zebrafish (Danio rerio) and toxic effects in liver. Environ Sci Technol 50:4054–4060. https://doi.org/10.1021/acs.est.6b00183

Lu L, Wan Z, Luo T, Fu Z, Jin Y (2018) Polystyrene microplastics induce gut microbiota dysbiosis and hepatic lipid metabolism disorder in mice. Sci Total Environ 631–632:449–458. https://doi.org/10.1016/j.scitotenv.2018.03.051

Lu L, Luo T, Zhao Y, Cai C, Fu Z, Jin Y (2019) Interaction between microplastics and microorganism as well as gut microbiota: a consideration on environmental animal and human health. Sci Total Environ 667:94–100. https://doi.org/10.1016/j.scitotenv.2019.02.380

Lyche JL, Rosseland C, Berge G, Polder A (2015) Human health risk associated with brominated flame-retardants (BFRs). Environ Int 74:170–180. https://doi.org/10.1016/j.envint.2014.09.006

Ma, H. et al. (2019). Correlation analysis of intestinal redox state with the gut microbiota reveals the positive intervention of tea polyphenols on hyperlipidemia in high fat diet fed mice. J Agric Food Chem. https://pubs.acs.org/doi/abs/10.1021/acs.jafc.9b02211. Accessed 11 Dec 2020

Mao X, Yang Q, Chen D, Yu B, He J (2019) Benzoic acid used as food and feed additives can regulate gut functions. Biomed Res Int 2019:5721585. https://doi.org/10.1155/2019/5721585

Maranduba, CMC, et al. (2015). Intestinal microbiota as modulators of the immune system and neuroimmune system: impact on the host health and homeostasis. J Immunol Res. https://www.hindawi.com/journals/jir/2015/931574/. Accessed 11 Dec 2020

Mardinoglu A, Shoaie S, Bergentall M, Ghaffari P, Zhang C, Larsson E, et al. The gut microbiota modulates host amino acid and glutathione metabolism in mice. Mol Syst Biol 2015;11:834. https://doi.org/10.15252/msb.20156487

Martin, R., Nauta, A., Ben Amor, K., Knippels, L., Knol, J., Garssen, J. (2010). Early life: gut microbiota and immune development in infancy. Benefic Microbes. https://www.ingentaconnect.com/content/wagac/bm/2010/00000001/00000004/art00008. Accessed 11 Dec 2020

Mason S, Garneau D, Sutton R, Chu Y, Ehmann K, Barnes J et al (1987) Microplastic pollution is widely detected in US municipal wastewater treatment plant effluent. Environ Pollut Barking Essex 2016:218. https://doi.org/10.1016/j.envpol.2016.08.056

McCubrey JA, LaHair MM, Franklin RA (2006) Reactive oxygen species-induced activation of the MAP kinase signaling pathways. Antioxid Redox Signal 8:1775–1789. https://doi.org/10.1089/ars.2006.8.1775

J Michels, A Stippkugel, M Lenz, K Wirtz, A Engel. (2018). Rapid aggregation of biofilm-covered microplastics with marine biogenic particles. Proc Biol Sci. https://pubmed.ncbi.nlm.nih.gov/30158309/. Accessed 12 Dec 2020

Milani, C. et al. (2017). The first microbial colonizers of the human gut: composition, activities, and health implications of the infant gut microbiota. Microbiol Mol Biol Rev. https://mmbr.asm.org/content/81/4/e00036–17.short. Accessed 11 Dec 2020

Moore CJ (2008) Synthetic polymers in the marine environment: a rapidly increasing, long-term threat. Environ Res 108:131–139. https://doi.org/10.1016/j.envres.2008.07.025

Neuman H, Debelius JW, Knight R, Koren O (2015) Microbial endocrinology: the interplay between the microbiota and the endocrine system. FEMS Microbiol Rev 39:509–521. https://doi.org/10.1093/femsre/fuu010

Park JC, Han J, Lee M-C, Kang H-M, Jeong C-B, Hwang D-S et al (2017) Adverse effects of BDE-47 on life cycle parameters, antioxidant system, and activation of MAPK signaling pathway in the rotifer Brachionus koreanus. Aquat Toxicol 186:105–112. https://doi.org/10.1016/j.aquatox.2017.02.025

Paul-Pont I, Lacroix C, González Fernández C, Hégaret H, Lambert C, Le Goïc N et al (2016) Exposure of marine mussels *Mytilus* spp. to polystyrene microplastics: toxicity and influence on fluoranthene bioaccumulation. Environ Pollut 216:724–737. https://doi.org/10.1016/j.envpol.2016.06.039

JM Pickard, MY Zeng, R Caruso, G Núñez. (2017). Gut microbiota: role in pathogen colonization, immune responses, and inflammatory disease. Immunol Rev. https://onlinelibrary.wiley.com/doi/abs/10.1111/imr.12567. Accessed 11 Dec 2020

Pohlmann AR, Fonseca FN, Paese K, Detoni CB, Coradini K, Beck RC et al (2013) Poly (ϵ-caprolactone) microcapsules and nanocapsules in drug delivery. Expert Opin Drug Deliv 10:623–638. https://doi.org/10.1517/17425247.2013.769956
Polymer Solutions News Team. Heat stabilizers: the coolest polymer additive. 2015. https://www.polymersolutions.com/blog/heat-stabilizer-analysis/. Accessed 17 Dec 2020
Qiao Y et al (2013) Alterations of the gut microbiota in high-fat diet mice is strongly linked to oxidative stress. Appl Microbiol Biotechnol 97:1689–1697
Qiao R, Sheng C, Lu Y, Zhang Y, Ren H, Lemos B (2019) Microplastics induce intestinal inflammation, oxidative stress, and disorders of metabolome and microbiome in zebrafish. Sci Total Environ 662:246–253. https://doi.org/10.1016/j.scitotenv.2019.01.245
Rakesh SS, Ramasamy M, Ramesh PT, Maheswari M, Shri Rangasami SR, Yuvaraj M (2020) Nanoplastics in 21st Century. Biot Res Today 2:356–358
Raval U, Harary JM, Zeng E, Pasinetti GM (2020) The dichotomous role of the gut microbiome in exacerbating and ameliorating neurodegenerative disorders. Expert Rev Neurother 20:673–686. https://doi.org/10.1080/14737175.2020.1775585
JF Rawls, et al. (2006). Reciprocal gut microbiota transplants from Zebrafish and mice to germ-free recipients reveal host habitat selection. Cell. https://www.sciencedirect.com/science/article/pii/S0092867406012268. Accessed 11 Dec 2020
Reddivari L, Veeramachaneni DNR, Walters WA, Lozupone C, Palmer J, Hewage MKK et al (2017) Perinatal bisphenol A exposure induces chronic inflammation in rabbit offspring via modulation of gut bacteria and their metabolites. MSystems 2. https://doi.org/10.1128/mSystems.00093-17
Revel M, Châtel A, Mouneyrac C (2018) Micro(nano)plastics: a threat to human health? Curr Opin Environ Sci Health 1:17–23. https://doi.org/10.1016/j.coesh.2017.10.003
Ribeiro CFA, Silveira GG, de OS, Cândido E d S, Cardoso MH, Espínola Carvalho CM, Franco OL (2020) Effects of antibiotic treatment on gut microbiota and how to overcome its negative impacts on human health. ACS Infect Dis 6:2544–2559. https://doi.org/10.1021/acsinfecdis.0c00036
Roman P, Cardona D, Sempere L, Carvajal F (2019) Microbiota and organophosphates. Neurotoxicology 75:200–208. https://doi.org/10.1016/j.neuro.2019.09.013
Rubio L, Marcos R, Hernández A (2020) Potential adverse health effects of ingested micro- and nanoplastics on humans. Lessons learned from in vivo and in vitro mammalian models. J Toxicol Environ Health Part B 23:51–68. https://doi.org/10.1080/10937404.2019.1700598
Sadri SS, Thompson RC (2014) On the quantity and composition of floating plastic debris entering and leaving the Tamar Estuary, Southwest England. Mar Pollut Bull 81:55–60. https://doi.org/10.1016/j.marpolbul.2014.02.020
Schirinzi GF, Pérez-Pomeda I, Sanchís J, Rossini C, Farré M, Barceló D (2017) Cytotoxic effects of commonly used nanomaterials and microplastics on cerebral and epithelial human cells. Environ Res 159:579–587. https://doi.org/10.1016/j.envres.2017.08.043
Schubert J (1972) A program to abolish harmful chemicals. Ambio 1:79–89
ScienceDirect Topics. n.d. Gut-Brain Axis - an overview
Shen M, Zhang Y, Zhu Y, Song B, Zeng G, Hu D et al (2019) Recent advances in toxicological research of nanoplastics in the environment: a review. Environ Pollut 252:511–521. https://doi.org/10.1016/j.envpol.2019.05.102
Shi X, Zhou B (2010) The Role of Nrf2 and MAPK Pathways in PFOS-induced oxidative stress in Zebrafish embryos. Toxicol Sci 115:391–400. https://doi.org/10.1093/toxsci/kfq066
Sies H, Stahl W (1998) Lycopene: antioxidant and biological effects and its bioavailability in the human. Proc Soc Exp Biol Med Soc Exp Biol Med N Y N 218:121–124. https://doi.org/10.3181/00379727-218-44285a
F Sommer, F Bäckhed. (2013). The gut microbiota — masters of host development and physiology. Nat Rev Microbiol. https://www.nature.com/articles/nrmicro2974?report=reader. Accessed 11 Dec 2020

SM Soucy, J Huang, JP Gogarten (2015). Horizontal gene transfer: building the web of life. Nat Rev Genet. https://www.nature.com/articles/nrg3962. Accessed 18 Dec 2020

Stephens B, Azimi P, El Orch Z, Ramos T (2013) Ultrafine particle emissions from desktop 3D printers. Atmos Environ 79:334–339. https://doi.org/10.1016/j.atmosenv.2013.06.050

Sung M-K, Yeon J-Y, Park S-Y, Park JHY, Choi M-S (2011) Obesity-induced metabolic stresses in breast and colon cancer. Ann N Y Acad Sci 1229:61–68. https://doi.org/10.1111/j.1749-6632.2011.06094.x

Sykes EA, Dai Q, Tsoi KM, Hwang DM, Chan WCW (2014) Nanoparticle exposure in animals can be visualized in the skin and analysed via skin biopsy. Nat Commun 5:3796. https://doi.org/10.1038/ncomms4796

Teleanu, DM, et al. (2018). Impact of nanoparticles on brain health: an up to date overview. J Clin Med. https://www.mdpi.com/2077–0383/7/12/490. Accessed 24 Dec 2020

Teleanu DM, Chircov C, Grumezescu AM, Teleanu RI (2019) Neurotoxicity of nanomaterials: an Up-to-Date Overview. Nano 9. https://doi.org/10.3390/nano9010096

Thursby E, Juge N (2017) Introduction to the human gut microbiota. Biochem J 474:1823–1836. https://doi.org/10.1042/BCJ20160510

Tilocca B, Pieroni L, Soggiu A, Britti D, Bonizzi L, Roncada P et al (2020a) Gut–brain axis and neurodegeneration: state-of-the-art of meta-omics sciences for microbiota characterization. Int J Mol Sci 21:4045. https://doi.org/10.3390/ijms21114045

Tilocca, B, et al. (2020b). Gut–brain axis and neurodegeneration: state-of-the-art of meta-omics sciences for microbiota characterization. Int J Mol Sci. https://www.mdpi.com/1422–0067/21/11/4045. Accessed 30 Dec 2020)

Tosin M, Weber M, Siotto M, Lott C, Degli-Innocenti F (2012) Laboratory test methods to determine the degradation of plastics in marine environmental conditions. Front Microbiol 3. https://doi.org/10.3389/fmicb.2012.00225

Tripathi AK, Ray AK, Mishra SK (2022) Molecular and pharmacological aspects of piperine as a potential molecule for disease prevention and management: evidence from clinical trials. Beni-Suef Univ J Basic Appl Sci 11:16. https://doi.org/10.1186/s43088-022-00196-1

Ventura M, O'Flaherty S, Claesson MJ, Turroni F, Klaenhammer TR, van Sinderen D et al (2009) Genome-scale analyses of health-promoting bacteria: probiogenomics. Nat Rev Microbiol 7:61–71. https://doi.org/10.1038/nrmicro2047

von Moos N, Burkhardt-Holm P, Köhler A (2012) Uptake and effects of microplastics on cells and tissue of the blue mussel *Mytilus edulis* L. after an experimental exposure. Environ Sci Technol 46:11327–11335. https://doi.org/10.1021/es302332w

Watts A, Urbina M, Goodhead R, Moger J, Lewis C, Galloway T (2016) Effect of Microplastic on the Gills of the Shore Crab Carcinus maenas. Environ Sci Technol 50. https://doi.org/10.1021/acs.est.6b01187

Wen B, Jin S-R, Chen Z-Z, Gao J-Z, Liu Y-N, Liu J-H et al (2018) Single and combined effects of microplastics and cadmium on the cadmium accumulation, antioxidant defence and innate immunity of the discus fish (Symphysodon aequifasciatus). Environ Pollut 243:462–471. https://doi.org/10.1016/j.envpol.2018.09.029

Wright SL, Kelly FJ (2017) Plastic and human health: a micro issue? Environ Sci Technol 51:6634–6647. https://doi.org/10.1021/acs.est.7b00423

Yang Y, Bazhin AV, Werner J, Karakhanova S (2013) Reactive oxygen species in the immune system. Int Rev Immunol 32:249–270. https://doi.org/10.3109/08830185.2012.755176

Yong CQY, Valiyaveetill S, Tang BL (2020) Toxicity of microplastics and nanoplastics in mammalian systems. Int J Environ Res Public Health 17:1509. https://doi.org/10.3390/ijerph17051509

Yousif E, Haddad R (2013) Photodegradation and photostabilization of polymers, especially polystyrene: review. Springerplus 2:398. https://doi.org/10.1186/2193-1801-2-398

Yu F, Yang C, Zhu Z, Bai X, Ma J (2019) Adsorption behavior of organic pollutants and metals on micro/nanoplastics in the aquatic environment. Sci Total Environ 694:133643. https://doi.org/10.1016/j.scitotenv.2019.133643

Zalasiewicz J, Waters CN, Ivar do Sul JA, Corcoran PL, Barnosky AD, Cearreta A et al (2016) The geological cycle of plastics and their use as a stratigraphic indicator of the Anthropocene. Anthropocene 13:4–17. https://doi.org/10.1016/j.ancene.2016.01.002
Zaneveld JR, McMinds R, Vega TR (2017) Stress and stability: applying the Anna Karenina principle to animal microbiomes. Nat Microbiol 2:1–8. https://doi.org/10.1038/nmicrobiol.2017.121
Zbyszewski M, Corcoran PL, Hockin A (2014) Comparison of the distribution and degradation of plastic debris along shorelines of the Great Lakes, North America. J Gt Lakes Res 40:288–299. https://doi.org/10.1016/j.jglr.2014.02.012
Zhang H, Kuo Y-Y, Gerecke AC, Wang J (2012) Co-release of hexabromocyclododecane (HBCD) and nano- and microparticles from thermal cutting of polystyrene foams. Environ Sci Technol 46:10990–10996. https://doi.org/10.1021/es302559v
Zhao Y, Li M-Z, Shen Y, Lin J, Wang H-R, Talukder M et al (2020) Lycopene prevents DEHP-induced leydig cell damage with the Nrf2 antioxidant signaling pathway in mice. J Agric Food Chem 68:2031–2040. https://doi.org/10.1021/acs.jafc.9b06882
B-K Zhu, Y-M Fang, D Zhu, P Christie, X Ke, Y-G Zhu. (2018). Exposure to nanoplastics disturbs the gut microbiome in the soil oligochaete *Enchytraeus crypticus*. Environ Pollut. https://www.sciencedirect.com/science/article/abs/pii/S0269749117347073. Accessed 11 Dec 2020

Chapter 14
Gut Microbiome, COVID-19, and Neurological Impairment

Richa Das, Riya Singh, and Amit Kumar Tripathi

Abstract The pandemic of COVID-19 caused by SARS-CoV-2 has created havoc worldwide. It causes lung infection through binding of ACE2 receptors that are present on the alveolar epithelial cells. The respiratory infection can cause the alteration of gut-microbiota. COVID-19 infections also manifest diarrhea which is an indication of bacterial infection and it can be modulated by dietary interventions. Dietary pattern, environmental factors, and genetics play important role in shaping the gut-microbiota which can influence the immunity. The plant-based rich fiber diet, which happens to be consumed by a majority of the Indian population, appears to be advantageous during the lockdown. Fiber diet includes beneficial microorganisms within the gut of the humans, thereby leading to a symbiotic association. It is delivering various health benefits to the host, including enhanced immunity and prevention of infection. Non-pharmacological measures such as implementation of lockdown seem to increase the consumption of home cooked food, thereby providing us with all the essential nutrients. The potential role of gut-microbiota in lung diseases has been evaluated, which is referred to as gut-lung axis (GLA). GLA is important for developing protection against COVID-19 infection. Inclusion of healthy vegetarian and fiber-rich food in our diet can help modulate GLA. The present chapter focuses on impact of western and healthy vegetarian diet on COVID-19. We will also explore the effects of dietary nutrients, like macronutrients, micronutrients, vitamin B_6, C, D and COVID-19 prevention and management. In addition to this, we will also focus on impact of improving gut microbiome in neurological impaired COVID-19 patients and personalized nutritional intervention for treating COVID-19.

R. Das
Parul Institute of Applied Science, Parul University, Vadodara, India

R. Singh · A. K. Tripathi (✉)
Molecular Biology Unit, Institute of Medical Science, Banaras Hindu University, Varanasi, India

A. K. Tripathi, M. Kotak (eds.), *Gut Microbiome in Neurological Health and Disorders*, Nutritional Neurosciences,
https://doi.org/10.1007/978-981-19-4530-4_14

Keywords Healthy diet · COVID-19 · Personalized medicine · Neurological Impairment · Gut-brain axis

Abbreviations

ACE	Angiotensin-converting enzyme
ARDS	Acute respiratory distress syndrome
CDC	Centre of disease control
COVID-19	Coronavirus disease
FOS	Fructo-oligosaccharides
GI	Gastrointestinal tract
GLA	Gut-lung axis
GOS	Galacto-oligosaccharides
RSV	Respiratory viral infection
SARS-COV	Severe acute respiratory syndrome coronavirus
ssRNA	Single stranded ribonucleic acid
TH17	T helper 17
Treg cell	T-regulatory cell
WHO	World Health Organization

14.1 Introduction

Microbiome is the branch of science that deals with the study of microorganisms residing in human body. It consists of abundant microorganisms such as bacteria, protozoa, fungi, and viruses. Microbiota disparity or discordance which is known as microbiota dysbiosis leads to significant human disease (Tungland 2018). Microbiome is an integral part of human life. It has the capability to expand the mechanism of drawing out all unavoidable nutrients from the food we eat (Alverdy and Krezalek 2017). It helps in maintaining homeostasis. It has been studied that microbiota disparity or residue discordance plays a major role in causing several chronic diseases, which may become life-threatening if not taken care. The diseases include cardiometabolic disorder, diabetes mellitus, obesity, inflammatory bowel disease, and cardiovascular disease (Mörkl et al. 2020). Healthy diet promotes healthy gut, which in turn improves brain health, as well as intestinal health, as they absorb all necessary nutrients from healthy food (Gill et al. 2006). Several beneficial microorganisms reside within different organs of the human body, such as gut, lungs, and liver (Carding et al. 2015). Gut-microbiota, also known as gut-flora or gut microbiome, comprises of all necessary and beneficial microorganisms for control and coordination of essential organs such as brain, kidney, liver, and lungs (Gut Flora - an overview 2006). Within a healthy gut of an individual more than thousands of diverse bacterial species have been found when compared to other parts of the body (Albillos et al. 2019; Kaushik et al. 2020). The virus SARS-CoV 2 is the causative agent of COVID-19 detected in natural host (Holmes 2003; Ge et al. 2013). COVID-19 is formed from positive enfolding of single stranded RNA virus (+ssRNA) which may range from 26 to 32 kb in size. COVID-19 is a three-layered capsulated virus. The layers are membrane, spike, and envelope (Murphy et al. 2012). It is associated mainly with acute respiratory distress syndrome (ARDS) which is a leading cause of death. Apart from this it also results in fever, pneumonia, cough, and indigestion. In humans, SARS-COV-2 binds with the ACE-2 (angiotensin-converting enzyme-2) receptors. ACE-2 is an enzyme that gives rise to small proteins by fragmenting up the larger angiotensin protein which regulates cell functions (Li et al. 2003). It is a crown shaped virus whose surface consists of spike-like structures. These spikes are made up of protein and bind to host cell receptor, ACE-2 in humans (Sriram et al. 2020). Its diameter ranges from approximately 60–140 nm. SARS-CoV-2 which belongs to the family of corona virus are basically subdivided into four categories of genera that is alpha-, beta-, gamma-, and delta (CDC 2020a). Various studies have reflected that there is a strong impact of healthy gut-microbiota in COVID-19. Gut-microbiota plays a vital role in enhancement of immune system (Calder 2020).

14.2 Human Diet and COVID-19

During recent studies and clinical trials with respect to COVID-19, it has been found that people with healthier immune system are able to fight back, which can only be achieved by introducing balanced and nutrition rich diet with proper daily hydration (Raje et al. 2020). Almost 60–70% of human health depends on balanced diet, which has also been put forward by World Health Organization (WHO) globally during COVID-19.

Weak immune system promotes unhealthy gut, which further leads to distinct pathogenic diseases such as type-2 diabetes mellitus, obesity, and stress (Pradhan et al. 2020). It has also been studied that consuming high fat diet is not the only reason for COVID-19 infection, malnutrition is also one of the reasons (Mackowiak 2013). SARS-CoV-2 uses the ACE-2 receptor to enter the lungs and cause ARDS (Ni et al. 2020). However, healthy dietary lifestyle can play a crucial role in influencing the levels of ACE (Sommerstein et al. 2020).

Different types of diet are being followed globally. Consumption of high calorie junk food and drinks which include western diet pattern, are low in fiber and essential nutrients, and work as a stimulant for ACE level (Clapp 2016). Whereas, uptake of diet rich in fruits, vegetables, whole grains, and dairy products act as an ACE-inhibitor (Cryan et al. 2019). Another type of diet, known as Mediterranean diet, rich in plant-based food products, eggs, seafood, whole grain and very moderate amount of meat, helps against COVID-19. Mediterranean diet comprises of both vegetarian and animal-based food in a balanced format (Angelidi et al. 2021) as shown in Fig. 14.1.

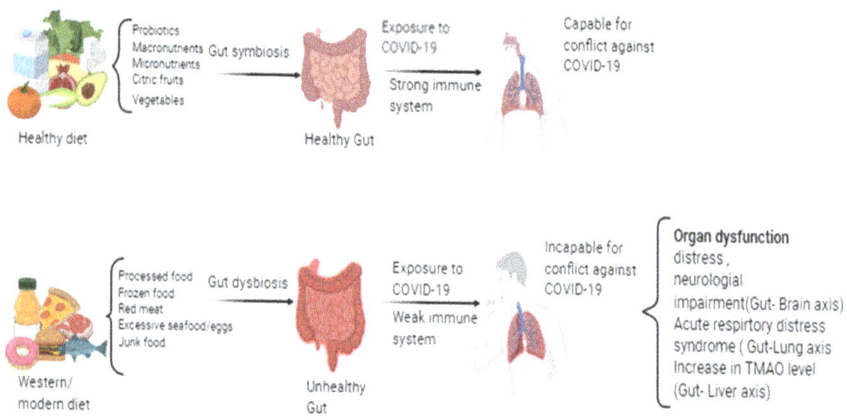

Fig. 14.1 Diagrammatic representation of the impact of healthy as well as western/modern dietary lifestyle on gut-microbiota, and its influence on immune system; in case of strong immune system individual was able to fight against COVID-19, while the weak immune system was unable to fight against COVID-19 after the exposure; ultimately leading to several disorders

To fight against COVID-19 certain balanced diet parameters should be taken care of which includes consumption of probiotics and prebiotics. Probiotics consists of living microorganisms such as bacteria and yeast, which are beneficial for human digestive system and are capable to protect against diarrhea. Probiotics are widely known for the presence of good bacteria. Yogurt, sourdough bread, kimchi, pickles, and cheese are some of the probiotic foods (Parvez et al. 2006). Prebiotics have characteristics that can alter the composition of microorganisms within the gut microbiome. Prebiotics consists of banana, garlic, onion, and asparagus (Pourabedin and Zhao 2015).

Including appropriate amount of micronutrients such as zinc, cobalt, magnesium, and polyphenols is also crucial. Macronutrients such as dietary fibers, carbohydrates, proteins, fats and vitamins, especially vitamin D (meat, seafood, eggs), vitamin C (lemon, spinach, broccoli), vitamin E (almonds, spinach, olive oil), Vitamin B_6 (bananas, peanuts, oats) are also required to fight against COVID-19. Giloy, citric fruits, along with Ayurveda herbs are a must to consume to boost immune system and to fight against the virus (Arshad et al. 2020). Few supplements have been found to modulate beneficial microbiota such as *Bifidobacteria*, *Lactobacillus* and microbes of the genus *Roseburia* (Tojo et al. 2014). Adapting good diet not only helps against COVID-19 but also plays a vital role in maintaining homeostasis within the human gut which helps to improve gut-microbiota and prevent several diseases (Lin and Zhang 2017).

14.3 Gut-Lung Axis

During the study of SARS-COV-2, it was found that the virus is not only limited in causing malfunctioning of the lungs, but also to different organs. Immune system can play an important role against COVID-19, which is possible only with healthy diet and gut (Mattioli et al. 2020). Human gut consists of unhealthy diet, such as processed food and drinks, which can cause disparity or discordance within the gut, termed as gut-dysbiosis. It leads to several diseases such as type-2 diabetes mellitus, cardiovascular disease, neurological impairment, and immunology bowel disease (WHO EMRO, n.d.). Patients infected with COVID-19 also present symptoms of diarrhea which reflects an interlinking between the gut and COVID-19. This linkage is termed as gut-lung axis (CDC 2022). The human gut-microbiota basically comprises of four bacterial families, *Firmicutes*, *Proteobacteria*, *Actinobacteria*, and *Bacteroidetes* (Quigley et al. 2013). Likewise, lungs also provide residence to microorganisms belonging to the family of *Bacteroidetes*, *Firmicutes*, and *Proteobacteria* (O'Dwyer et al. 2016). This similarity of microorganisms works as a connection between gut-microbiota and lung-microbiota, which then gives rise to gut-lung axis. The microbial metabolites present within the gut cause pulmonary-related diseases via blood circulation. Inflammation in the lungs can influence the gut and vice versa (Enaud et al. 2020). It has also been found that a remarkable shift in lipid metabolism and fecal metabolome, together with respiratory viral infection (RSV), shows inappropriate impact in gut-microbiota. Therefore, it can be said that COVID-19 plays a significant role in gut-microbiota dysbiosis, as evidenced by the relationship between respiratory viral infection and gut-microbiota (Matthay and Wick 2020), shown in Fig. 14.2. Healthy gut-microbiota helps to maintain pro- and anti-inflammatory responses, and promote homeostasis within the immune system (Notz et al. 2020).

14.4 Diet and Gut-microbiota in the Population of Developed and Developing Countries

Human gut-microbiota plays a crucial role in dealing with several pathogenic diseases, which are associated with different organs of the human body, such as kidney, liver, lungs, and brain (Riccio and Rossano 2020). Microbiota deals with immune system and other related diseases in our body. Its function comes into role because of several microorganisms present within the gut. Human gut consists of species that usually ranges from 500 to 1000 in number which belongs to the population of more than hundred trillions of microorganisms (Thursby and Juge 2017). Both small and large intestines are the residences for major population of microorganisms. Human gastrointestinal tract involves the presence of different phyla. These are categorized as bacteria which plays a vital role in starch degradation and metabolism of carbohydrates. They are called as *Bacteroidetes (Bacteroidetes*

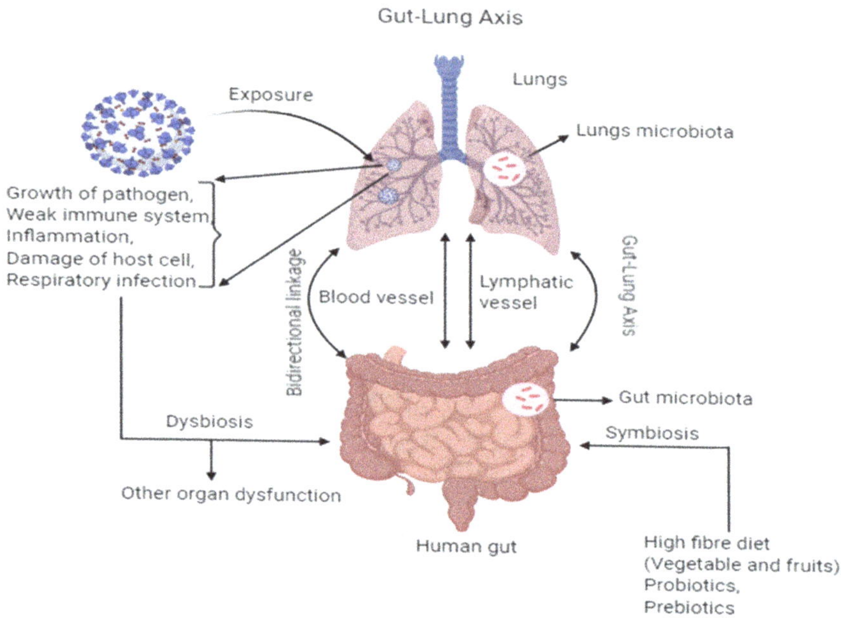

Fig. 14.2 Diagrammatic representation of gut-lung axis, indicating the bidirectional interrelationship between the lung-microbiota and gut-microbiota via the respective lymphatic vessel and blood vessels; exposure to COVID-19 affects the lung-microbiota hence leading to dysbiosis of gut-microbiota and vice versa

fragilis) (Eckburg et al. 2005). Second category is *Firmicutes*, which includes *Lactobacillus*. It has anti-inflammatory properties and also prevents diarrhea in children. Another example of *Firmicutes* is *Eubacterium*, whose function is to carry bile juice and steroid throughout the intestine (Quigley et al. 2013). Third category includes *Bifidobacteria (Bifidobacterium bifidum)* member of *Actinobacteria.* They inhibit harmful pathogens and bacteria, and help in maintaining healthy mucosal gut barrier (G. E. Team 2012). Fourth category is *Proteobacteria* which includes *Desulfovibrio*. They are known as sulfur-reducing and hydrogen-consuming bacteria (Valdes et al. 2018). Apart from these, bacteria, namely *Peptococcus clostridium* and *Verrucomicrobium* are vital for gut-microbiota. In human small intestine, colon provides residence to *Bacteroidetes* and *Firmicutes*, which comprises of more than 90% of the bacterial population inhabiting within the colon. Maintenance of microbial health is crucial to sustain homeostasis within the human gut (Kastl et al. 2020). Several experiments have been performed, on how different types of diet consumed by different countries contribute to variation in human gut-microbiota. Different types of diets have been categorized, like vegetarian diet, modern or western diet, and animal-based diet (WHO, n.d.-a). It has been found that people who belong to western countries and consume modern or western food have less number of *Bacteroidetes* within their gut as compared to people belonging to India (Clarke et al. 2012). People with vegetarian diet have high populations of *Bacteroidetes* within their gut, whereas people with animal-based diet are found highly populated with *Firmicutes* (Tomova et al. 2019). A case study was conducted to compare the diet and diet-associated influence on gut-microbiota among people belonging to different countries. Indian diet was found to be involved in providing a much healthier gut-microbiota (WHO, n.d.-b). Chinese and Japanese population has much less amount of bacterial genera as compared to Indian population in which the gut is populated with *Prevotella, Lactobacillus,* and *Carnobacterium* (Chitguppi 2020). It has also been found that migration from India to the United states modernize the gut microbiome (Vangay et al. 2018). Moreover, the gut of people belonging Africa are filled with a decent population of *Bacteroidetes* because of high fiber diet, as compared to those of European population (MEB et al. 2019). According to several reports it has been studied that with such a large and huge population compared to other countries across the world, India is not only said to be diverse in culture but also in diet and lifestyle. Indian population reflects 80% of *Prevotella,* 55% of *Bacteroidetes, Megasphaera,* and *Roseburia* in their gut (Gut Microbiota for Health, n.d.). Rural and urban India both are very different in their bacterial genera. The *Bacteroides* species are highly found in rural areas (Das et al. 2018).

14.5 Effect of Microbiota on COVID-19 Cases During Lockdown

During the pandemic of COVID-19 most of the countries, including India, were under lockdown, following various non-pharmacological protection like masking, self-hygiene, and social distancing (Alvarez et al. 2020). This not only affected human mental and physical health, but also led to the huge downfall in global economic data because of restrictions on outdoor work (Andrade 2020). However, this lockdown turned out to be more effective against the spreading of this deadly virus at the time of almost no availability of proper treatment (Góis 2020; Sardar et al. 2020). Interestingly, it has also been found that mortality rate in India was less than other countries (western) worldwide (Jain et al. 2020). In India affected people were broadly categorized into two types, symptomatic and asymptomatic, in which asymptomatic people comprised of more infected population of approximately 80%, as these are found to be unintentional carriers of the virus (Zens et al. 2020). During lockdown Indians were provided with proper supply of fruits and vegetables, which also plays an essential role against COVID-19. Whereas, in western countries the supply of fruits and vegetables were prohibited which forced them to consume processed food. This caused disparity of gut-microbiota in countries like Brazil, the USA, and European countries (Kumari et al. 2020). Non-pharmacological measures and consumption of home cooked food are beneficial for healthy microbiota (Singh et al. 2017).

Indian dietary lifestyle is known as one of the richest across the globe. During COVID-19 almost all the countries across the world were under lockdown and were following all the non-pharmacological measures (Lam et al. 2020). During the period of lockdown, people were more attracted towards food as a stress releasing source (Ashritha and Somasundaram 2020). Indian diet comprises of whole grain that is rich in oligosaccharides, dietary fiber, and starch. These results in the production of short-chain fatty acids, and work as the source of energy for the colonocytes (Hati et al. 2019; Slavin 2007). Consumption of whole grain also acts as an antioxidant for disease prevention (CABI, n.d.). During lockdown less access to animal-based diet by Indians, and consumption of more plant-based diet enhanced the microbiota (Van Loo et al. 2017). The homemade ayurvedic drink called kadha that consists of cinnamon, bae-leaf, black pepper, ginger, tulsi, turmeric, mint leaves, cloves, and lemon grass, proved to be an excellent immunity booster, and moreover, turned out to be beneficial for gut-microbiota (Maurya and Sharma 2022).

14.6 Diet Induced Dysbiosis, Inflammation, and Commodity

Gut health is established as a powerhouse for several organs against different diseases including COVID-19. The gastrointestinal tract, the most significant immunological organ, serves as a defense mechanism against harmful diseases (Lane et al. 2010). Any sort of disturbance within the gut can cause dysbiosis and lead to inflammation which can strongly affect the body when infected with COVID-19 (Buttó and Haller 2016). To maintain symbiosis across GI tract healthy diet must be adapted (Zhukova 2019). In contrast to different food products, it has been found that consumption of vegetables, nuts, and fruits turn out to be beneficial for generating anti-inflammatory responses which lead to more steady microbiota (Benenden Health, n.d.). As vegetarian diet is high in fiber which generates short-chain fatty acids, it can help in promoting anti-inflammatory pathway leading to the enhancement of reactive oxygen species-mediated killing coupled with enhanced phagocytosis (Haas 2020). Whereas, consumption of western diet works as a promoter for dysbiosis of microbiota which provokes pro-inflammatory responses leading to chronic diseases (Prana et al. 2019). Degradation of strong junctions within the gut leads to gut leakage by different proteases. This causes chronic inflammation, and the dispersal of cells and pathogenic microorganisms into the bloodstream, where they can sometimes be undetected for years and work as a deadly intermediary known as commodities (Singh and Misra 2020). It has been found that people severely infected with COVID-19 that are already going through some commodities such as obesity, stress, and diabetes mellitus, do not receive adequate therapy which eventually leads to death (International Diabetes Federation 2021). This is being examined by the Centre of Disease Control and Prevention (CDC) by studying the medical history of COVID-19 infected patient among different countries (CDC 2020b).

14.7 Personalized Nutritional Invention for Treating COVID-19

Diet has a significant impact on health and reflects how well a human body can fight deadly diseases such as COVID-19 and maintain a healthy gut-microbiota (CDC 2020c). Several research in discovering novel personalized nutrition through supplements have been conducted to improve current regular treatment. This can be accomplished by altering the individual's lifestyle (EIT Food 2020). For the improvement of gut-microbiota in such patients, inclusion of essential symbiotic, probiotics and prebiotics, like fructo-oligosaccharides (FOS), galacto-oligosaccharides (GOS), along with different *Lactobacillus* strains is important (Errasfa 2021). These supplementing measures aid in improving the immune system that is weakened by SARS-COV-2 (Fig. 14.3), allowing patients to receive proper treatment and promote a faster and better recovery rate, particularly among the elderly people (Rastogi et al. 2022).

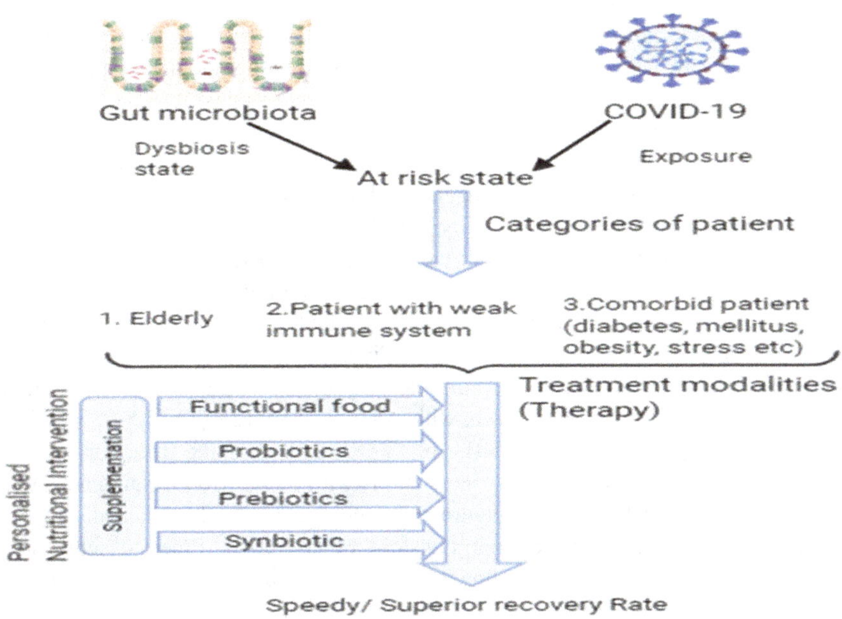

Fig. 14.3 Diagrammatic representation of personalized nutritional invention through treatment modalities (therapy) and supplementations; hence beneficial for several categories of vulnerable individuals when exposed to COVID-19

14.8 Molecular Mechanism of Microbiota–Virus Interaction

Among several pathogenic diseases, it has been found that viral infections are the most fatal worldwide, among other bacterial pathogens (Murphy 2020). As viruses have the capability of infecting prokaryotic as well as eukaryotic organisms at the very same instance, this reflects the concept of studying the molecular mechanism of microbiota–virus interaction (Koonin et al. 2015). Microbiota plays a vital role in defeating pathogenic viruses by suppressing its infection (Zheng and Perlman 2018). Binding of viral receptors to the epithelial cells surface includes various mechanisms to invade the host and this is achieved by disrupting several barriers, such as tissue, body temperature, and mucus membrane (Chua et al. 2020).

14.8.1 Piperine as a Repurposing Molecule for Reversing the COVID-19 Pandemic

COVID-19 patients with a healthy gut-microbiota have a stronger immune system. Microbial metabolites have the ability to cross the blood–brain barrier (BBB) and alter the gene expression in the brain and other organs, resulting in pleiotropic effects. Using Fecal Microbiota Transplantation (FMT) technology to identify healthy gut microbiomes in stool samples of COVID-19 patients may be a better technique for precision therapy. In addition to its immunomodulatory properties, consumption of black pepper may directly aid in the fight against SARS-CoV-2 with the help of its antiviral properties (Choudhary et al. 2020). Piperine has recently been found to possess the ability to bind with the spike glycoprotein and the ACE2 receptor. By forming a single hydrogen bond with each amino acid residue, the interactions of hydrogen bonds with Gly399, His401, Glu402, Arg514, and Arg518 were determined to be important (Maurya et al. 2020). Piperine forms one hydrogen bond with His41 and interacts with the protease having a docking score of −90.95 and binding energy of −78.10 kcal mol^{-1}. Other stable interactions are π-sulfur, π–σ, π–π T-shaped, and alkyl interactions. Piperine forms hydrogen bond with GLY164 and GLY170, having a binding affinity of −6.4 kcal mol^{-1}. In addition to hydrogen bonds, piperine interacts with ARG71, TYR121 (TYR453), TYR163 (TYR495), and ASN169 (ASN501) of SARS-CoV-2 spike receptor-binding domain. Piperine's main stabilizing interactions with SARS-CoV-2 spike receptor-binding domain are by covalent hydrogen bonding, π–π T-shaped, and van der Waals force of interactions (Alagu Lakshmi et al. 2020). Piperine suppresses SARS-CoV-2 replication by binding to the Nsp15 viral protein (Kumar et al. 2020a,b; Tripathi et al. 2022).

14.8.2 Interplay Between Gut Microbiome, COVID-19, and Neurological Impairment

There is relationship between gut microbiome, COVID-19, and neurological impairment (Villapol 2020). COVID-19 is associated with several neurological disorders such as headache, sleep-associated disorder, stroke, motor neuron disease (MDN), Alzheimer disease (AD), Parkinson disease (PD), multiple sclerosis (MS), and Creutzfeldt–Jakob disease (CJD) (Follmer 2020). Bilateral and long-lasting headache is most common in COVID-19 patients (Kocasoy Orhan et al., n.d.). The major courses of headache characteristics during the pandemic are the duration, severity, and frequency. Some patients with COVID-19 exhibit significantly lower microbial diversity, with increased abundance of opportunistic pathogens and a decreased population of protective bacteria (Gu et al. 2020).

14.9 Conclusion and Future Perspectives

The impact of COVID-19 worldwide has put everyone in danger. Each country is fighting day and night to generate suitable vaccines for all categories of people. To ensure safety measures against this deadly virus, it has become compelling for every individual to know that what is COVID-19 and what all precautions need to be taken to avoid getting infected, and also following all safety rules at the time of infection. Several research reports about SARS-CoV-2 convey that it is not only limited to infect lungs but it can also infect several other organs. The symptoms of COVID-19, such as fever, cold, acute respiratory distress syndrome, and diarrhea, reflect that COVID-19 can have a major impact on our gut which can lead to gut-microbiota disparity or dysbiosis. Therefore, good gut-microbiota health is important. Dietary fiber intake, which increases short-chain fatty acids in the blood, protects the lungs from allergic inflammation. In addition to fiber, probiotics such as wheat bran, FOS, and GOS are also involved in treating asthma, cystic fibrosis, and inflammation. By looking at the current scenario, it is very much clear that corona virus is going to stay with us even after the introduction of proper vaccines. Also, it might be possible that anytime a novel variant might evolve ahead in future. This makes it important for us to adapt healthy dietary lifestyle. Balanced diet is crucial, but environmental health should also be a top priority for everyone around the world, as there is a strong correlation between human and environmental health. As a result of lockdown, the environment was able to heal from global warming because people were restricted from leaving their homes. This resulted in less pollution. Water bodies were also healed, as there was much less water pollution. However, the pandemic of COVID-19 has raised much awareness and provided opportunity for people to grasp the value of living a healthy lifestyle, which should not only benefit today and tomorrow, but also future generations. Healthy diet promotes healthy immune system, and healthy immunity contributes in making human as a warrior against deadly diseases.

References

Alagu Lakshmi S, Shafreen RMB, Priya A, Shunmugiah KP (2020) Ethnomedicines of Indian origin for combating COVID-19 infection by hampering the viral replication: using structure-based drug discovery approach. J Biomol Struct Dyn 39(13):4594–4609. https://doi.org/10.1080/07391102.2020.1778537

Albillos A, de Gottardi A, Rescigno M (2019) The gut-liver axis in liver disease: pathophysiological basis for therapy. J Hepatol. https://www.journal-of-hepatology.eu/article/S0168–8278(19)30604-X/fulltext. Accessed 16 Jan 2021

Alvarez FE, Argente D, Lippi F (2020) A simple planning problem for COVID-19 Lockdown. NBER. https://www.nber.org/papers/w26981. Accessed 2 Jan 2021

Alverdy JC, Krezalek MA (2017) Collapse of the microbiome, emergence of the pathobiome and the immunopathology of sepsis. Crit Care Med 45:337–347. https://www.ncbi.nlm.nih.gov/pmc/articles/PMC5245179/. Accessed 8 Dec 2020

Andrade C (2020) COVID-19: humanitarian and health care crisis in a third world country. J Clin Psychiatry 81(3). https://doi.org/10.4088/JCP.20com13383

Angelidi AM, Kokkinos A, Katechaki E, Ros E, Mantzoros CS (2021) Mediterranean diet as a nutritional approach for COVID-19. Metab Clin Exp 114. https://doi.org/10.1016/j.metabol.2020.154407

Arshad MS et al (2020) Coronavirus disease (COVID-19) and immunity booster green foods: a mini review. Food Sci Nutr. https://onlinelibrary.wiley.com/doi/full/10.1002/fsn3.1719. Accessed 1 Jan 2021

Ashritha M, Somasundaram J (2020) Change in the dietary habits of people during their normal lifestyle vs lockdown period – a questionnaire survey. Eur J Mol Clin Med 7:709–722. https://ejmcm.com/article_2239.html. Accessed 2 Jan 2021

Benenden Health (n.d.) Gut food – 15 foods for good gut health. https://www.benenden.co.uk/be-healthy/nutrition/gut-food-15-foods-for-good-gut-health/. Accessed 2 Jan 2021

Buttó LF, Haller D (2016) Dysbiosis in intestinal inflammation: cause or consequence. Int J Med Microbiol 306(5):302–309. https://doi.org/10.1016/j.ijmm.2016.02.010

CABI (n.d.) Whole-grain products and antioxidants. https://www.cabdirect.org/cabdirect/abstract/20001412230. Accessed 2 Jan 2021

Calder, PC. (2020). Nutrition, immunity and COVID-19. BMJ Nutr Prev Health. 3. https://nutrition.bmj.com/content/3/1/74. Accessed 2 Jan 2021

Carding S, Verbeke K, Vipond DT, Corfe BM, Owen LJ (2015) Dysbiosis of the gut microbiota in disease. Microb Ecol Health Dis 26(s2):26191. https://doi.org/10.3402/mehd.v26.26191

CDC (2020a) Coronavirus Disease 2019 (COVID-19). Centers for Disease Control and Prevention, Feb. 11. https://www.cdc.gov/coronavirus/2019-ncov/index.html. Accessed 2 Jan 2021

CDC 2020b. CDC Works 24/7. *Centers for disease control and prevention*, Dec. 31. https://www.cdc.gov/index.htm. Accessed 2 Jan 2021

CDC (2020c). COVID-19 and your health. *Centers for Disease Control and Prevention*, Feb. 11. https://www.cdc.gov/coronavirus/2019-ncov/need-extra-precautions/people-with-medical-conditions.html. Accessed 2 Jan 2021

CDC (2022) Symptoms of coronavirus. https://www.cdc.gov/coronavirus/2019-ncov/symptoms-testing/symptoms.html. Accessed 1 Jan 2021

Chitguppi, R. (2020) *Prevotella* & endemic TB might explain low COVID 19 deaths in India-Sandeep Chakraborty (Interview). *Dental Tribune South Asia*. https://in.dental-tribune.com/news/*Prevotella*-endemic-tb-might-explain-low-covid-19-deaths-in-india-sandeep-chakraborty-interview/. Accessed 2 Jan 2021

Choudhary P, Chakdar H, Singh D, Selvaraj C, Singh SK, Kumar S, Saxena AK (2020) Computational studies reveal piperine, the predominant oleoresin of black pepper (Piper nigrum) as a potential inhibitor of SARS-CoV-2 (COVID-19). Curr Sci 119:1333–1342

Chua RL et al (2020) COVID-19 severity correlates with airway epithelium–immune cell interactions identified by single-cell analysis. Nat Biotechnol 38(8). https://doi.org/10.1038/s41587-020-0602-4

Clapp S. (2016) Fixing the Food System: Changing How We Produce and Consume Food. https://books.google.co.in/books?hl=en&lr=&id=wbZCDQAAQBAJ&oi=fnd&pg=PP1&dq=+consumption+of+high+calories+junk+food+drinks++are+low+in+fibre+and+essential+nutrients+and+work+as+a+growing+ladder+for+ACE+level+which+also+include+consuming&ots=r7sLYDNPtB&sig=6Ayg2jan_itGPbvxLzG0NLf1BaI&redir_esc=y#v=onepage&q&f=false. Accessed 1 Jan 2021

Clarke SF, et al. (2012). "The gut microbiota and its relationship to diet and obesity." Gut Microbes. https://www.ncbi.nlm.nih.gov/pmc/articles/PMC3427212/. Accessed 2 Jan 2021

Cryan JF et al (2019) The microbiota-gut-brain axis. Physiol Rev 99(4):1877–2013. https://doi.org/10.1152/physrev.00018.2018

Das, B., et al. (2018). Analysis of the gut microbiome of rural and urban healthy Indians living in sea level and high altitude areas. Sci Rep. https://www.nature.com/articles/s41598–018–28550-3. Accessed 2 Jan 2021

Eckburg PB et al (2005) Diversity of the human intestinal microbial flora. Science 308(5728):1635–1638. https://doi.org/10.1126/science.1110591

EIT Food. (2020). Personalised nutrition to help reduce the effect of diseases such as COVID-19, according to scientists at the 'EIT Food Innovation Forum. https://www.eitfood.eu/news/post/personalised-nutrition-to-help-reduce-the-effect-of-diseases-such-as-covid-19-according-to-scientists-at-the-eit-food-innovation-forum. Accessed 2 Jan 2021

Enaud R, et al. (2020) The gut-lung axis in health and respiratory diseases: a place for inter-organ and inter-kingdom crosstalks. Front Cell Infect Microbiol. https://www.frontiersin.org/articles/10.3389/fcimb.2020.00009/full. Accessed 1 Jan 2021

Errasfa, M. (2021). Milk oligosaccharides and lectins as candidates for clinical trials against Covid-19. Curr Nutr Food Sci. 16. https://www.researchgate.net/publication/343831543_Milk_Oligosaccharides_and_Lectins_as_Candidates_for_Clinical_Trials_Against_Covid-19. Accessed 2 Jan 2021

Follmer C (2020) Gut microbiome imbalance and neuroinflammation: impact of covid-19 on Parkinson's disease. Mov Disord 1

G. E. Team (2012) Information about gut microbiota Gut Microbiota for Health https://www.gutmicrobiotaforhealth.com/about-gut-microbiota-info/. Accessed 2 Jan 2021

Ge X-Y et al (2013) Isolation and characterization of a bat SARS-like coronavirus that uses the ACE2 receptor. Nature 503:535–538. https://doi.org/10.1038/nature12711

Gill SR et al (2006) Metagenomic analysis of the human distal gut microbiome. Science 312(5778):1355–1359. https://doi.org/10.1126/science.1124234

Góis AN. (2020) Lockdown as an intervention measure to mitigate the spread of COVID-19: a modeling study. Rev Soc Bras Med Trop. 53. https://www.scielo.br/scielo.php?pid=S0037–86822020000100368&script=sci_arttext. Accessed 2 Jan 2021

Gu S, Chen Y, Wu Z, Chen Y, Gao H, Lv L, Guo F, Zhang X, Luo R, Huang C, Lu H (2020) Alterations of the gut microbiota in patients with coronavirus disease 2019 or H1N1 influenza. Clin Infect Dis 71(10):2669–2678

Gut Flora – an overview. 2006. ScienceDirect. https://www.sciencedirect.com/topics/immunology-and-microbiology/gut-flora. Accessed 16 Jan 2021

Gut Microbiota for Health. (n.d.). The largest gut microbiome study of the Indian population reveals microbial composition is both region- and diet-specific. https://www.gutmicrobiotaforhealth.com/the-largest-gut-microbiome-study-of-the-indian-population-reveals-microbial-composition-is-both-region-and-diet-specific/. Accessed 2 Jan 2021

Haas, S. (2020). Anti-inflammatory vegan meal plan. EatingWell. http://www.eatingwell.com/article/7823682/anti-inflammatory-vegan-meal-plan/. Accessed 2 Jan 2021

Hati, S., Patel, M., Mishra, BK and Das, S. (2019). Short-chain fatty acid and vitamin production potentials of *Lactobacillus* isolated from fermented foods of Khasi Tribes, Meghalaya, India.

Ann Microbiol. 69. 1191–1199. https://link.springer.com/article/10.1007/s13213-019-01500-8. Accessed 2 Jan 2021
Holmes, K.V. (2003). SARS-associated coronavirus. N Engl J Med. https://www.nejm.org/doi/full/10.1056/NEJMp030078. Accessed 1 Jan 2021
International Diabetes Federation 2021. COVID-19 and diabetes. https://www.idf.org/aboutdiabetes/what-is-diabetes/covid-19-and-diabetes/1-covid-19-and-diabetes.html. Accessed 2 Jan 2021
Jain VK, Iyengar K, Vaish A, Vaishya R (2020) Differential mortality in COVID-19 patients from India and western countries. Diabetes Metab Syndr Clin Res Rev 14(5):1037–1041. https://doi.org/10.1016/j.dsx.2020.06.067
Kastl AJ, Terry NA, Wu GD, Albenberg LG (2020) The structure and function of the human small intestinal microbiota: current understanding and future directions. Cell Mol Gastroenterol Hepatol 9(1):33–45. https://doi.org/10.1016/j.jcmgh.2019.07.006
Kaushik, A., Gupta, S., and Sood, M. (2020) COVID-19 pandemic in India: what lies ahead-letter to the editor. Indian J Clin Biochem. https://www.ncbi.nlm.nih.gov/pmc/articles/PMC7286206/. Accessed 1 Jan 2021
Kocasoy Orhan E, Ekizoglu E, ÖZGE A, Cagatay AA, Baykan B, Uygun O, Ertas M, Bolay H. Headache characteristics in COVID-19 pandemic-a survey study
Koonin, E.V., Dolj, V.V., and Krupovic, M. (2015). Origins and evolution of viruses of eukaryotes: the ultimate modularity. Virology. 479, 2–25. https://www.ncbi.nlm.nih.gov/pmc/articles/PMC5898234/. Accessed 2 Jan 2021
Kumar G, Kumar D, Singh NP (2020a) Therapeutic approach against 2019-nCoV by inhibition of ACE-2 receptor. Drug Res (Stuttg) 71(04):213–218. https://doi.org/10.1055/a-1275-0228
Kumar S, Kashyap P, Chowdhury S, Kumar S, Panwar A, Kumar A (2020b) Identification of phytochemicals as potential therapeutic agents that binds to Nsp15 protein target of coronavirus (SARS-CoV-2) that are capable of inhibiting virus replication. Phytomedicine 85:0944–7113. https://doi.org/10.1016/j.phymed.2020.153317
Kumari A et al (2020) A short questionnaire to assess changes in lifestyle-related behaviour during COVID 19 pandemic. Diabetes Metab Syndr Clin Res Rev 14(6):1697–1701. https://doi.org/10.1016/j.dsx.2020.08.020
Lam, S., Lombardi, A., Ouanounou, A. (2020). COVID-19: a review of the proposed pharmacological treatments. Eur J Pharmacol. 886. https://www.sciencedirect.com/science/article/abs/pii/S0014299920305434. Accessed 2 Jan 2021
Lane, JA et al. 2010. The food glycome: a source of protection against pathogen colonization in the gastrointestinal tract. Int J Food Microbiol. 142, 1–2, 1–13. https://www.sciencedirect.com/science/article/pii/S0168160510003223?casa_token=Z9ZlAgz9TjUAAAAA:Mdg_3K-PQD892naHV__MSKwr2Li-sWMFCTuPJ7w0QZF5fpjqCWl7BM4cJ1iVpdAts4AxchoEBQ. Accessed 2 Jan 2021
Li, W, et al. (2003). Angiotensin-converting enzyme 2 is a functional receptor for the SARS coronavirus. Nature. 426:450–454. https://www.nature.com/articles/nature02145. Accessed 1 Jan 2021
Lin, L., Zhang, J. (2017). Role of intestinal microbiota and metabolites on gut homeostasis and human diseases. BMC Immunol. 18. https://bmcimmunol.biomedcentral.com/articles/10.1186/s12865-016-0187-3 Accessed 1 Jan 2021
P. A. Mackowiak, Recycling Metchnikoff: probiotics, the intestinal microbiome and the quest for long life. Front Public Health, vol. 1, 2013, doi: 10.3389/fpubh.2013.00052
Matthay MA, Wick KD (2020). JCI – corticosteroids, COVID-19 pneumonia, and acute respiratory distress syndrome. https://www.jci.org/articles/view/143331. Accessed 1 Jan 2021
Mattioli AV, Sciomer S, Cocchi C, Maffei S, Gallina S (2020) Quarantine during COVID-19 outbreak: changes in diet and physical activity increase the risk of cardiovascular disease. Nutr Metab Cardiovasc Dis 30(9):1409–1417. https://doi.org/10.1016/j.numecd.2020.05.020
Maurya, D.K., Sharma D. (2022). Evaluation of traditional ayurvedic Kadha for prevention and management of the novel Coronavirus (SARS-CoV-2) using in silico approach. J Biomol Struct

Dyn. 40:3949–3964. https://www.tandfonline.com/doi/full/10.1080/07391102.2020.1852119. Accessed 2 Jan 2021

Maurya VK, Kumar S, Prasad AK, Bhatt MLB, Saxena SK (2020) Structure-based drug designing for potential antiviral activity of selected natural products from ayurveda against SARS-CoV-2 spike glycoprotein and its cellular receptor. VirusDisease 31:179–193. https://doi.org/10.1007/s13337-020-00598-8

Hansen MEB, et al. (2019) Population structure of human gut bacteria in a diverse cohort from rural Tanzania and Botswana. Genome Biol. https://genomebiology.biomedcentral.com/articles/10.1186/s13059-018-1616-9. Accessed 2 Jan 2021

Mörkl S, et al. The role of nutrition and the gut-brain axis in psychiatry: a review of the literature. Neuropsychobiology 2020, Vol. 79, No. 1. https://www.karger.com/Article/Fulltext/492834#ref7. Accessed 8 Dec 2020

Murphy, J. (2020). 5 viruses more dangerous than the new coronavirus. *MDLinx*. https://www.mdlinx.com/article/5-viruses-more-dangerous-than-the-new-coronavirus/7wQzXjJudXjqIZVb8eU6LI. Accessed 2 Jan 2021

F. A. Murphy *et al.*, Virus Taxonomy: Classification and Nomenclature of Viruses. Springer Science & Business Media, 2012

Ni, W. et al. (2020). Role of angiotensin-converting enzyme 2 (ACE2) in COVID-19. Crit Care. https://ccforum.biomedcentral.com/articles/10.1186/s13054-020-03120-0. Accessed 1 Jan 2021

Notz Q et al (2020) Pro- and anti-inflammatory responses in severe COVID-19-induced acute respiratory distress syndrome—an observational pilot study. Front Immunol 11. https://doi.org/10.3389/fimmu.2020.581338

O'Dwyer DN, Dickson RP, Moore BB (2016) The lung microbiome, immunity and the pathogenesis of chronic lung disease. J Immunol Baltim Md 1950 196(12):4839–4847. https://doi.org/10.4049/jimmunol.1600279

Parvez, S. et al. 2006. Probiotics and their fermented food products are beneficial for health. J Appl Microbiol. https://sfamjournals.onlinelibrary.wiley.com/doi/full/10.1111/j.1365-2672.2006.02963.x. Accessed 1 Jan 2021

Pourabedin M, Zhao X. Prebiotics and gut microbiota in chickens. FEMS Microbiol Lett. 2015; 362. https://academic.oup.com/femsle/article/362/15/fnv122/550322?login=true. Accessed 1 Jan 2021

Pradhan D, Biswasroy P, Kumar Naik P, Ghosh G, Rath G (2020) a review of current interventions for COVID-19 prevention. Arch Med Res 51(5):363–374. https://doi.org/10.1016/j.arcmed.2020.04.020

V. Prana, P. Tieri, M. C. Palumbo, E. Mancini, and F. Castiglione, "Modeling the effect of high calorie diet on the interplay between adipose tissue, inflammation, and diabetes," Comput Math Methods Med, 2019. https://www.hindawi.com/journals/cmmm/2019/7525834/. Accessed 2 Jan 2021)

Quigley, EMM, et al. "Gut bacteria in health and disease. 2013; 9. https://www.ncbi.nlm.nih.gov/pmc/articles/PMC3983973/. Accessed 1 Jan 2021

Raje, D.V., Bajaj, A., Chakraborty, M., Purohit, H.J. (2020). Cumulative active and recovery rates based criterion for gradual lockdown exit: a global observation of SARS Cov-2 management. medRxiv. https://www.medrxiv.org/content/10.1101/2020.06.05.20123364v1. Accessed 1 Jan 2021

Rastogi, A., et al. (2022). Short term, high-dose vitamin D supplementation for COVID-19 disease: a randomised, placebo-controlled, study (SHADE study). Postgrad Med J. https://pmj.bmj.com/content/early/2020/11/12/postgradmedj-2020-139065. Accessed 2 Jan 2021

Riccio, P., Rossano, R., The human gut microbiota is neither an organ nor a commensal. 2020; 594. FEBS Lett. https://febs.onlinelibrary.wiley.com/doi/full/10.1002/1873-3468.13946. Accessed 1 Jan 2021

Sardar T, Nadim SS, Rana S, Chattopadhyay J (2020) Assessment of lockdown effect in some states and overall India: a predictive mathematical study on COVID-19 outbreak. Chaos Solitons Fractals 139:110078. https://doi.org/10.1016/j.chaos.2020.110078

Singh AK, Misra A (2020) Impact of COVID-19 and comorbidities on health and economics: focus on developing countries and India. Diabetes Metab Syndr 14(6):1625–1630. https://doi.org/10.1016/j.dsx.2020.08.032

R. K. Singh *et al.*, "Influence of diet on the gut microbiome and implications for human health," J Transl Med., vol. 15, no. 1, p. 73, 2017, doi:10.1186/s12967-017-1175-y

Slavin, J. (2007). Why whole grains are protective: biological mechanisms. Proc Nutr Soc. https://www.cambridge.org/core/journals/proceedings-of-the-nutrition-society/article/why-whole-grains-are-protective-biological-mechanisms/25042F202584EDAE68C7CBBFCDBB5471. Accessed 2 Jan 2021

Sommerstein, R., Kochen, M.M., Messerli, F.H., and Gräni, C. (2020) Coronavirus disease 2019 (COVID-19): do angiotensin-converting enzyme inhibitors/angiotensin receptor blockers have a biphasic effect? J Am Heart Assoc. https://www.ahajournals.org/doi/10.1161/JAHA.120.016509. Accessed 1 Jan 2021

Sriram, K., Insel, P., Loomba, R. (2020). What is the ACE2 receptor, how is it connected to coronavirus and why might it be key to treating COVID-19? The experts explain. *The Conversation*. https://theconversation.com/what-is-the-ace2-receptor-how-is-it-connected-to-coronavirus-and-why-might-it-be-key-to-treating-covid-19-the-experts-explain-136928. Accessed 2 Jan 2021

Thursby E, Juge N (2017) Introduction to the human gut microbiota. Biochem J 474(11):1823–1836. https://doi.org/10.1042/BCJ20160510

Tojo, R., et al. (2014). Intestinal microbiota in health and disease: role of bifidobacteria in gut homeostasis. World J Gastroenterol. 20: 15163–15176. https://www.ncbi.nlm.nih.gov/pmc/articles/PMC4223251/. Accessed 1 Jan 2021

Tomova A et al (2019) The effects of vegetarian and vegan diets on gut microbiota. Front Nutr 6. https://doi.org/10.3389/fnut.2019.00047

Tripathi AK, Ray AK, Mishra SK (2022) Molecular and pharmacological aspects of piperine as a potential molecule for disease prevention and management: evidence from clinical trials. Beni-Suef Univ J Basic Appl Sci 11:16. https://doi.org/10.1186/s43088-022-00196-1

Tungland, B. (2018) The human microbiota in health and disease. ScienceDirect. https://www.sciencedirect.com/science/article/pii/S2095809917301492. Accessed 8 Dec 2020

Valdes, AM, et al. Role of the gut microbiota in nutrition and health. BMJ. 2018;361. https://www.bmj.com/content/361/bmj.k2179. Accessed 2 Jan 2021

Van Loo EJ, Hoefkens C, Verbeke W (2017) Healthy, sustainable and plant-based eating: perceived (mis)match and involvement-based consumer segments as targets for future policy. Food Policy 69:46–57. https://doi.org/10.1016/j.foodpol.2017.03.001

Vangay P, et al. (2018). U.S. immigration westernizes the human gut microbiome. Cell. https://www.ncbi.nlm.nih.gov/pmc/articles/PMC6498444/. Accessed 2 Jan 2021

Villapol S (2020) Gastrointestinal symptoms associated with COVID-19: impact on the gut microbiome. Transl Res 226:57–69

WHO. (n.d.-a). Food and nutrition tips during self-quarantine. https://www.euro.who.int/en/health-topics/health-emergencies/coronavirus-covid-19/publications-and-technical-guidance/food-and-nutrition-tips-during-self-quarantine. Accessed 2 Jan 2021

WHO (n.d.-b) Coronavirus disease (COVID-19): Food safety and nutrition. https://www.who.int/emergencies/diseases/novel-coronavirus-2019/question-and-answers-hub/q-a-detail/coronavirus-disease-covid-19-food-safety-and-nutrition. Accessed 2 Jan 2021

WHO EMRO. (n.d.). Nutrition advice for adults during the COVID-19 outbreak. Nutrition-infocus. Nutrition. http://www.emro.who.int/nutrition/nutrition-infocus/nutrition-advice-for-adults-during-the-covid-19-outbreak.html. Accessed 1 Jan 2021

Zens M, Brammertz A, Herpich J, Südkamp N, Hinterseer M. App-based tracking of self-reported COVID-19 Symptoms: analysis of questionnaire data. J Med Internet Res. 2020. 22(9):e21956. https://www.jmir.org/2020/9/e21956/. Accessed 2 Jan 2021

Zheng J, Perlman S (2018) Immune responses in influenza A virus and human coronavirus infections: an ongoing battle between the virus and host. Curr Opin Virol 28:43–52. https://doi.org/10.1016/j.coviro.2017.11.002

Zhukova NV (2019) Fatty acids of marine mollusks: impact of diet, bacterial symbiosis and biosynthetic potential. Biomol Ther 9(12). https://doi.org/10.3390/biom9120857

Chapter 15
Tools to Study Gut Microbiome

K. S. Sreevatshan, Veena G. Nair, C. S. Srinandan, and Ganesh Babu Malli Mohan

Abstract Human microbiome consists of collective genomes of microbiota predominantly of bacteria, which are living in various parts of the human body, out of these the gut harbors more. The most important properties of the gut microbiome are not clearly understood due to the lack of scientific tools for isolating non-cultivable microbes. The existence of gut microbiota as a biofilm mode of growth helps them to withstand stress. Gripping enroute to human microbiome, gut microbiome is roaring in current decades because of quick fostering technologies especially molecular sequencing tools and techniques. Even though these newly developed techniques are at their initial stage, through which the functional properties of the extremely complicated gut microbiota abide less explored, a handful of positive findings have been uncovered and cataloged. These findings disclosed an exceptional future on the road to reorganizing and reforming the disease pathology and medicaments. New scientific approaches such as metagenomic tools by next generation sequencing have paved the way for novel therapies from gut microbiome. In this chapter we have provided an overview of the lifestyle of gut microbiota as a biofilm form and their role in human health. We elaborated on the gut metagenomics experimental tools and their study design. Towards the end, we had described some of the challenges faced during the analysis.

Keywords Human microbiome · Metagenomics · Next-generation sequencing · Biofilm · Experimental design · Gut microbiota · Bioinformatics tools · Amplicon sequencing

---

Dr. Srinandan was an inspirational biofilm biologist and extraordinary human being to students, colleagues, friends, and family. ***He will be with us and always live in our heart***. We dedicate the book chapter to beloved late Dr. Srinandan.

K. S. Sreevatshan · G. B. Malli Mohan (✉)
Microbial Omics Lab, Centre for Research in Infectious Diseases, School of Chemical and Biotechnology, SASTRA Deemed to be University, Thanjavur, India

V. G. Nair · C. S. Srinandan
Biofilm Biology Lab, Centre for Research in Infectious Diseases, School of Chemical and Biotechnology, SASTRA Deemed to be University, Thanjavur, India

A. K. Tripathi, M. Kotak (eds.), *Gut Microbiome in Neurological Health and Disorders*, Nutritional Neurosciences,
https://doi.org/10.1007/978-981-19-4530-4_15

15.1 Introduction

15.1.1 Gut Microbes Are Highly Abundant

The number of microorganisms present in the earth is rip-roaring and outnumbers the count of macroscopic organisms (Flemming and Wuertz 2019). The human microbiota itself comprises approximately 10^{13} to 10^{14} microbial cells (Sender et al. 2016). The gastrointestinal microbiota is the highest, making up around 10^{11} bacteria per gram of wet weight, which predominantly comprises Firmicutes, Bacteroidetes, and Actinobacteria. However, the microbiota in the gut is known to serve as a regulatory as well as the functional role in host physiology and metabolism.

15.1.2 The Great Plate Count Anomaly

The first multispecies microbial communities were observed more than 300 years ago by Antonie van Leeuwenhoek when studying his own teeth (Dobell 1932). However, the development of microbiology languished for the next 200 years after Leeuwenhoek's observations of tiny animalcules. Little progress was made because microscopic observations of microorganisms do not give clear information to understand the science behind them. For the entire discipline to develop, techniques for isolating and culturing microbes in the laboratory were needed. The culturing of microorganisms using growth media is the gold standard procedure used by many microbiologists from centuries to the present day to isolate microbes from samples. The addition and elimination of physical as well as chemical parameters which nourish the microbes made this technique popular. But it failed to meet the expectation of cultivating all the possible microorganisms from the original sample, which is known as "The Great plate count anomaly." We have a limited understanding of culturing these unculturable little wonders, as it is said, "to really know them, you have to grow them" (Charnock et al. 2017). Due to the limited knowledge and understanding about the unculturable organisms, the characteristics and behavior of gut microbiota in relationship to the host is unknown.

15.1.3 From Microscope to Genoscope

The study of microorganisms has transitioned from microscopy to molecular study. Metagenomics and other high-throughput studies burgeoned the microbiology field and have provided new knowledge in the structure and function of the human microbiome (Qin et al. 2010; Huttenhower et al. 2012). However, the intestinal tract contains the most densely colonized microbial ecosystem of the human body and keeps beneficial interactions with its host (Sommer and Bäckhed 2013). The complexity of gut microbiomes is unique and novel even though their importance in the host immune system and influence in host development is recognized. The colonization of microbes in the gut commences from birth as the infant comes across its environment. In the uterine environment, placenta, and amniotic fluid consisting of microbes and amniotic biofilms leads to microbial programming (Romero et al. 2008). During breastfeeding too, the microbial community is transferred to infants (Rautava et al. 2012).

15.1.4 Microbiota Establishment in the Gut by Forming Biofilms

The microbial communities are inhabited over the intestinal mucosa of the gut as biofilm. Bacterial communities enclosed in exopolymeric coat are called biofilms. However, the gut bacteria may adhere to the mucus in the colon, exploit the host mucus as a matrix, and establish biofilm. Biofilm microbiota can in turn disperse to free swimming planktonic bacteria in the intestinal lumen. The microbiota that established itself on the gut mucosa by forming biofilm, displays colonization resistance, where they prevent the colonization of pathogens. Colonization resistance occurs by the mechanism of niche exclusion, modulation of pH, production of antibiotics, or suppressing the virulence of pathogens. It helps the gut biofilms to hold water, safeguard against antibiotics and other unfavorable conditions as well as helps in the horizontal gene transfer (Dunne 2002). The gut microbiota uses surface adhesins such as serine-rich repeat proteins for the formation of biofilm and it also helps for the selection of niches inside the gut. Genes that encode motility, pili, matrix, and adhesin proteins are very essential commodities for biofilm formation (Buret et al. 2019).

In this chapter, we provide a glimpse of the experimental tools to study gut microbiota including their study design and some of the challenges faced during the analysis.

15.2 Gut Metagenomics Experimental Tool

15.2.1 Experimental/Study Design

A virtuous experiment and/or study design helps to mitigate erroneous and inconsistent outcomes which are commonly observed in metagenomics investigations. At large, any biological question should be principally intensive by pilot/small-scale studies and careful literature survey which is necessary to circumvent ambiguity (Bharti and Grimm 2021). Streamlined metagenomic studies include sample numbers, case-controls, randomized-controlled trials (RCT), cross-sectional and longitudinal samples, metadata which positively help the scientific community to advance big data processing and eliminate confounding effects. A significant number of samples consisting of both biological and technical replicates, preserved fixed sample size, and unaltered during the study are key factors for final results, which majorly plays a critical role in microbial population diversity interpretation under similar conditions (Kadam and Bhalerao 2010; Goodrich et al. 2014). An appropriate control sample is necessary to distinguish between the real microbial signature and an artifact from the collected samples. Remarkably, it is very challenging to obtain suitable controls, particularly in clinical trials where the microbial burden is much influenced by gender, age, geographical locations, ethnicity, diet, and lifestyle (Laukens et al. 2016). Additionally, microbiome investigation could be impacted by experimental factors including nucleic acids extraction kits, sampling approaches, contaminations, and sequencing methods, which could be mitigated by counting on positive and negative control samples. However, a good study should design suitably to control and document as many possible features as metadata which will help to avoid false understanding of results and highlights the actual size of individual factors. In addition, a less complex cross-sectional metagenomic study integrates comparisons between two groups, e.g., healthy vs disease condition and/or placebo vs treatment (Martin et al. 2018) (Fig. 15.1). Still, it is well known that the environmental factors that influence the microbiome population may lead to various additive or multiplicative effects. Hence, one should design longitudinal studies to satisfy a statistical standpoint, where the same sample collection at various time points can help to avoid biases (Caruana et al. 2015).

15.2.2 Sample Types, Collection, Handling, and Processing

The choice of sample type for microbial community analysis will be driven by fundamental research questions being answered. For example, the majority of human gut metagenomic studies focused on fecal samples which are simple procedures to collect and can be used to analyze for longitudinal studies. Alternatively, biopsy samples are more beneficial for understanding the host–microbiota interactions (Tang et al. 2020). Hence, it is crucial to collect samples from a fixed site and adequate for the final outcome. Further, subsequent parameters such as contamination, transportation, storage, and safety should be considered during sample collection and handling. The sample preservation and storage should be tailored to the sample type and study design. It is suggested that the sample should be frozen and preserved at −20 °C within 15 mins of sample collection, then transport within 24 hrs to the lab on dry ice and stored at −80 °C until further procedure (Jenkins et al. 2018). The choice of nucleic acid (DNA/RNA) extraction procedures is important for the quality and completeness of metagenomic big data analysis of any microbial population. Effective recovery of DNA from all types of microbes should be employed. Particularly, gram-positive and spore-forming bacteria are rigid

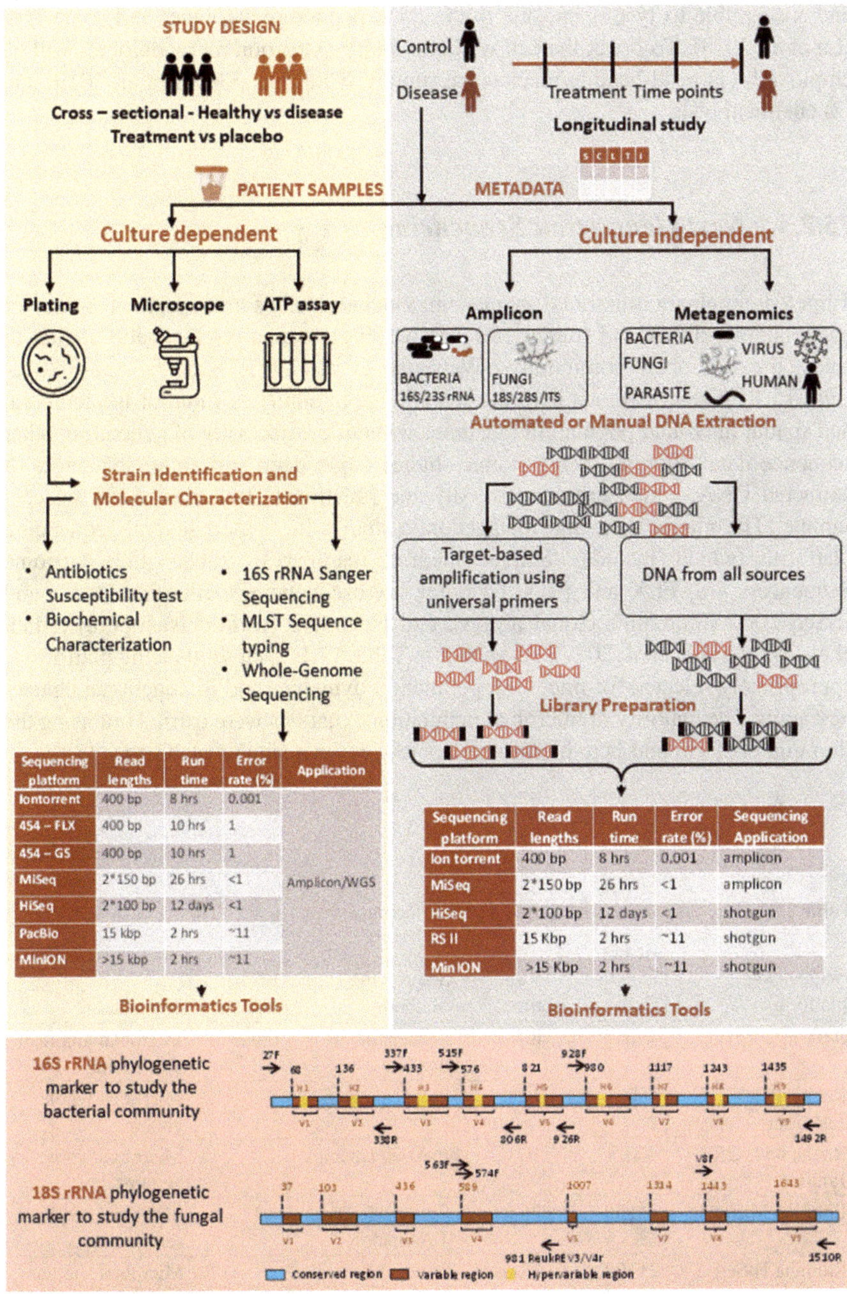

Fig. 15.1 Schematic representation of experimental design and sequencing methodologies to investigate the human gut metagenomics

and susceptible to lyse open, due to the thick peptidoglycan layer and spore coat (Lu et al. 2015). To break the cell wall two major extraction methodologies will be employed; (1) bead-beating/mechanical rupture (ChecinskaSielaff et al. 2019) and (2) chemical lysis.

15.2.3 Next-Generation Sequencing

Targeted amplicon microbial sequencing includes ribosomal small subunits of bacterial 16S rDNA and fungal 18S rDNA/ITS gene as a phylogenetic marker to study the microbial community (Weisburg et al. 1991; Schoch et al. 2012) (Fig. 15.1). These methods help the scientific community to monitor the temporal and spatial microbial population dynamics without the necessity of generating other sequence data. Shotgun metagenomic studies target deep and entire sequences of extracted DNA, which provide diversity and functional properties of the selected sample. The microbiome and metagenomic studies have used several sequencing platforms which includes Sanger method (capillary), Roche 454, Genome Sequencers GS, FLX and FLX Titanium (Pyrosequencing), Illumina GAIIx and HiSeq 2000 (Illumina's clonal arrays), and Nanopore (Oxford Nanopore) (Allali et al. 2017; Malla et al. 2019). As shown in Table 15.1, each sequencing platform is appropriately selected for microbial population dynamics and its functional characterization. The majority of microbial population dynamics were carried out using the Illumina platform and here bioinformatics tool cover a similar platform.

Table 15.1 Microorganism's sequencing platform comparison

Sequencing platform	Read lengths (bp)	Run time	Sequencing Methods	Error rate (%)	Application
ABI–Iontorrent	400	8 h	Sequencing by ligation	0.001	Microbial amplicon sequencing
Roche 454—FLX Titanium	400	10 h	Pyrosequencing	1	Microbial whole genome
Roche 454–GS Junior	400	10 h	Pyrosequencing	1	Microbial whole genome
Illumina MiSeq	2*150	26 h	Reversible Dye Terminators	<1	Microbial amplicon sequencing
Illumina HiSeq	2*100	12 days	Reversible Dye Terminators	<1	Microbial shotgun metagenomics
PacBio	8500	2 h	ZMW-Single molecule	~11	Microbial shotgun metagenomics
NanoporeMiniION	15,000	72 h	–	5–15	Microbial shotgun metagenomics

15.3 Gut Metagenomics Bioinformatics Tool

15.3.1 Preprocessing of Raw Reads

Preprocessing is a crucial prerequisite step of metagenome sequencing analysis, which involves quality trimming and contamination removal (He et al. 2020). Reads can be either single-end, mate pairs, or paired-end reads constructed on the prime of adapter ligation. The computational tools for quality trimming detect and efficiently remove the following sequence details of raw reads including low-quality, adapters, host-associated sequence contamination. In general, sequences that are suspected of a high level of errors are removed in this quality trimming. Parameters to identify true DNA fragments and remove the sequencing artifacts include average quality score, number of homopolymers length, number of primer mismatches, and length of sequence tested. The FASTQC tool (Bioinformatics, n.d.) is employed to check the quality of raw reads which includes identification of the sequence distribution, primer dimers, GC contents, and the presence of adapter sequences. To trim the low-quality reads and adapter sequences from the raw file, sickle, cutadapt, and AdapterRemoval (GitHub n.d.-a; Martin 2011; Schubert et al. 2016) tools were developed and used since 2012. Furthermore, to improve the quality of raw reads, various algorithms were developed and widely used tools which include Trimmomatic, seqtk, ea-utils, FASTX-toolkit, BBTools (GitHub n.d.-b; GitHub n. d.-c; FASTX n.d.; BBMap n.d.; Aronesty 2013; Bolger et al. 2014), Knead-data—a pipeline tool which had integrated Trimmomatic for low-quality read removal and bowtie2 and Burrows-Wheeler Aligner (BWA) for mapping and removing the host contamination (GitHub n.d.-c). In general, a good quality score is recommended greater than 25 to be adapted for the trimming of reads. For amplicon libraries, performing the trimming along the 5′ side using the defined quality score is recommended. For shotgun libraries, before quality trimming, removal of the host contamination either by mapping them through bowtie2 or Burrows-Wheeler Alignment (BWA) is highly recommended (Li and Durbin 2009; Langmead and Salzberg 2012). After DNA sequences are obtained from the quality check, they must be sequentially analyzed and interpreted. The big sequence data requires sophisticated bioinformatics analysis tools; here we differentiate the computation analysis of targeted amplicon and shotgun metagenomics data, which is shown in Fig. 15.2.

15.3.2 Amplicon Analysis

One of the thoughtful concerns for targeted amplicon sequence analysis is to discern the real from read errors sequences. To overcome the difficulties major tools were widely used. Firstly, QIIME, Mothur, and VAMPS tools were developed and utilized with predefined identity threshold clustering the reads (97% similarity-Operational Taxonomic Unit (OTU)). These tools allow researchers to compare

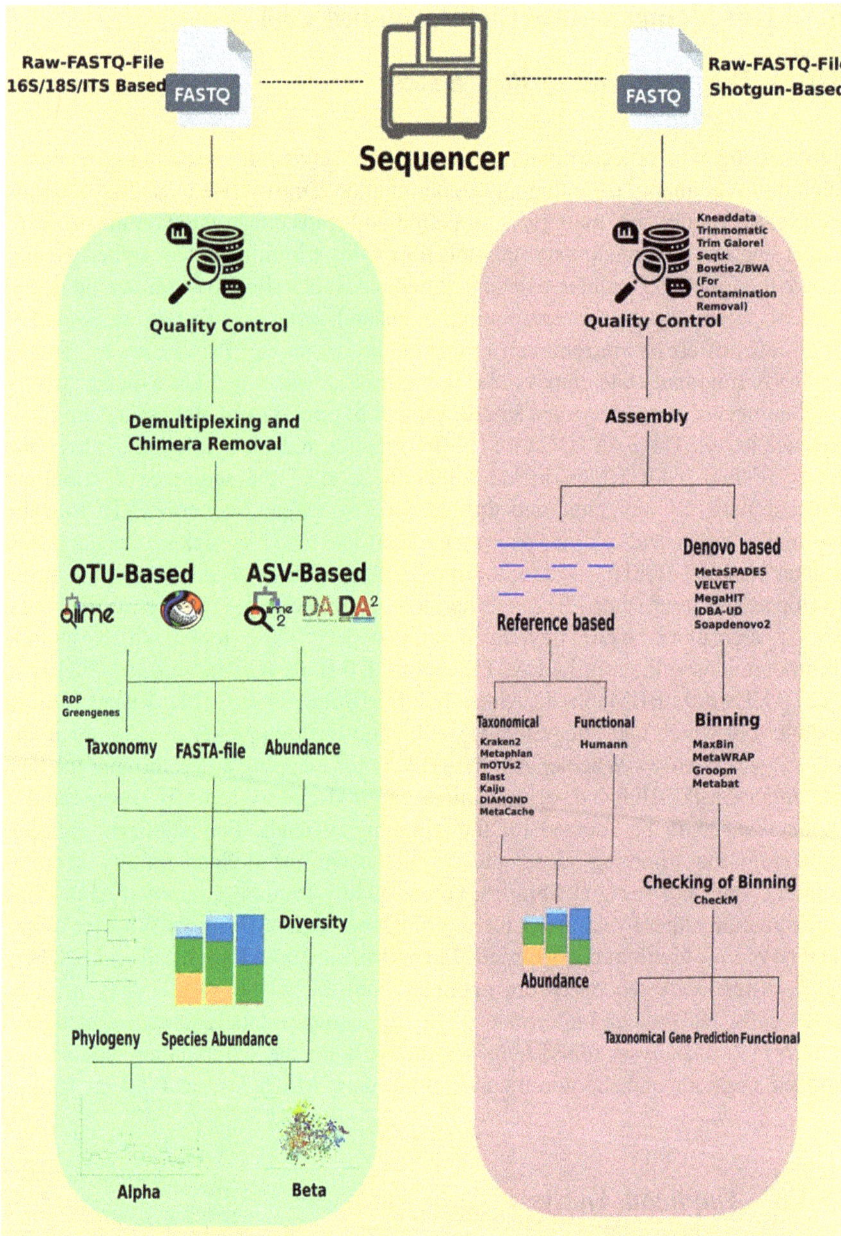

Fig. 15.2 Schematic representation of bioinformatics tools to study human gut metagenomics

and analyze microbial population dynamics using a large set of DNA sequence big data. Alternatively, QIIME2, DADA2, Deblur, MED, and UNOISE tools operate a biological sequence denoising steps before the amplification and sequencing errors

Table 15.2 List of preprocessing scripts

S.no	Pipeline/tool	Script/function
1	qiime	Split_libraries_fastq.py (n.d.)
2	mothur	Make.contigs (Mothur, n.d.)
3	qiime2	Q2-demux (QIIME, n.d.-a)

and provide the amplicon sequence variant (ASV) (Schloss et al. 2009; Caporaso et al. 2010; Bolyen et al. 2019; Schloss 2020). All aforementioned pipeline (or) tools are available both as Graphical User Interface (GUI) in Galaxy (Afgan et al. 2018), Qiita (n.d.) and as Command-Line Interface (CLI).

Selecting representative sequences for species profiling is the key step in amplicon analysis. The operational taxonomic unit was developed in the early 2000s and earlier tools based on this approach are DOTUR and SONS (Schloss and Handelsman 2005, 2006). From time to time many algorithms have been developed for exploring various microbial communities. The most popular OTU-clustering algorithms are based upon greedy heuristic methods (Liu et al. 2021). The primary advantage over the OTU approach is that it needs less computational power for clustering over 97% similarity and the disadvantage is that clustering is based on 97% (or) 99%, hence this method identifies lesser biological variation and rare taxa identification. To overcome this, another method was introduced in 2013 called "Oligotyping" where the sequences change by one nucleotide can be identified (Eren et al. 2013). Further, it was developed and renamed as Amplicon Sequence Variant (ASV) (or) Exact Sequence Variant (ESV) and preceded the OTU-based method. The ASV approach is based upon generating an error model for every sequencing run and using that to predict the original biological variation and error sequences generated during amplification and sequencing. The ASV approach is able to distinguish the sequences even at the single nucleotide level. Popular tools based on this algorithm are DADA2, Deblur, and pipelines such as qiime2 (Callahan et al. 2016; Amir et al. 2017; Bolyen et al. 2019). Whatever the approach may be, the final output will be a feature table (or) abundance table which will contain the frequency of feature sequences in particular samples. This feature table can be linked further to the taxonomy for finding the abundance generally from Kingdom to Species level.

Amplicon analysis begins with removing chimeras and demultiplexing after preprocessing. Demultiplexing is a step where the reads are grouped based on their barcodes. Various functions/scripts used to demultiplex the sequences are given in Table 15.2.

Chimeras are mismatched and unwanted sequences obtained during Polymerase Chain Reaction (PCR) and often occurs while using mixed templates. Hence chimeric sequences need to be removed before the annotation process or they may show a wrong diversity or mis-identify as taxa that are not originally available in the particular nature. Taking this into consideration most of the pipelines and tools had incorporated their own script for detection and removal of chimeric reads (Table 15.3).

Table 15.3 List of scripts for removal of chimeric reads

S. no	Pipeline	Script/function used
1	qiime	Parallel_identify_chimeric_seqs.py (or)Identify_chimeric_seqs.py (QIIME n.d.-b, c)
2	mothur	chimera.vsearch (or) chimera.uclust (or) chimera.persaus (Chimera Detection Commands n.d.)
3	qiime2[a] and DADA2[a]	–

[a]In the ASV-based pipeline, the tools used have incorporated the chimeric removal step in the denoising script directly. Hence there is no separate script/function for chimeric removal

Generally, amplicon sequences are used to check the microbial diversity and composition. However, many tools were developed and used to predict the functional composition. One such tool is PiCrust (Langille et al. 2013), which was developed based on the OTU table of greengenes (DeSantis et al. 2006) database which could predict the functional composition through KEGG (Kanehisa et al. 2021) pathways. Newly developed and improved version Picrust2 (Douglas et al. 2020) has a similar background but both ASV and OTU tables can be used. Tax4Fun (Aßhauer et al. 2015) is an R package which could predict the functional pathways through KEGG (Kanehisa et al. 2021) and SILVA (Quast et al. 2013) databases.

15.3.3 Shotgun Analysis

Compared to amplicon-based sequence analysis, shotgun metagenomic analysis can provide much more functional information and deeper taxonomic resolution. Nevertheless, analysis requires higher computational power to complete due to the huge datasets, and the majority of the tools implemented under Linux/Mac OS environments. After the clean raw reads from the preprocessing step, the important step of shotgun analysis is to generate a taxonomic and functional table using the read-based (or) assembly-based method. In read-based, MetaPhLan (Beghini et al. 2021) tool is predominately employed for taxonomic annotation which is based on clade-specific marker genes. MetaPhLan is available both in CLI and GUI in Galaxy (Afgan et al. 2018). It is possible to construct user-defined taxonomic databases and use alignment tools such as DIAMOND (Buchfink et al. 2014), Bowtie2 (Langmead and Salzberg 2012), BWA (Li and Durbin 2009), Blast+ (Camacho et al. 2009), and kASA (Weging et al. 2021). Further, it is also feasible to use tools such as Metalign (Lapierre et al. 2020) which is based upon the min hash approach, and MetaCache (Müller et al. 2017) which is based on the k-mer approach for annotation.

15.3.4 Assembly

Assembly is a process of stitching reads into longer fragments, which denotes contig and rebuilding the genes. Remarkably, these algorithms were developed for whole-genome assemblies, however, later extended for broader application. The choice of the assembly algorithm remains critical for further analysis. For a typical metagenomic sequence assembly regularly used algorithms include Velvet, IDBA-UD, MegaHIT, METASpades, RayMETA, MetaVelvet, SOAPDenovo2, and Omega (Zerbino 2010; Boisvert et al. 2012; Peng et al. 2012; Liu et al. 2014; Nurk et al. 2017; Luo et al., n.d.). All these metagenome assembler algorithms are based on De Bruijn graphs and are available as open source. Binning is employed after assembly of the reads where assembled contigs are assigned to an individual group of the microbial genome. There are two kinds of methods for binning, first is taxonomic binning where assembled contigs are mapped to reference databases using alignment tools such as Bowtie2, BWA, Blast+, and kASA (Camacho et al. 2009; Li and Durbin 2009; Langmead and Salzberg 2012; Weging et al. 2021). Another one is genome binning which is based on machine learning methods without any help from a reference database. From DNA sequences to microbial species diversity analysis majorly obtained based on genus/species-based or OTU-based ecological metrics. These can be achieved by clustering the sequences using external information either denovo OTU or reference-based OTU picking (greengenes or by SILVA) approaches.

Genome binning has three classifications, namely sequence-based, differential-abundance-based, and hybrid method which was explained well in Yue et al. (Yue et al. 2020). Currently used binning tools such as MaxBin2, GroopM, and Metabat2 are based upon hybrid methods (Imelfort et al. 2014; Wu et al. 2016; Kang et al. 2019). Preceding the taxonomic annotation, the binned files are checked using CheckM which assesses the quality of the file using the marker database (Parks et al. 2015). After that, we can deploy various taxonomic annotation tools such as prokka, Kraken, and Kraken2 which is based upon an algorithm that links k-mer and lowest common ancestor (LCA), CAT & BAT which is also based on LCA and Prodigal (Hyatt et al. 2010; Seemann 2014; Wood and Salzberg 2014; Von Meijenfeldt et al. 2019; Wood et al. 2019). MEGAN which is GUI can be used both for taxonomic and functional annotation of the reads (Huson et al. 2016). Functional annotation is where the clean raw reads are mapped with databases such as KEGG and Metacyc. Humann3 tool can be used for annotation. Further, if de novo sequencing was carried out, it is possible to run with prokka (Beghini et al. 2021; Seemann 2014).

15.3.5 Diversity Measures

Diversity measures are one of the downstream analyses where it tells how well the samples are diverse. It is sub-categorized as alpha and beta diversity. Alpha diversity is defined as diversity within a sample that depends upon richness and evenness. Richness is defined as the total number of species present in a sample. Simplest index available is Observed, other indices such as ACE and Chao1 are also available for measuring richness. Evenness mainly focuses upon the species' abundance. Diversity indices such as Shannon and Pielou's evenness are available for measuring evenness. Phylogenetic-based alpha diversity is also available where the measure depends upon the phylogeny relationship between the species. Faith PD is one of the best examples for the phylogeny-based alpha diversity where it sums the branch length of all species available in the sample (Hughes et al. 2001; Kim et al. 2017; Luz Calle 2019; Liu et al. 2021).

Another diversity measure available is beta diversity which tells the difference between the samples. The most predominantly used diversity indices are Bray-Curtis Dissimilarity index which is used for compositional information, Jaccard index which is based upon presence and absence data, Aitchison distance which is based upon Euclidean distance calculation along with clr (Central Log Ratio) transformation, and Unifrac distances which are based upon the phylogenetic tree (Lozupone and Knight 2005; Luz Calle 2019). Irrespective of the beta diversity indices, the final output is the distance matrix table which needs to be plotted in an ordination plot for visualization purposes. Ordination is used to reduce the dimensionality of the dataset which can be used further for the visualization purpose. For visualizing the most commonly used plots are Principal Component Analysis (PCA), Metric Dimensional Scaling (MDS), Non-metric Dimensional Scaling (NDMS), and Principal Coordinates Analysis (PCoA) (Chengsong and Jianming 2009; Jolliffe 2011; Gower 2015; Abdi, n.d.). Recently developed ordination plots t-SNE and UMAP can be also used for the microbiome analysis (van der Maaten and Hinton 2008; McInnes et al. 2018). For amplicon libraries, the qiime2 pipeline has its own tools for both alpha and beta diversity analysis. Further, it is also possible to import the qiime2 files to R using the package "qiime2R" and use packages such as "vegan," "phyloseq," and "microbiome" for both distance matrices and ordination plot generation. For the shotgun libraries, especially for reference-based analysis, various in-house R script/python modules have been generated and published for importing the data as phyloseq object further generated the diversity indices. Commonly used R packages for diversity analysis are "phyloseq," "vegan," "microbiome," and "mia" (GitHub n. d.-d; Dixon 2003; McMurdie and Holmes 2013; Lahti et al. 2017; Ernst et al. 2021).

15.3.6 Challenges

The first challenge faced by the beginners is working on the command line as most of the tools are available as command-line versions. But many initiatives on courses and workshops are being taken to teach the basics of working on the command line. As a next step, many tool developers are trying to build GUI for their tools, hence even non-coding researchers can work with the tools and have new findings. Galaxy and qiime2 are the best examples of running GUI-based tools. Another challenge faced by every researcher is computational power and storage. The most predominant challenge is reproducibility where the researchers must share their accession no. of the raw reads, metadata in which they must note the primers, adapters, barcodes, DNA extraction kit, location, time, sample type, in-house script which had been developed to get the result along with the parameters/settings used for the analysis.

References

Abdi H (n.d.). Metric multidimensional scaling (MDS): analyzing distance matrices

Aßhauer KP, Wemheuer B, Daniel R, Meinicke P (2015) Tax4Fun. Bioinformatics 31:2882–2884

Afgan E, Baker D, Batut B, Van Den Beek M, Bouvier D, Ech M, Chilton J, Clements D, Coraor N, Grüning BA, Guerler A, Hillman-Jackson J, Hiltemann S, Jalili V, Rasche H, Soranzo N, Goecks J, Taylor J, Nekrutenko A, Blankenberg D (2018) The Galaxy platform for accessible, reproducible and collaborative biomedical analyses: 2018 update. Nucleic Acids Res 46:W537–W544

Allali I, Arnold JW, Roach J, Cadenas MB, Butz N, Hassan HM, Koci M, Ballou A, Mendoza M, Ali R, Azcarate-Peril MA (2017) A comparison of sequencing platforms and bioinformatics pipelines for compositional analysis of the gut microbiome. BMC Microbiol 17(1):194

Amir A, McDonald D, Navas-Molina JA, Kopylova E, Morton JT, Zech Xu Z, Kightley EP, Thompson LR, Hyde ER, Gonzalez A, Knight R (2017) Deblur rapidly resolves single-nucleotide community sequence patterns. mSystems 2

Aronesty E (2013) Comparison of sequencing utility programs. Open Bioinforma J 7:1–8

BBMap. n.d. BBMap – Bushnell B. - sourceforge.net/projects/bbmap/

Beghini F, McIver LJ, Blanco-Míguez A, Dubois L, Asnicar F, Maharjan S, Mailyan A, Manghi P, Scholz M, Thomas AM, Valles-Colomer M, Weingart G, Zhang Y, Zolfo M, Huttenhower C, Franzosa EA, Segata N (2021) Integrating taxonomic, functional, and strain-level profiling of diverse microbial communities with biobakery 3. Elife 10

Bharti R, Grimm DG (2021) Current challenges and best-practice protocols for microbiome analysis. Brief Bioinform 22(1):178–193

Bioinformatics. (n.d.). Babraham Bioinformatics - FastQC A Quality Control tool for High Throughput Sequence Data. https://www.bioinformatics.babraham.ac.uk/projects/fastqc/. Accessed 3 Jan 2022

Boisvert S, Raymond F, Godzaridis É, Laviolette F, Corbeil J (2012) Ray meta: scalable de novo metagenome assembly and profiling. Genome Biol 13:1–13

Bolger AM, Lohse M, Usadel B (2014) Trimmomatic: a flexible trimmer for illumina sequence data. Bioinformatics 30:2114–2120

Bolyen E, Rideout JR, Dillon MR, Bokulich NA, Abnet CC, Al-Ghalith GA, Alexander H, Alm EJ, Arumugam M, Asnicar F, Bai Y, Bisanz JE, Bittinger K, Brejnrod A, Brislawn CJ, Brown CT,

Callahan BJ, Caraballo-Rodríguez AM, Chase J, Cope EK, Da Silva R, Diener C, Dorrestein PC, Douglas GM, Durall DM, Duvallet C, Edwardson CF, Ernst M, Estaki M, Fouquier J, Gauglitz JM, Gibbons SM, Gibson DL, Gonzalez A, Gorlick K, Guo J, Hillmann B, Holmes S, Holste H, Huttenhower C, Huttley GA, Janssen S, Jarmusch AK, Jiang L, Kaehler BD, Kang KB, Keefe CR, Keim P, Kelley ST, Knights D, Koester I, Kosciolek T, Kreps J, MGI L, Lee J, Ley R, Liu YX, Loftfield E, Lozupone C, Maher M, Marotz C, Martin BD, McDonald D, LJ MI, Melnik AV, Metcalf JL, Morgan SC, Morton JT, Naimey AT, Navas-Molina JA, Nothias LF, Orchanian SB, Pearson T, Peoples SL, Petras D, Preuss ML, Pruesse E, Rasmussen LB, Rivers A, Robeson MS, Rosenthal P, Segata N, Shaffer M, Shiffer A, Sinha R, Song SJ, Spear JR, Swafford AD, Thompson LR, Torres PJ, Trinh P, Tripathi A, Turnbaugh PJ, Ul-Hasan S, van der Hooft JJJ, Vargas F, Vázquez-Baeza Y, Vogtmann E, von Hippel M, Walters W, Wan Y, Wang M, Warren J, Weber KC, CHD W, Willis AD, Xu ZZ, Zaneveld JR, Zhang Y, Zhu Q, Knight R, Caporaso JG (2019) Reproducible, interactive, scalable and extensible microbiome data science using QIIME 2. Nat Biotechnol 37:852–857

Buchfink B, Xie C, Huson DH (2014) Fast and sensitive protein alignment using DIAMOND. Nat Methods 12(12):59–60

Buret, A.G., Motta, JP., Allain, T. *et al* (2019). Pathobiont release from dysbiotic gut microbiota biofilms in intestinal inflammatory diseases: a role for iron? J Biomed Sci 26, 1

Caruana EJ, Roman M, Hernández-Sánchez J, Solli P (2015) Longitudinal studies. J Thorac Dis 7(11). https://doi.org/10.3978/j.issn.2072-1439.2015.10.63

Charnock C, Tunsjø H, Hjeltnes B (2017) To really know them you have to grow them. Bioingenioren 8:18–23

ChecinskaSielaff A, Urbaniak C, Mohan GBM, Stepanov VG, Tran Q, Wood JM, Minich J, McDonald D, Mayer T, Knight R, Karouia F, Fox GE, Venkateswaran K (2019) Characterization of the total and viable bacterial and fungal communities associated with the International Space Station surfaces. Microbiome 7(1):50

Chengsong Z, Jianming Y (2009) Nonmetric multidimensional scaling corrects for population structure in association mapping with different sample types. Genetics 182:875

Chimera Detection Commands. (n.d.). https://mothur.org/wiki/chimera_detection_commands/. Accessed 3 Jan 2022

Callahan BJ, McMurdie PJ, Rosen MJ, Han AW, Johnson AJA, Holmes SP (2016) DADA2: high-resolution sample inference from Illumina amplicon data. Nat Methods 13(7):581–583

Camacho C, Coulouris G, Avagyan V, Ma N, Papadopoulos J, Bealer K, Madden TL (2009) BLAST+: architecture and applications. BMC Bioinformatics 10:1–9

Caporaso JG, Kuczynski J, Stombaugh J, Bittinger K, Bushman FD, Costello EK, Fierer N, P⍰a AG, Goodrich JK, Gordon JI, Huttley GA, Kelley ST, Knights D, Koenig JE, Ley RE, Lozupone CA, McDonald D, Muegge BD, Pirrung M, Reeder J, Sevinsky JR, Turnbaugh PJ, Walters WA, Widmann J, Yatsunenko T, Zaneveld J, Knight R (2010) QIIME allows analysis of high-throughput community sequencing data. Nat Methods 7:335–336

DeSantis TZ, Hugenholtz P, Larsen N, Rojas M, Brodie EL, Keller K, Huber T, Dalevi D, Hu P, Andersen GL (2006) Greengenes, a chimera-checked 16S rRNA gene database and workbench compatible with ARB. Appl Environ Microbiol 72:5069–5072

Dixon P (2003) VEGAN, a package of R functions for community ecology. J Veg Sci 14:927–930

Dobell C (1932) Antony Van Leeuwenhoek and his little animals. Harcourt Brace & Company, New York

Douglas GM, Maffei VJ, Zaneveld JR, Yurgel SN, Brown JR, Taylor CM, Huttenhower C, Langille MGI (2020) PICRUSt2 for prediction of metagenome functions. Nat Biotechnol 38(6):685–688

Dunne WM (2002) Bacterial adhesion: seen any good biofilms lately? Clin Microbiol Rev 15(2): 155–166

Eren AM, Maignien L, Sul WJ, Murphy LG, Grim SL, Morrison HG, Sogin ML (2013) Oligotyping: differentiating between closely related microbial taxa using 16S rRNA gene data. Methods Ecol Evol 4:1111–1119

Ernst F, Shetty S, Borman T, Lahti L (2021). mia: Microbiome analysis. R package version 1.2.3, https://github.com/microbiome/mia
FASTX. n.d. FASTX-Toolkit. http://hannonlab.cshl.edu/fastx_toolkit/. Accessed 3 Jan 2022
Flemming HC, Wuertz S (2019) Bacteria and archaea on Earth and their abundance in biofilms. Nat Rev Microbiol 17:247–260
GitHub. (n.d.-a). GitHub - najoshi/sickle: windowed Adaptive Trimming for fastq files using quality. https://github.com/najoshi/sickle. Accessed 3 Jan 2022
GitHub. n.d.-b GitHub – lh3/seqtk: toolkit for processing sequences in FASTA/Q formats. https://github.com/lh3/seqtk. Accessed 3 Jan 2022
GitHub. n.d.-c GitHub – biobakery/kneaddata: quality control tool on metagenomic and metatranscriptomic sequencing data, especially data from microbiome experiments. https://github.com/biobakery/kneaddata. Accessed 3 Jan 2022
GitHub (n.d.-d). GitHub - jbisanz/qiime2R: Import qiime2 artifacts to R. https://github.com/jbisanz/qiime2R. Accessed 4 Jan 2022
Goodrich JK, Di Rienzi SC, Poole AC, Koren O, Walters WA, Caporaso JG, Knight R, Ley RE (2014) Conducting a microbiome study. Cell 158(2):250–262
Gower JC (2015) Principal coordinates analysis. Wiley Stats Ref Stat Ref Online:1–7
He B, Zhu R, Yang H, Lu Q, Wang W, Song L, Sun X, Zhang G, Li S, Yang J, Tian G, Bing P, Lang J (2020) Assessing the impact of data preprocessing on analyzing next generation sequencing data. Front Bioeng Biotechnol 8:817. https://doi.org/10.3389/fbioe.2020.00817
Hughes JB, Hellmann JJ, Ricketts TH, Bohannan BJM (2001) Counting the uncountable: statistical approaches to estimating microbial diversity. Appl Environ Microbiol 67:4399
Huson DH, Beier S, Flade I, Górska A, El-Hadidi M, Mitra S, Ruscheweyh H-J, Tappu R (2016) MEGAN community edition - interactive exploration and analysis of large-scale microbiome sequencing data. PLoS Comput Biol 12:e1004957
Huttenhower C, Gevers D, Knight R, Abubucker S, Badger JH, Chinwalla AT et al (2012) Structure, function and diversity of the healthy human microbiome. Nature 486:207–214
Hyatt D, Chen GL, LoCascio PF, Land ML, Larimer FW, Hauser LJ (2010) Prodigal: prokaryotic gene recognition and translation initiation site identification. BMC Bioinformatics 11:1–11
Imelfort M, Parks D, Woodcroft BJ, Dennis P, Hugenholtz P, Tyson GW (2014) GroopM: an automated tool for the recovery of population genomes from related metagenomes. PeerJ 2014
Jenkins SV, Vang KB, Gies A, Griffin RJ, Jun S-R, Nookaew I, Dings RPM (2018) Sample storage conditions induce post-collection biases in microbiome profiles. BMC Microbiol 18(1):227
Jolliffe I (2011) Principal component analysis. Int Encycl Stat Sci:1094–1096
Kadam P, Bhalerao S (2010) Sample size calculation. Int J Ayurveda Res 1(1):55–57
Kanehisa M, Furumichi M, Sato Y, Ishiguro-Watanabe M, Tanabe M (2021) KEGG: integrating viruses and cellular organisms. Nucleic Acids Res 49:D545–D551
Kang DD, Li F, Kirton E, Thomas A, Egan R, An H, Wang Z (2019) MetaBAT 2: an adaptive binning algorithm for robust and efficient genome reconstruction from metagenome assemblies. PeerJ 2019
Kim B-R, Shin J, Guevarra RB, Lee JH, Kim DW, Seol K-H, Lee J-H, Kim HB, Isaacson RE (2017) Deciphering diversity indices for a better understanding of microbial communities. J Microbiol Biotechnol 27:2089–2093
Langille MGI, Zaneveld J, Caporaso JG, McDonald D, Knights D, Reyes JA, Clemente JC, Burkepile DE, Vega Thurber RL, Knight R, Beiko RG, Huttenhower C (2013) Predictive functional profiling of microbial communities using 16S rRNA marker gene sequences. Nat Biotechnol 319(31):814–821
Langmead B, Salzberg SL (2012) Fast gapped-read alignment with Bowtie 2. Nat Methods 9:357
Leo Lahti, Sudarshan Shetty et al. (2017). Tools for microbiome analysis in R. Version. URL: http://microbiome.github.com/microbiome
Lapierre N, Alser M, Eskin E, Koslicki D, Mangul S (2020) Metalign: efficient alignment-based metagenomic profiling via containment min hash. Genome Biol 21:1–15

Laukens D, Brinkman BM, Raes J, De Vos M, Vandenabeele P (2016) Heterogeneity of the gut microbiome in mice: guidelines for optimizing experimental design. FEMS Microbiol Rev 40(1):117–132
Li H, Durbin R (2009) Fast and accurate short read alignment with Burrows–Wheeler transform. Bioinformatics 25:1754–1760
Liu YX, Qin Y, Chen T, Lu M, Qian X, Guo X, Bai Y (2021) A practical guide to amplicon and metagenomic analysis of microbiome data. Protein Cell 12:315–330
Liu C-M, Luo R, Lam T-W (2014) GPU-accelerated BWT construction for large collection of short reads. Bioinformatics 31:1674–1676
Lozupone C, Knight R (2005) unifrac: a new phylogenetic method for comparing microbial communities. Appl Environ Microbiol 71:8228
Lu Y, Hugenholtz P, Batstone DJ (2015) Evaluating DNA extraction methods for community profiling of pig hindgut microbial community. PLoS One 10(11):e0142720
Luo R et al SOAPdenovo2: an empirically improved memory-efficient short-read de novo assembler. Gigascience 1:18
Luz Calle M (2019) Statistical analysis of metagenomics data. Genomics Inform 17
van der Maaten L, Hinton G (2008) Visualizing data using t-SNE. J Mach Learn Res 9:2579–2605
Malla MA, Dubey A, Kumar A, Yadav S, Hashem A, Allah EF (2019) Exploring the human microbiome: the potential future role of next-generation sequencing in disease diagnosis and treatment. Front Immunol 9
Martin M (2011) Cutadapt removes adapter sequences from high-throughput sequencing reads. EMBnet J 17:10
Martin I, Djuardi Y, Sartono E, Rosa BA, Supali T, Mitreva M, Houwing-Duistermaat JJ, Yazdanbakhsh M (2018) Dynamic changes in human-gut microbiome in relation to a placebo-controlled anthelminthic trial in Indonesia. PLoS Negl Trop Dis 12(8):e0006620
McInnes L, Healy J, Melville J (2018) UMAP: Uniform Manifold Approximation and Projection for Dimension Reduction
McMurdie PJ, Holmes S (2013) phyloseq: an R Package for Reproducible Interactive Analysis and Graphics of Microbiome Census Data. PLoS One 8:e61217
Mothur. (n.d.). make.contigs. https://mothur.org/wiki/make.contigs/. Accessed 3 Jan 2022
Müller A, Hundt C, Hildebrandt A, Hankeln T, Schmidt B (2017) MetaCache: context-aware classification of metagenomic reads using minhashing. Bioinformatics 33:3740–3748
Nurk S, Meleshko D, Korobeynikov A, Pevzner PA (2017) MetaSPAdes: a new versatile metagenomic assembler. Genome Res 27:824–834
Parks DH, Imelfort M, Skennerton CT, Hugenholtz P, Tyson GW (2015) CheckM: assessing the quality of microbial genomes recovered from isolates, single cells, and metagenomes. Genome Res 25:1043
Peng Y, Leung HCM, Yiu SM, Chin FYL (2012) IDBA-UD: a de novo assembler for single-cell and metagenomic sequencing data with highly uneven depth. Bioinformatics 28:1420–1428
QIIME. (n.d.-a). demux — QIIME 2 2021.11.0 documentation. https://docs.qiime2.org/2021.11/plugins/available/demux/?highlight=demux. Accessed 3 Jan 2022
QIIME. (n.d.-b). parallel_identify_chimeric_seqs.py – Parallel chimera detection — Homepage. http://qiime.org/scripts/parallel_identify_chimeric_seqs.html. Accessed 3 Jan 2022
QIIME. (n.d.-c). identify_chimeric_seqs.py – Identify chimeric sequences in input FASTA file — Homepage. http://qiime.org/scripts/identify_chimeric_seqs.html. Accessed 3 Jan 2022
Qiita. (n.d.) https://qiita.ucsd.edu/. Accessed 3 Jan 2022
Qin J, Li R, Raes J, Arumugam M, Burgdorf KS, Manichanh C et al (2010) A human gut microbial gene catalogue established by metagenomic sequencing. Nature 464:59–65
Quast C, Pruesse E, Yilmaz P, Gerken J, Schweer T, Yarza P, Peplies J, Glöckner FO (2013) The SILVA ribosomal RNA gene database project: improved data processing and web-based tools. Nucleic Acids Res 41:D590–D596
Rautava S, Luoto R, Salminen S, Isolauri E (2012) Microbial contact during pregnancy, intestinal colonization and human disease. Nat Rev Gastroenterol Hepatol 9:565–576

Romero R, Schaudinn C, Kusanovic JP, Gorur A, Gotsch F, Webster P et al (2008) Detection of a microbial biofilm in intraamniotic infection. Am J Obstet Gynecol 198:135e1–135e5

Schloss PD (2020) Reintroducing mothur: 10 years later. Appl Environ Microbiol 86:e02343–e02319

Schloss PD, Handelsman J (2005) Introducing DOTUR, a computer program for defining operational taxonomic units and estimating species richness. Appl Environ Microbiol 71:1501

Schloss PD, Handelsman J (2006) Introducing SONS, a tool for operational taxonomic unit-based comparisons of microbial community memberships and structures. Appl Environ Microbiol 72: 6773

Schloss PD, Westcott SL, Ryabin T, Hall JR, Hartmann M, Hollister EB, Lesniewski RA, Oakley BB, Parks DH, Robinson CJ, Sahl JW, Stres B, Thallinger GG, Van Horn DJ, Weber CF (2009) Introducing mothur: Open-source, platform-independent, community-supported software for describing and comparing microbial communities. Appl Environ Microbiol 75:7537–7541

Schoch CL, Seifert KA, Huhndorf S, Robert V, Spouge JL, Levesque CA, Chen W, Consortium FB, List FBCA, Bolchacova E, Voigt K, Crous PW, Miller AN, Wingfield MJ, Aime MC, An K-D, Bai F-Y, Barreto RW, Begerow D, Bergeron M-J, Blackwell M, Boekhout T, Bogale M, Boonyuen N, Burgaz AR, Buyck B, Cai L, Cai Q, Cardinali G, Chaverri P, Coppins BJ, Crespo A, Cubas P, Cummings C, Damm U, de Beer ZW, de Hoog GS, Del-Prado R, Dentinger B, Diéguez-Uribeondo J, Divakar PK, Douglas B, Dueñas M, Duong TA, Eberhardt U, Edwards JE, Elshahed MS, Fliegerova K, Furtado M, García MA, Ge Z-W, Griffith GW, Griffiths K, Groenewald JZ, Groenewald M, Grube M, Gryzenhout M, Guo L-D, Hagen F, Hambleton S, Hamelin RC, Hansen K, Harrold P, Heller G, Herrera C, Hirayama K, Hirooka Y, Ho H-M, Hoffmann K, Hofstetter V, Högnabba F, Hollingsworth PM, Hong S-B, Hosaka K, Houbraken J, Hughes K, Huhtinen S, Hyde KD, James T, Johnson EM, Johnson JE, Johnston PR, Jones EBG, Kelly LJ, Kirk PM, Knapp DG, Kõljalg U, Kovács GM, Kurtzman CP, Landvik S, Leavitt SD, Liggenstoffer AS, Liimatainen K, Lombard L, Luangsa-ard JJ, Lumbsch HT, Maganti H, Maharachchikumbura SSN, Martin MP, May TW, McTaggart AR, Methven AS, Meyer W, Moncalvo J-M, Mongkolsamrit S, Nagy LG, Nilsson RH, Niskanen T, Nyilasi I, Okada G, Okane I, Olariaga I, Otte J, Papp T, Park D, Petkovits T, Pino-Bodas R, Quaedvlieg W, Raja HA, Redecker D, Rintoul TL, Ruibal C, Sarmiento-Ramírez JM, Schmitt I, Schüßler A, Shearer C, Sotome K, Stefani FOP, Stenroos S, Stielow B, Stockinger H, Suetrong S, Suh S-O, Sung G-H, Suzuki M, Tanaka K, Tedersoo L, Telleria MT, Tretter E, Untereiner WA, Urbina H, Vágvölgyi C, Vialle A, Vu TD, Walther G, Wang Q-M, Wang Y, Weir BS, Weiß M, White MM, Xu J, Yahr R, Yang ZL, Yurkov A, Zamora J-C, Zhang N, Zhuang W-Y, Schindel D (2012) Nuclear ribosomal internal transcribed spacer (ITS) region as a universal DNA barcode marker for Fungi. Proc Natl Acad Sci U S A 109(16): 6241–6246

Schubert M, Lindgreen S, Orlando L (2016) AdapterRemoval v2: rapid adapter trimming, identification, and read merging. BMC Res Notes 9

Seemann T (2014) Prokka: rapid prokaryotic genome annotation. Bioinformatics 30:2068–2069

Sender R, Fuchs S, Milo R (2016) Revised estimates for the number of human and bacteria cells in the body. PLoS Biol 14:e1002533

Sommer F, Bäckhed F (2013) The gut microbiota-masters of host development and physiology. Nat Rev Microbiol 11:227–238

split_libraries_fastq.py. (n.d.) split_libraries_fastq.py – this script performs demultiplexing of Fastq sequence data where barcodes and sequences are contained in two separate fastq files (common on Illumina runs). — Homepage. http://qiime.org/scripts/split_libraries_fastq.html. Accessed 3 Jan 2022

Tang Q, Jin G, Wang G, Liu T, Liu X, Wang B, Cao H (2020) Current sampling methods for gut microbiota: a call for more precise devices. Front Cell Infect Microbiol 10

Von Meijenfeldt FAB, Arkhipova K, Cambuy DD, Coutinho FH, Dutilh BE (2019) Robust taxonomic classification of uncharted microbial sequences and bins with CAT and BAT. Genome Biol 20:1–14

Weging S, Gogol-Döring A, Grosse I (2021) Taxonomic analysis of metagenomic data with kASA. Nucleic Acids Res 49:e68–e68

Weisburg WG, Barns SM, Pelletier DA, Lane DJ (1991) 16S ribosomal DNA amplification for phylogenetic study. J Bacteriol 173(2):697–703

Wood DE, Lu J, Langmead B (2019) Improved metagenomic analysis with Kraken 2. Genome Biol 20:1–13

Wood DE, Salzberg SL (2014) Kraken: Ultrafast metagenomic sequence classification using exact alignments. Genome Biol 15:1–12

Wu YW, Simmons BA, Singer SW (2016) MaxBin 2.0: an automated binning algorithm to recover genomes from multiple metagenomic datasets. Bioinformatics 32:605–607

Yue Y, Huang H, Qi Z, Dou HM, Liu XY, Han TF, Chen Y, Song XJ, Zhang YH, Tu J (2020) Evaluating metagenomics tools for genome binning with real metagenomic datasets and CAMI datasets. BMC Bioinformatics 21:1–15

Zerbino DR (2010) Using the Velvet de novo assembler for short-read sequencing technologies. Curr Protoc Bioinformatics CHAPTER:Unit

Chapter 16
Germ-free Mice Technology: Opportunity for Future Research

Ashish Jain and Anand Maurya

Abstract The most popular approach to measure key functions of any living entity is to remove it and then study the consequences of its removal. Microorganisms influence their host in several manners and their role can be studied by eliminating them from their host and observe the host's response, in their absence. Numerous studies have justified the vital role of microbiota in human health and disease development. Germ-free (GF) animal models are useful tools to improve our understanding of the host–microbiota relationship in vivo. Although different animal models, lacking microbiota (partially or completely) have been extensively used in research but germ-free (GF) mice are the most widely used rodent model in human research due to its close proximity to humans. In modern research, GF technology is one of the most attractive and informative tools for getting insights into host's microbial community. Each body part harbors unique microorganisms with unique functions. Because of the advancement of microbial characterization techniques, the human microbiota community is expanding day by day. GF mice model can efficiently reveal the role of these valuable partners of humans. In spite of its high cost and obligation of skilled experts, GF research is a hot field for investigators and has a huge possibility for future applications. The present book chapter is a summary of the basics of GF technology and its main applications with future prospects.

Keywords Germ-free · Gnotobiotics · Microbiota · Microbiome · Mice

16.1 Introduction

All multicellular organisms including humans live in close association with the microbial communities (Mendes and Raaijmakers 2015). All the microorganisms that live in or on the human body are collectively known as microbiota (Human Microbiome Project Consortium 2012). The members of this microbial community

A. Jain (✉) · A. Maurya
Department of Microbiology, Smt. CHM College, Thane, India

A. K. Tripathi, M. Kotak (eds.), *Gut Microbiome in Neurological Health and Disorders*, Nutritional Neurosciences,
https://doi.org/10.1007/978-981-19-4530-4_16

have been shown to interact with each other and also with their host in different manners. It is a huge task to define the function of each individual microorganism of this complex community and their relation with the host. Germ-free animal research has been done extensively on three different animal models: Germ-free (GF) animals, gnotobiotic (GN) animals, and specific pathogen-free (SPF) animals. Germ-free or axenic animals are completely devoid of any kind of living microorganisms. Gnotobiotic animals are initially germ-free and subsequently inoculated with either one strain (monoxenics) or two different strains (dixenics) of microorganisms and so on. These animals thus have a defined and known microbiota. Specific pathogen-free (SPF) animals are those from which a defined set of microorganisms, usually pathogenic organisms are excluded (Festing and Blackmore 1971). Since the use of humans in such studies is not ethically possible, use of an animal model having several similarities with humans is the only left option. Mouse is the most common germ-free model in human studies (Luczynski et al. 2016a). However, rats, guinea pigs, chickens, piglets, and calves have also been used (Luczynski et al. 2016a, b). Experimental germ-free (GF) mice models are valuable tools for establishing the impact of microbiota on host metabolism, physiology and to explore interactions between microbiota and with their host (Fiebiger et al. 2016; Bhattarai et al. 2017). Recent studies performed on germ-free mice have proved the specific role of specific microbial communities for their host (Braniste et al. 2014; Jourová et al. 2017; Kaden-Volynets et al. 2019). Although germ-free mice technology has broadened our knowledge of microbe–microbe and host–microbe interactions but, due to the huge cost and expertise associated with the GF facilities, this field has not been explored as per the expectations and possibilities. This book chapter covers the basics of GF mice technology, its applications and future prospects.

Axenic: free of all detectable microorganisms
Monoxenic: a culture in which one organism is grown with only one other organism
Dixenic: a mixed culture of one organism together with two other organisms
Germ Free (GF): historical term same as axenic but continues to remain the more popular than axenic
Gnotobiotic (GN) animal: an animal with known and defined microorganisms
Pathogen-free (PF) animal: an animal free of all known pathogens
Specific pathogen-free (SPF) animal: an animal from which a defined set of microorganisms, usually pathogenic are excluded
Conventional (CV) animal: an animal maintained under accepted husbandry practices
Altered Schaedler flora (ASF): a model community of eight cultivable microorganisms derived from mice and used for establishing stable GI colonization in the GF mouse

16.2 Germ-Free Mice Technology

16.2.1 History

Louis Pasteur (1885) conceptualized the idea of a germ-free animal more than a century ago with the remark that bacteria-free existence is impossible (Pasteur 1885). Ten years later in 1895, Nuttall and Thierfelder produced the first GF guinea pigs, which survived for two weeks (Nuttall and Thierfelder 1895). Due to the technology developed by James Arthur Reyniers and Philip C. Trexler, the hypothesis of germ-free life came into reality in the 1950s (Reyniers 1957). By the late 1950s, researchers successfully developed GF mice, rats, guinea pigs, and chicks inside sterile stainless steel and plastic housings (Reyniers 1959a). The first GF mice were successfully developed by Pleasants in 1959 (Pleasants 1959).

16.2.2 GF Technology

Today's methodology for keeping GF animals has not changed much since 1959. To start a germ-free colony, pups must be delivered from the mother's womb through a careful cesarean section to protect them from exposer of microorganisms that inhabit on the mother's vagina and skin (Gustafsson 1959a; Reyniers 1959b).Then, the newly born pups are introduced to the GF foster mother and raised in an aseptic isolator and only exposed to food, water, and other equipment that has also been sterilized. GF animals can also be produced via embryo transfer, in an isolator by the implantation of cleaned embryos into GF female in well-controlled conditions. A recipient female normally delivers and caress the pups assuming them her own offsprings, hence enhancing the survival rate of pups. These mice regularly monitored in order to guarantee GF status by analyzing the presence of any kind of microorganism in their feces using cultural and sequencing techniques (Smith et al. 2007). Once GF animals are produced next lineage can be generated by crossing GF individuals (Gustafsson 1959a; Reyniers 1959b). Then GF mice colonies can be shipped in a sterile container for different purposes including GF research. Isolators maintain a sterile environment for GF animals. A typical isolator has an air supply, air inlet and outlet, transfer port, and arm-length gloves, as well as a special tank filled with disinfectant and used for the transfer of mice in and out. Bedding, food, water, and equipment, including cages, must first be sterilized before putting them into the isolator through the sterile lock. Sterilization of entire steel isolators is accomplished by autoclaving the whole isolator, as well as with portable vacuum and steam equipment. Plastic isolators are sterilized by steam accomplished with germicidal vapor (2% peracetic acid and chlorine dioxide). Air sterilization is ensured upon entry and exhaust by mechanical air filtration under positive pressure.

Autoclave jars are used for transferring of animals in and out of the isolator. Maintenance of the GF status of rodents during the execution of the entire experiment is technically challenging. The probability of their contamination is always high. The experimentation cost of the GF facility is extremely high since multiple mouse strains and multiple inoculation groups are housed in separate isolators. Recently the use of positive-pressure isocages has been increased in short duration experiments since they offer low cost and space effective (Hecht et al. 2014). Future research is needed to optimize these isocages to apply them in long-term experiments.

16.2.3 Customized Flora and Control Group for Experiments

In 1965, Schaedler and Dubos characterized the bacterial population of the gastrointestinal (GI) tract of conventional mice (Schaedler et al. 1965). One of the most pronounced phenotypes observed in almost all GF rodents is unusually enlarged cecum that is of normal length in conventional rodents (Wostmann and Bruckner-Kardoss 1959). Surprisingly a reduction in cecum size was observed upon colonization of normal gut microbiota. On the basis of these findings, Schaedler colonized GF animals with a mixed bacterial population of Bacteroides, lactobacilli, an anaerobic Streptococcus, and a slow lactose-fermenting coliform (Schaedler et al. 1965). This flora is subsequently known as the "Schaedler flora" and has been used globally as an essential tool for the standardization of experimental animals' microbiota. In 2015, extremely oxygen-sensitive (EOS) bacteria were included in Schaedler's flora. This altered flora is now known as Altered Schaedler Flora (ASF) (Orcutt et al. 1987). Recently a synthetic bacterial community is created for experimental GF mice known as "Oligo-Mouse-Microbiota" (OMM^{12}) which consists of 12 sequenced and easily available bacterial strains isolated from mice (Brugiroux et al. 2016; Lagkouvardos et al. 2016). The specific pathogen-free (SPF) mice which are free from particular pathogens are treated as a control group for GF mice research (Smith et al. 2007). SPF mice are generally exposed to the defined colonization of modified Schaedler flora or other customized flora (Wymore Brand et al. 2015; Brugiroux et al. 2016). Broader categories of mice models are summarized in Fig. 16.1.

16.3 Why Mice Model?

Although many GF animals have been used in investigations, GF mice is the most acceptable GF model in human studies for long period (Haldane 1928). Over 95% of animal studies have been conducted on the *Mus musculus* model because of almost

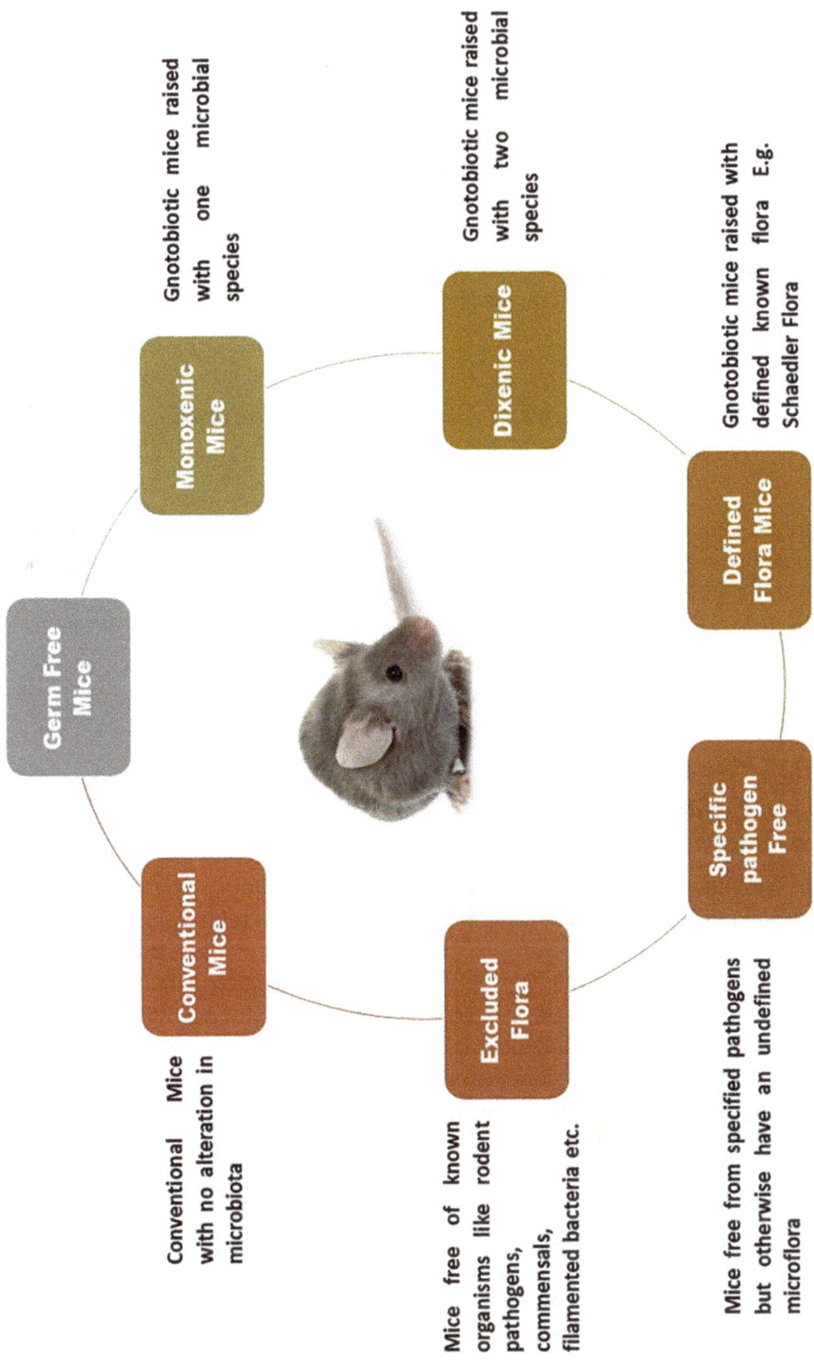

Fig. 16.1 Different types of mice models used in gnotobiotic technology

99% genetic homology with the human genome, availability of good genetic/molecular technologies for mice research, and replicability of many human conditions in mice (Gregory et al. 2002; Mouse Genome Sequencing Consortium et al. 2002). Investigators prefer mice model for various other reasons including their small size, easy to maintain, and adaptability in altered conditions. Because of quick reproduction and a short life span of two to three years, many generations can be observed in a short period of time. They are relatively inexpensive and can be bought in large quantities from commercial producers. Their mild-tempered and docile nature makes them easy for researchers to handle (Canales and Walz 2019). Due to constrain of working space inside the GF isolator and extremely higher maintenance cost mice offer more experimental units in a relatively smaller area in a cost-effective manner.

16.3.1 Differences Between Germ-free and Conventionally Raised Animals

Germ-free animals exhibited several physiological and functional alterations, not associated with conventionally raised animals. Some crucial differences are summed up in Table 16.1. These dissimilarities have created the research-ground for gnotobiotics, to assess the effect of microbiota in postnatal development and metabolism of host.

16.3.2 GF mice Technology: Applications and Future Guideline

Germ-free animal models are essential tools of investigators to explore the complexity and functions of host's microbiota. GF mice provide researchers with a better way to get insights of host–microbe and microbe–microbe interactions. GF mice have been widely used in the field of nutrition, metabolism, drug response, neuroscience, immune response, cancer biology, and infectious diseases and some of the important fields are also highlighted in Fig. 16.2. Germ-free mice technology has revealed the crucial role of commensal microbiota in normal aging, and normal functioning and development of immune system, GI system, and nervous system (Grenham et al. 2011).

Table 16.1 Key differences between GF animals and conventionally raised animals

	GF Animal (Comparison with conventional animals)
Gastrointestinal tract physiology	
Weight of the small intestine	Lighter (Gordon and Bruckner-Kardoss 1961)
Mucosa of the small intestine	Thinner (Gordon and Bruckner-Kardoss 1961)
Mucosal surface area, Lamina propria	Reduced by approximately 30% (Gordon and Bruckner-Kardoss 1961)
Digestion and absorption	More efficient (Phillips and Smith 1959; Heneghan 1963)
Passage time through the small intestine	Increased (Abrams and Bishop 1967)
Cecal	Excessively enlarged (Wostmann and Bruckner-Kardoss 1959)
Mitosis indexes of epithelial cells	Lower
Renewal rates of intestinal epithelium	Reduced (Abrams et al. 1963)
Cells, tissues, and organs of immune system	
Expression of certain TLRs	Decreased or absent (Shanahan 2002; Grenham et al. 2011)
IgA secretion	Decreased (Abrams and Bishop 1961; Wostmann et al. 1970)
Peyer's patches, lymphoid follicles in the intestine	Fewer and smaller (Abrams and Bishop 1961; Wostmann et al. 1970)
Lymphoid tissue and lymph nodes	Undeveloped and smaller (Abrams and Bishop 1961)
γ-globulin-bearing plasma cells in mesenteric lymph nodes	Absent (Hobby et al. 1968)
Antibody-producing cells of lymph-node	Reduced (one-twelfth) (Olson and Wostmann 1966)
Antibody-producing cells after Ag challenge	Increased (Olson and Wostmann 1966)
Total white blood cell count	Lower (Reyniers et al. 1960)
B- and γ-globulins in the serum	Reduced (Gustafsson and Laurell 1958)
Plasma cells synthesizing IgA	Less (10%) (Crabbé et al. 1968)
Lysozyme in saliva	Absent (Makulu and Wagner 1967)
Thymus	Smaller in size (Wilson et al. 1965)
Nutrition, digestion, and metabolism	
Nutrient requirements	Consume more food (Wostmann et al. 1983)
Diet-induced obesity	Not observed (Bäckhed et al. 2007)
Water intake	Higher (Coates 1973)
Lipid: requirement for essential fatty acids	Lower (Coates 1973)
Cholesterol absorption	Absorb up to 50% more
Protein:	
Fecal nitrogen	More excreted (Levenson and Tenivant 1963)
Urinary nitrogen	Less excretion (Reddy et al. 1969)
Starving conditions	Lesser survival rate (Loesche 1969)
Minerals: urinary calcium excretion	Higher (5X higher) (Gustafsson and Norman 1962)

(continued)

Table 16.1 (continued)

	GF Animal (Comparison with conventional animals)
Vitamin B & K in diet	Required (Gustafsson 1959b; Sumi et al. 1977)
Trypsin and chymotrypsin in feces	Higher (Borgstrom et al. 1959)
Serum cholesterol levels	Higher (Danielsson and Gustafsson 1959)
Cardiovascular system	
Heart weight, total blood volume, and cardiac output	Reduced (Gordon et al. 1963)
RBC count, Hematocrit values	Higher (Gordon et al. 1963)
Age	Live significantly longer (Gordon et al. 1966; Tazume et al. 1991)

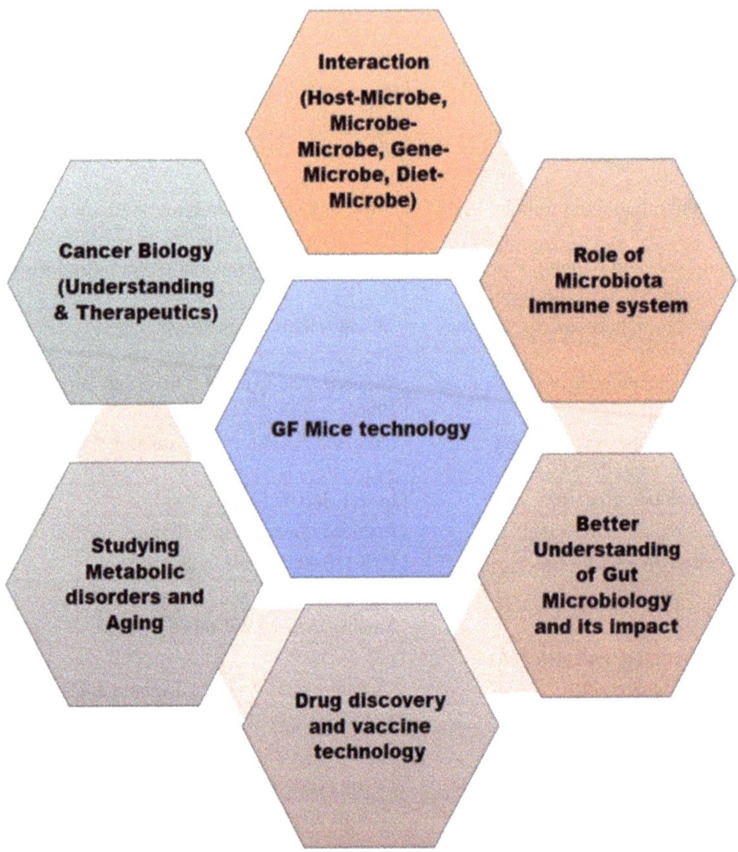

Fig. 16.2 Key applications of germ-free mice technology

16.4 Metabolic Disorders

Our generation is facing many metabolic disorders including obesity, heart disease, stroke, type 2 diabetes, hyperglycemia and hyperlipidemia, phenyl-ketonuria (PKU), and a number of other hepatic disorders. Studies performed on GF mice have established the very fact that the presence or absence of specific gut microbiota is correlated with certain metabolic disorders (Karlsson et al. 2013). It is found that the composition of gut microbiota is unstable and far sensitive for alteration but surprisingly found relatively stable during obesity, which suggests the possible role of gut microbiota in obesity. Jeff Gordon et al. showed that the transfer of gut microbiota of obese humans or mice to GF mice with no change to mouse diet, resulted in weight gain relative to GF mice that had received microbiota transplants from lean donors (Turnbaugh et al. 2006; Ridaura et al. 2013). Transfer of microbiota from third-trimester pregnant mothers to GF mice promoted low-grade inflammation, increased adiposity, and insulin resistance relative to GF mice receiving microbiota from first trimester pregnancies (Koren et al. 2012). As gut microbiota synthesizes additional essential nutrients, its absence affects the process of absorption and digestion (Sekirov et al. 2010; Grenham et al. 2011). GF mice have reduced production of short-chain fatty acids, which are beneficial to host metabolism and are produced when dietary fiber is fermented by gut bacteria (Høverstad and Midtvedt 1986; denBesten et al. 2013). GF rats have deficient thiamine absorption: when these animals are fed radio-labelled thiamine, large quantities of the nutrient are found within the feces but little is found within the tissue (Wostmann et al. 1962). A recent study proposed that fecal microbiota transplantation is the recent technology used for treating various neurological diseases (Tripathi et al. 2022). This suggests that the gut microbiome plays an important role in influencing metabolism and adiposity. Phenylketonuria (PKU) is a genetic disorder related to an inability to metabolize phenylalanine (Phe), which may end in neurotoxicity. Recently a genetically engineered *Escherichia coli* strain administration to PKU mouse model showed significantly reduced blood phenylalanine concentration independent of dietary protein intake (Isabella et al. 2018). Collectively, such studies certainly prove that not only the commensal microbiota but genetically modified microorganisms can also be used to target genetic disorders along with altered metabolism. Gut microbiota is the most abundant microbiota and also termed as "neglected endocrine organ" (Clarke et al. 2014). Gut microbiome contains hundred-times more genes than the human genome (Qin et al. 2010). This microbiome comprises an enormous possibility of future investigation because every member of this massive microbial community may have a unique function for its host that can be assessed efficiently with the assistance of GF technology. Once the individual role of every member of microbiota revealed, specific microorganisms can be selectively administered or removed from the host to treat several metabolic disorders. Initial findings with microbiota research might be milestones for future investigations during which GF animals are going to be considered as an indispensable tool for exploring the role of microorganisms to study the complexity of microbiota-gut-brain axis and metabolic disorders.

16.5 Inflammatory Bowel Disease (IBD)

Inflammatory Bowel Disease (IBD) is a set of chronic inflammatory conditions of the gastrointestinal tract (Belkaid and Hand 2014). IBD features a complex etiology and is influenced by the genetic factors, host immune system, and external factors like the microbiota (Maloy and Powrie 2011). IL-17-producing Th17 cells are correlated with IBD while RORγt$^+$ Treg cells adversely affect IBD by maintaining homeostasis at the mucosal barrier. Studies performed on mice showed the elevated number of IL-17-producing Th17 cells upon colonization of mice with anaerobic, uncultivatable segmented filamentous bacteria while the introduction of commensals showed an adverse effect (Nakanishi and Tamai 2015). In another study transfer of microbiotas from IBD donors into germ-free mice increased numbers of intestinal Th17 cells and Th2 cells and decreased numbers of RORγt$^+$ Treg cells while microbiota from healthy donors exhibited an adverse effect (Britton et al. 2019). Administration of fecal microbiota transplants from patients with IBS to germ-free mice induced alterations in GI motility, also as hypersensitivity to colonic distension (Crouzet et al. 2013; De Palma et al. 2014). The absence of IBD in germ-free animals is that the classical evidence that microorganisms are crucial for the development of IBD. *We can't neglect the colonic microbiota and a heavier metagenome and its functions in several physiological conditions including IBD. Future attempts should be focused on manipulating the amount and composition of commensal or altered microbiota to alleviate the severity of IBS either by oral administration of probiotic formulations, fecal microbiota transplants from healthy donors, and/or diet modifications. Germ-free models might be a milestone in the collaborative attempts to cure IBD.*

16.5.1 Host Immune Response

The role of microbiota in the development and regulation of the immune system has been extensively studied with the assistance of GF mice. Factually the exposure to microbes early in life is essential for the proper development and performance of the immune system (Blümer et al. 2005; Douwes et al. 2008; Kaplan et al. 2011). The host's immune system and gut microbiota have a mutualistic relationship. The microbiota helps in shaping our immune system, and the later shape the composition of host microbiota (Nicholson et al. 2012). These hypotheses further get strength by the fact that about 80% of the host's immune cells are located in or around the gut (Abbas et al. 2017). Commensal microbiota of host is crucial for proper intestinal immune response, protection from pathogens, and suppression of detrimental inflammatory reactions. GF mice show many abnormal features generally absent in conventionally raised mice, including the presence of fewer and smaller Peyer's patches, a reduced number of $CD4^{+}T$ cells, low level of IgA producing plasma cells, under-developed gut-associated lymphoid tissues (GALT), fewer intraepithelial lymphocytes as well as reduced production of antimicrobial peptides (Wostmann and Bruckner-Kardoss 1959; Gordon and Bruckner-Kardoss 1961; Shanahan 2002; Round and Mazmanian 2009). Upon administration of normal microbiota, most of those altered structures and immunological functions of GF mice are corrected and restored. These inducible structures normally develop in conventional animals exposed naturally to diverse populations of microorganism, suggesting a complex relationship between the host's immune response and its commensal microbiota. *Microbes produce a variety of known and unidentified metabolites that can modulate the host's metabolic pathways in a complex manner. Class, quantity, and roles of metabolites are influenced by the composition of the host's microbiota. Despite advanced molecular characterization technology most of the commensal microorganisms, their different metabolites and function of every metabolite are yet to be defined. There is an enormous challenge and also an opportunity for the investigators to address this issue and define the functionality of each single metabolite of this complex immune response regulated by the host and its microbiota.*

16.5.2 Vaccine Response

More recently, mice treated with a cocktail of antibiotics, exhibited impaired IgG responses upon systemic Ag ova challenge, the same could be restored after the colonization with a mixed bacterial population (Lamousé-Smith et al. 2011). Pre-antibiotic-treated mice showed enhanced antibody response upon oral administration of Rota-virus (Uchiyama et al. 2014). These findings suggest that gut flora can enhance systemic vaccine responses but can suppress oral vaccine responses. A study on seasonal influenza vaccine showed that after vaccination, in germ-free or antibiotic-treated mice, IgG and IgM antibody responses were significantly impaired (Oh et al. 2014). It is also reported that microbial metabolites can modulate various immune cell types including Mfs, DCs, T cells, and B-cells (Dorrestein et al. 2014). *However, further investigations are desirable to conclude whether these effects occur in all situations or only observed in some special circumstances. Currently, the knowledge of various microbial metabolites and their role in vaccine response is in its infancy. Future research can reveal the potential mechanisms by which the gut microbiome modulate vaccine response in various populations. The study of gut microbiota for a successful vaccination strategy may open a new area for investigators with unlimited possibilities and GF mice may serve as a valuable tool in this cause.*

16.5.3 Host–microbe and Microbe–Microbe Interaction

Germ-free mouse facilitates to introduce microbes individually or sequentially in its different body parts to assess the role of a single bacterium or known consortia of bacteria on host function in vivo (Reigstad et al. 2015). Germ-free mice offer to study microbe–microbe interaction, to better understand how the introduction of a new microbial member affects the whole microbial community and host functions. Germ-free models provide insights into the host processes regulated by the presence and/or composition of the microbiota in health and disease (Reyniers 1959b; Smith et al. 2007). Based on their interaction with the host, members of the microbiota can be classified as beneficial species (Commensals) including probiotic bacteria, like *Bifidobacterium* and benign organisms such as members of the defined "altered Schaedler flora," or pathogenic species, including pathobionts such as *Helicobacter pylori* and opportunistic pathogens (Fanning et al. 2012; Biggs et al. 2017). The development of gnotobiotic animal models provides an opportunity to compare them with conventionally raised animals but also the ability to introduce one or few bacterial species at a time to understand host–microbe interactions in a simplified environment (Williams 2014).

16.5.4 Host–pathogen Interaction

Very soon after the invention of germ-free technology, gnotobiotic animals became a key tool to study the host–microbiota interaction and later used to investigate the host immune responses to pathogens. Initial studies performed on the mono-associated animal to assess the resisting power of host towards infections demonstrated that lack of an intestinal microbiota impairs early innate immunity. Mono-associated animals showed higher sensitivity towards *Listeria monocytogenes* infection while di-associated mice having commensal flora, remained unaffected from the pathogen (Zachar and Savage 1979; Czuprynski and Balish 1981). A similar conclusion concerning the significance of native flora on the host's immunity was made in other studies conducted with *Salmonella typhimurium* and *Vibrio cholera* infections (Nardi et al. 1991; Butterton et al. 1996). A commensal microbiota competes for space, nutrients and mediates the production of antimicrobial metabolites, which subsequently prevent growth and colonization of numerous pathogenic bacteria (Mack et al. 1999, 2003; Srikanth and McCormick 2008). GF mice models in combination with mono-associated and di-associated gnotobionts will be obligatory tools in future investigations, aimed to recognize the pathogenesis and treatment of newly emerging infections.

16.5.5 Reproductive Health

Germ-free mice are considered reproductively inferior to their conventional counterparts. Surprisingly, when germ-free female mice di-associated with *B. distasonis* and *C. perfringens* displayed a normalized estrous cycle, and increase rates of copulation and implantation (Shimizu et al. 1998). The lower reproductive tract of the female mouse is anatomically almost like that of humans (Leppi 1964). During a recent study, GF mice vaginally inoculated with *Prevotella bivia* displayed increased numbers of mucosal activated $CCR5^+$ $CD4^+$ T cells (HIV target cells) within the female genital tract compared to mice inoculated with *Lactobacillus crispatus*. Hence colonization of altered bacteria is often associated with increased HIV risk and other STIs (Gosmann et al. 2017). In future investigations, germ-free models can be exploited to address different aspects of reproductive health correlation with host microbiota.

16.5.6 Cancer Biology

Cancer development in GF rodents can partially be associated with the absence of commensal flora (Pollard and Teah 1963; Walburg Jr 1973; Pollard et al. 1985). Experimental cancer yields were found to be lower in GF rodents when the carcinogens tested required enzymatic metabolic activation (Weisburger et al. 1975). Generally, the oncogenic potential is that the same as in conventional rats, but tumor-related changes are more clearly defined in GF animals (Pollard et al. 1968). GF rodents with either spontaneous or induced tumors have higher numbers of plasma cells but haven't any germinal zones in their lymph nodes (Pollard et al. 1968). Gnotobiotic animals are particularly suitable for testing candidate viral carcinogens, since derivation by hysterectomy and gnotobiotic maintenance has been found to eliminate all known viruses from GF rodents (Luckey 1963; Pleasants 1974). Cycasin from cycad bean flour is carcinogenic for conventional rats because the microbiome present in them converts it into a carcinogen, whereas it doesn't induce tumors in GF rats (Laqueur et al. 1967; Luckey 1968). Spontaneous colon adenomas are twice as prevalent in GF rats (Weisburger et al. 1975). The foremost frequent spontaneous tumors in aged GF rats involve the mammary and pituitary glands (Pittermann and Deerberg 1975).

16.5.7 Aging

GF mice tend to live longer than their conventionally colonized counterpart animals (Reyniers and Sacksteder 1958; Gordon et al. 1966; Tazume et al. 1991). There is growing evidence indicating that gut microbiota influences the aging process. As GF mice are raised in sterile conditions, their longer life span is likely due to the absence of pathological infections. Premature mortality of GF mice is mainly due to infection or by environmental factors. Delayed morbidity in 2- to 3-year-old GF rodents is a common observation, which shows them to be virtually free of the age-related kidney, heart, and lung changes (Pollard and Kajima 1970; Pollard 1971). In humans, microbial diversity and stability decrease with age and are accompanied by a cognitive decline (O'Toole and Claesson 2010; Borre et al. 2014). These findings have prompted the thought that restoring microbial diversity within the elderly could improve general and mental health.

16.5.8 Drug Response and Xenobiotics

It is often assumed that gastrointestinal tract microbiota is probably the first which interacts with ingested xenobiotics. The gut microbiome can activate or deactivate pharmaceuticals and may alter their metabolic consequences. An altered microbiota also influences the outcome of various therapies, this proves the importance of intact microbiota in host immune responses (Pope et al. 2017). In experimental germ-free mice with induced tumors, immune cells poorly respond to immunotherapy that slows cancer growth and prolongs survival. These germ-free rodents hardly exhibited any response toward anticancer drugs like oxaliplatin and cisplatin (Viaud et al. 2015). Clinical use of anticancer drug cyclophosphamide (CTX) on tumor-bearing mice caused the translocation of some bacterial species into mesenteric lymph nodes and the spleen, where they stimulate a Th1 and Th17 immune response. Germ-free mice failed to generate the same response and were found immune to the CTX (Viaud et al. 2013). A study performed in National Cancer Institute (NCI) reported that in germ-free mice having subcutaneous tumors exhibited lower cytokine production and tumor necrosis after CpG-oligonucleotide treatment and deficient production of reactive oxygen species and cytotoxicity upon chemotherapy. This finding advocates the necessity of an intact microbiota for proper response to anticancer therapy (Iida et al. 2013). Recently in two parallel studies, the microbial population of fecal samples of melanoma patients was characterized prior to treatment with the anticancer drugs which block a T cell receptor PD-1. In both studies, certain bacterial species were reported in greater numbers in those patients who responded properly to the drug. When the same microbes were administered into the germ-free mice model, an anti-tumor immune response was observed (Gopalakrishnan et al. 2018; Gong et al. 2019). While some drugs get activated through bacterial metabolism, others can be inactivated due to microbial action. A single bacterium *Eggerthella lenta* inactivates the drug Digoxin, a treatment for heart failure by converting it into inactive form dihydrodigoxin (Lindenbaum et al. 1981). The microbiome also inactivates Parkinson's disease drug L-DOPA, initially by Enterococcus faecalis mediated decarboxylation and later *Eggerthella lenta* A2 mediated dihydroxylation. Treatment with broad-spectrum antibiotics can reverse this activity (Rekdal et al. 2019). Recently Klatt et al. reported that vaginal bacteria *Gardnerella vaginalis* could rapidly metabolize and breakdown the active form of "Tenofovir Microbicide" the drug for HIV treatment and thus results in a high HIV acquisition in those women. These findings highlight the contribution of intact microbiota and its poorly known factors in the therapy of cancer and other diseases (Klatt et al. 2017).

16.5.9 Gastrointestinal System and Enteric Nervous System

It is reported that gastrointestinal (GI) transit time was significantly faster in conventionally raised mice as compared to GF mice (Abrams and Bishop 1967). Subsequently, several investigators have shown the introduction of mouse-derived or human gut-derived bacteria into GF mice alters GI motility and transit time (Husebye et al. 2001; Kashyap et al. 2013). Various studies highlight the utility of GF mice as a model to understand host–microbe interaction and how microbes modulate GI motility and secretions (Husebye et al. 2001; Kashyap et al. 2013; Kaji et al. 2015; Reigstad et al. 2015; Yano et al. 2015). The microbiota was found essential for the postnatal development of the enteric sensory and motor neurons (Luczynski et al. 2016a, b).

The changes that have been reported in central nervous system development in GF mice are reflected during the maturation of the enteric nervous system (ENS) (Collins et al. 2014; Luczynski et al. 2016a). At postnatal day 3, the structure, neurochemical composition, and function of enteric neurons in the jejunum and ileum of GF mice were significantly altered, also in the small intestine, GF mice have decreased overall nerve density (Collins et al. 2014). The ganglia of intrinsic sensory neurons of the ENS are embedded in the gut wall and it has been established that the electrophysiological properties of afterhyperpolarization (AH) neurons are altered in the absence of colonizing bacteria (Forsythe and Kunze 2013; McVey Neufeld et al. 2013, 2015). As mentioned earlier, GF mice have altered intestinal motility and these sensory neurons synapse on enteric motor neurons controlling gut motility, so this may provide a possible explanation for the dysfunction. The AH sensory neurons also synapse, both anatomically and functionally, with vagal nerve endings in the gut and thus could represent a direct neural route whereby the intestinal bacterial status is transmitted to the brain (Powley et al. 2008; Perez-Burgos et al. 2014). All these studies carried out with help of GF mice provide us a crucial link between microbiota and development of the GI system and also how it alters the functioning of ENS when compared with conventional models.

16.6 Future Potentials of Germ-Free Technology

16.6.1 Technological Aspects

Introduction of automation, sterile room facilities and robotics could prove to be boon in the technology of in vivo mice models. Major challenges for this technology are scarcity of proficient technicians, and the time and space required to accommodate the bulky isolators along with huge cost associated with maintaining GF facilities is the major challenge for this technology (Mallapaty 2017). Innovative ideas such as the use of positive-pressure isocages for short-term experiments can be helpful to reduce the overall cost (Hecht et al. 2014). Availability of highly skilled technical staff will be another breakthrough in GF technology. Colonies of GF mice of specified genetic strains and novel minimal bacterial consortia should be established for the common laboratory species for target-based studies. Long-term rearing, development of breeding colonies of GF mice, and development of devices for shipping GF animals require extensive operations. In future, a centralized or region-based laboratory exclusively for GF mice technology can be established which not only will be center for resources but also skilled manpower.

16.6.2 Future Bio-therapeutic Agents and Pharmaceuticals Products

Due to increased incidences of antibiotic-resistance and therefore the side-effects of those drugs on host and off-target flora, alternate strategies should be developed to target pathogens (Meropol et al. 2008). Manipulation of the commensal microbiota and hence enabling its over-growth and competition with pathogens thus ultimately replace drug-resistance flora could also be a potent solution in the form of probiotics (Imperial and Ibana 2016). Germ-free mice colonies are often used as an experimental model to develop probiotics against antimicrobial-resistance pathogens. However, consistent monitoring of microbial load in germ-free models is critical for researchers to determine the load of contaminants or antibiotic-resistant microbes. These gnotobiotic models can also be potentially utilized in vaccine development program. Since it's evident that host microbiota can modulate vaccine response, hence it can be a decisive factor for a successful vaccination strategy (Wang et al. 2010; Cram et al. 2018). Microbiota features a capacity to alter the efficacy of any pharmaceutical formulation applied on its host. Different pharmaceutical products can be activated or inactivated by selective microorganisms (Iida et al. 2013; Klatt et al. 2017; Rekdal et al. 2019). Hence germ-free mice associated with such bacteria can be utilized in preclinical trials to develop stable and effective drugs (Fig. 16.3).

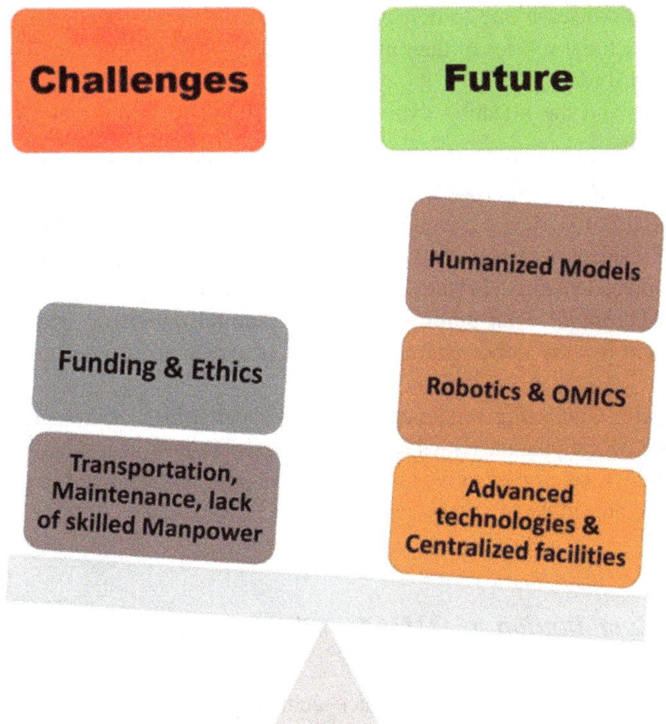

Fig. 16.3 Current challenges vs proposed future advancements in germ-free mice technology

16.7 Future Models

Humanized mouse model: GF mice can be colonized with human gut microbiota by using either a reductionist or holistic approach. In a reductionist approach, investigators seek the effect of known organisms, which affect the host, affected by the host or interact with one another. In the holistic approach, complex gut microbiota from diseased or healthy human donors is transferred in GF mice and their role in humans is predicted by profiling the alterations in recipient mice. The generation of "humanized mouse models" can support translational aspects of future research by creating human-like conditions within the mouse gut (Basic and Bleich 2019).

Knockout-gnotobiotic mouse model: Recently knockout-gnotobiotic animal models have been successfully developed and exploited to review the immune response modulated by a pathogen, in absence of crucial immune-modulatory genes (Balish et al. 1998; Yugo et al. 2018). Although the development of the knockout-gnotobiotic animal model is not an easy task, it will immensely help to understand the role of a specific bacterium in special circumstances.

1. In a genetic disease due to defective gene/gene product: to explore the role of various bacterial species or their metabolites that can perform an identical role to cure the disorder
2. To understand the infection dynamics of a pathogen in immune-compromised individuals

Model recombinant microorganisms: Engineering Human Microbiome is a novel concept (Kali 2015). Recombinant DNA technology can be employed to modify the genome of resident microflora and genetically modified microbes can be assessed to achieve unprecedented goals when associated with GF animals (Kayser et al. 2019).

Fecal Microbiota Transplantation (FMT): Administration of stool sample in solution from healthy donor to intestinal tract of recipient in order to change the gut health.

Cohousing: *Cohousing* recommendations for individual species are based, in part, on behavioral characteristics such as the desire to nest near a cage mate. In a study of male mice, *animals* given the choice to nest in an inhabited or empty cage preferred the proximity of another *animal*.

16.7.1 Combination of OMICS and GF Technology

The bioinformatics approaches in GF mice technology can prove to be indispensable in terms of applications in the future. When combined with approaches such as genomics, transcriptomics, metabolomics, and proteomics, GF mice technology can lead to the discovery of the exact functions and mechanisms of host colonization. It can also lead to a better understanding of the interaction and communication of specific microbiota representatives amongst each other and also with their respective hosts.

16.8 Conclusion

Here, we have summarized the foremost important outcomes of germ-free mice technology within the fields of health and allied sciences. These animal models offer immense advantages over other existing approaches for studying the role of varied microbial species and to know pathogenesis through host–microbe interaction, microbe–microbe interaction, gene–microbe interaction, diet–microbe interactions and senescence. GF animals are going to be the most important tool to investigate, how certain microorganisms are ready to colonize and survive within the host, while others can't. Germ-free mice have been extensively used for deciphering some mechanisms linked to metabolic diseases like Type II diabetes mellitus, behavioral functions at the gut-brain axis and autism, cardiovascular diseases, and cancer. Like every technology developed in the past few decades, germ-free mice technology is additionally into its evolving phase. The fact that technologies for detecting and characterizing microorganisms is continuously evolving, GF mice technology also needs to go in with pace.

References

Abbas AK, Lichtman AHH, Pillai S (2017) Cellular and Molecular Immunology E-Book, Elsevier Health Sciences

Abrams GD, Bauer H, Sprinz H (1963) Influence of the normal flora on mucosal morphology and cellular renewal in the ileum. A comparison of germ-free and conventional mice. Lab Investig 12:355–364

Abrams GD, Bishop JE (1961) Germfree techniques in experimental pathology: a survey of morphologic changes in, and the research potential of, the germfree guinea pig. Univ Mich Med Bull 27:136–147

Abrams GD, Bishop JE (1967) Effect of the normal microbial flora on gastrointestinal motility. Proc Soc Exp Biol Med 126:301–304

Bäckhed F, Manchester JK, Semenkovich CF, Gordon JI (2007) Mechanisms underlying the resistance to diet-induced obesity in germ-free mice. Proc Natl Acad Sci U S A 104(3): 979–984. https://doi.org/10.1073/pnas.0605374104

Balish E, Wagner RD, Vázquez-Torres A, Pierson C, Warner T (1998) Candidiasis in interferon-gamma knockout (IFN-gamma-/-) mice. J Infect Dis 178(2):478–487. https://doi.org/10.1086/51564

Basic M, Bleich A (2019) Gnotobiotics: past, present and future. Lab Anim 53(3):232–243. https://doi.org/10.1177/0023677219836715

Belkaid Y, Hand TW (2014) Role of the microbiota in immunity and inflammation. Cell 157(1): 121–141. https://doi.org/10.1016/j.cell.2014.03.011

Bhattarai Y, Muniz Pedrogo DA, Kashyap PC (2017) Irritable bowel syndrome: a gut microbiota-related disorder? Am J Physiol Gastrointest Liver Physiol 312(1):G52–G62. https://doi.org/10.1152/ajpgi.00338.2016

Biggs MB, Medlock GL, Moutinho TJ et al (2017) Systems-level metabolism of the altered Schaedler flora, a complete gut microbiota. ISME J 11(2):426–438. https://doi.org/10.1038/ismej.2016.130

Blümer N, Herz U, Wegmann M, Renz H (2005) Prenatal lipopolysaccharide-exposure prevents allergic sensitization and airway inflammation, but not airway responsiveness in a murine model of experimental asthma. Clin Exp Allergy 35(3):397–402. https://doi.org/10.1111/j.1365-2222.2005.02184.x

Borgstrom B, Dahlqvist A, Gustafsson BE, Lundh G, Malmquist J (1959) Trypsin, invertase and amylase content of feces of germfree rats. Proc Soc Exp Biol Med 102:154–155. https://doi.org/10.3181/00379727-102-25174

Borre YE, O'Keeffe GW, Clarke G, Stanton C, Dinan TG, Cryan JF (2014) Microbiota and neurodevelopmental windows: implications for brain disorders. Trends Mol Med 20(9): 509–518. https://doi.org/10.1016/j.molmed.2014.05.002

Braniste V, Al-Asmakh M, Kowal C et al (2014) The gut microbiota influences blood-brain barrier permeability in mice (published correction appears in SciTransl Med. 2014 Dec 10;6(266): 266er7. Guan, Ng Lai (corrected to Ng, Lai Guan)). Sci Transl Med 6(263):263ra158. https://doi.org/10.1126/scitranslmed.3009759

Britton GJ, Contijoch EJ, Mogno I et al (2019) Microbiotas from humans with inflammatory bowel disease alter the balance of Gut Th17 and RORγt$^+$ regulatory T Cells and EXACERBATE COLITIS IN MICE. Immunity 50(1):212–224.e4. https://doi.org/10.1016/j.immuni.2018.12.015

Brugiroux S, Beutler M, Pfann C et al (2016) Genome-guided design of a defined mouse microbiota that confers colonization resistance against Salmonella entericaserovarTyphimurium. Nat Microbiol 2:16215. https://doi.org/10.1038/nmicrobiol.2016.215

Butterton JR, Ryan ET, Shahin RA, Calderwood SB (1996) Development of a germfree mouse model of Vibrio cholerae infection. Infect Immun 64(10):4373–4377

Canales CP, Walz K (2019) The mouse, a model organism for biomedical research. In: Walz K, Young JI (eds) Cellular and animal models in human genomics research. Translational and Applied Genomics Series. Elsevier/Academic Press, San Diego, pp 119–140

Clarke G, Stilling RM, Kennedy PJ, Stanton C, Cryan JF, Dinan TG (2014) Minireview: gut microbiota: the neglected endocrine organ. MolEndocrinol 28(8):1221–1238. https://doi.org/10.1210/me.2014-1108

Coates ME (1973) Gnotobiotic animals in nutrition research. Proc Nutr Soc 32(2):53–58. https://doi.org/10.1079/pns19730015

Collins J, Borojevic R, Verdu EF, Huizinga JD, Ratcliffe EM (2014) Intestinal microbiota influence the early postnatal development of the enteric nervous system. Neurogastroenterol Motil 26(1): 98–107. https://doi.org/10.1111/nmo.12236

Crabbé PA, Bazin H, Eyssen H, Heremans JF (1968) The normal microbial flora as a major stimulus for proliferation of plasma cells synthesizing IgA in the gut. The germ-free intestinal tract. Int Arch Allergy ApplImmunol 34(4):362–375. https://doi.org/10.1159/000230130

Cram JA, Hager KW, Kublin JG (2018) Utilizing gnotobiotic models to inform the role of the microbiome in vaccine response heterogeneity. Curr Opin HIV AIDS 13(1):1–8. https://doi.org/10.1097/COH.0000000000000422

Crouzet L, Gaultier E, Del'Homme C et al (2013) The hypersensitivity to colonic distension of IBS patients can be transferred to rats through their fecal microbiota. Neurogastroenterol Motil 25(4):e272–e282. https://doi.org/10.1111/nmo.12103

Czuprynski CJ, Balish E (1981) Pathogenesis of Listeria monocytogenes for gnotobiotic rats. Infect Immun 32(1):323–331

Danielsson H, Gustafsson B (1959) On serum-cholesterol levels and neutral fecal sterols in germ-free rats; bile acids and steroids 59. Arch Biochem Biophys 83:482–485. https://doi.org/10.1016/0003-9861(59)90056-6

denBesten G, van Eunen K, Groen AK, Venema K, Reijngoud DJ, Bakker BM (2013) The role of short-chain fatty acids in the interplay between diet, gut microbiota, and host energy metabolism. J Lipid Res 54(9):2325–2340. https://doi.org/10.1194/jlr.R036012

De Palma G, Collins SM, Bercik P (2014) The microbiota-gut-brain axis in functional gastrointestinal disorders. Gut Microbes 5(3):419–429. https://doi.org/10.4161/gmic.29417

Dorrestein PC, Mazmanian SK, Knight R (2014) Finding the missing links among metabolites, microbes, and the host. Immunity 40:824–832

Douwes J, Cheng S, Travier N et al (2008) Farm exposure in utero may protect against asthma, hay fever and eczema. Eur Respir J 32(3):603–611. https://doi.org/10.1183/09031936.00033707

Fanning S, Hall LJ, Cronin M et al (2012) Bifidobacterial surface-exopolysaccharide facilitates commensal-host interaction through immune modulation and pathogen protection. Proc Natl Acad Sci U S A 109(6):2108–2113. https://doi.org/10.1073/pnas.1115621109

Festing MF, Blackmore DK (1971) Life span of specified-pathogen-free (MRC category 4) mice and rats. Lab Anim 5(2):179–192. https://doi.org/10.1258/002367771781006564

Fiebiger U, Bereswill S, Heimesaat MM (2016) Dissecting the interplay between intestinal microbiota and host immunity in health and disease: lessons learned from germfree and gnotobiotic animal models. Eur J Microbiol Immunol (Bp) 6(4):253–271. https://doi.org/10.1556/1886.2016.00036

Forsythe P, Kunze WA (2013) Voices from within: gut microbes and the CNS. Cell Mol Life Sci 70(1):55–69. https://doi.org/10.1007/s00018-012-1028-z

Gong J, Chehrazi-Raffle A, Placencio-Hickok V, Guan M, Hendifar A, Salgia R. The gut microbiome and response to immune checkpoint inhibitors: preclinical and clinical strategies.

Clin Transl Med. 2019;8(1):9. Published 2019 Mar 18. doi:https://doi.org/10.1186/s40169-019-0225-x

Gopalakrishnan V, Spencer CN, Nezi L et al (2018) Gut microbiome modulates response to anti-PD-1 immunotherapy in melanoma patients. Science 359(6371):97–103. https://doi.org/10.1126/science.aan4236

Gordon HA, Bruckner-Kardoss E (1961) Effect of normal microbial flora on intestinal surface area. Am J Phys 201:175–178. https://doi.org/10.1152/ajplegacy.1961.201.1.175

Gordon HA, Bruckner-Kardoss E, Wostmann BS (1966) Aging in germ-free mice: life tables and lesions observed at natural death. J Gerontol 21(3):380–387. https://doi.org/10.1093/geronj/21.3.380

Gordon HA, Wostmann BS, Bruckner-Kardoss E (1963) Effects of microbial flora on cardiac output and other elements of blood circulation. Proc Soc Exp Biol Med 114:301–304. https://doi.org/10.3181/00379727-114-28658

Gosmann C, Anahtar MN, Handley SA et al (2017) Lactobacillus-deficient cervicovaginal bacterial communities are associated with increased HIV acquisition in young South African women. Immunity 46(1):29–37. https://doi.org/10.1016/j.immuni.2016.12.013

Gregory SG, Sekhon M, Schein J et al (2002) A physical map of the mouse genome. Nature 418(6899):743–750. https://doi.org/10.1038/nature00957

Grenham S, Clarke G, Cryan JF, Dinan TG (2011) Brain-gut-microbe communication in health and disease. Front Physiol 2:94. https://doi.org/10.3389/fphys.2011.00094

Gustafsson BE (1959a) Lightweight stainless steel systems for rearing germfree animals. Ann N Y AcadSci 78:17–28

Gustafsson BE (1959b) Vitamin K deficiency in germfree rats. Ann N Y Acad Sci 78:166–174

Gustafsson BE, Laurell CB (1958) G Gamma globulins in germ-free rats. J Exp Med 108(2):251–258. https://doi.org/10.1084/jem.108.2.251

Gustafsson BE, Norman A (1962) Urinary calculi in germfree rats. J Exp Med 116(3):273–284. https://doi.org/10.1084/jem.116.3.273

Haldane JBS (1928) On being the right size Possible Worlds. Harper & Brothers, New York

Hecht G, Bar-Nathan C, Milite G, Alon I, Moshe Y, Greenfeld L et al (2014) A simple cage-autonomous method for the maintenance of the barrier status of germ-free mice during experimentation. Lab Anim 48:292–297

Heneghan JB (1963) Influence of microbial flora on xylose absorption in rats and mice. Am J Phys 205:417–420. https://doi.org/10.1152/ajplegacy.1963.205.3.417

Hobby GL, Lenert TF, Maier-Engallena J et al (1968) Further observations on the immunogenic effect of BCG in germfree mice. II. *Am Rev Respir Dis* 97(6):1095–1103. https://doi.org/10.1164/arrd.1968.97.6P1.1095

Høverstad T, Midtvedt T (1986) Short-chain fatty acids in germfree mice and rats. J Nutr 116(9):1772–1776. https://doi.org/10.1093/jn/116.9.1772

Human Microbiome Project Consortium (2012) A framework for human microbiome research. Nature 486(7402):215–221. https://doi.org/10.1038/nature11209

Husebye E, Hellström PM, Sundler F, Chen J, Midtvedt T (2001) Influence of microbial species on small intestinal myoelectric activity and transit in germ-free rats. Am J Physiol Gastrointest Liver Physiol 280:G368–G380

Iida N, Dzutsev A, Stewart CA et al (2013) Commensal bacteria control cancer response to therapy by modulating the tumor microenvironment. Science 342(6161):967–970. https://doi.org/10.1126/science.1240527

Imperial IC, Ibana JA. Addressing the antibiotic resistance problem with probiotics: reducing the risk of its double-edged sword effect. Front Microbiol. 2016;7:1983. Published 2016 Dec 15. doi:https://doi.org/10.3389/fmicb.2016.01983

Isabella VM, Ha BN, Castillo MJ et al (2018) Development of a synthetic live bacterial therapeutic for the human metabolic disease phenylketonuria. Nat Biotechnol 36(9):857–864. https://doi.org/10.1038/nbt.4222

Jourová L, Anzenbacher P, Lišková B et al (2017) Colonization by non-pathogenic bacteria alters mRNA expression of cytochromes P450 in originally germ-free mice. Folia Microbiol (Praha) 62(6):463–469. https://doi.org/10.1007/s12223-017-0517-8

Kaden-Volynets V, Basic M, Neumann U et al (2019) Lack of liver steatosis in germ-free mice following hypercaloric diets. Eur J Nutr 58(5):1933–1945. https://doi.org/10.1007/s00394-018-1748-4

Kaji I, Akiba Y, Said H, Narimatsu K, Kaunitz JD (2015) Luminal 5-HT stimulates colonic bicarbonate secretion in rats. Br J Pharmacol 172:4655–4670

Kali A (2015) Human microbiome engineering: the future and beyond. J ClinDiagn Res 9(9): DE01–DDE4. https://doi.org/10.7860/JCDR/2015/14946.6570

Kaplan JL, Shi HN, Walker WA (2011) The role of microbes in developmental immunologic programming. Pediatr Res 69(6):465–472. https://doi.org/10.1203/PDR.0b013e318217638a

Karlsson F, Tremaroli V, Nielsen J, Bäckhed F (2013) Assessing the human gut microbiota in metabolic diseases. Diabetes 62:3341–3349. https://doi.org/10.2337/db13-0844

Kashyap PC, Marcobal A, Ursell LK, Larauche M, Duboc H, Earle KA, Sonnenburg ED, Ferreyra JA, Higginbottom SK, Million M et al (2013) Complex interactions among diet, gastrointestinal transit, and gut microbiota in humanized mice. Gastroenterology 144:967–977

Kayser BD, Lhomme M, Prifti E et al (2019) Phosphatidylglycerols are induced by gut dysbiosis and inflammation, and favorably modulate adipose tissue remodeling in obesity. FASEB J 33(4):4741–4754. https://doi.org/10.1096/fj.201801897R

Klatt NR, Cheu R, Birse K et al (2017) Vaginal bacteria modify HIV tenofovirmicrobicide efficacy in African women. Science 356(6341):938–945. https://doi.org/10.1126/science.aai9383

Koren O, Goodrich JK, Cullender TC et al (2012) Host remodeling of the gut microbiome and metabolic changes during pregnancy. Cell 150(3):470–480. https://doi.org/10.1016/j.cell.2012.07.008

Lagkouvardos I, Pukall R, Abt B et al (2016) The Mouse Intestinal Bacterial Collection (miBC) provides host-specific insight into cultured diversity and functional potential of the gut microbiota (published correction appears in Nat Microbiol. 2016 Oct 17;1(11):16219). Nat Microbiol 1(10):16131. https://doi.org/10.1038/nmicrobiol.2016.131

Lamousé-Smith ES, Tzeng A, Starnbach MN (2011) The intestinal flora is required to support antibody responses to systemic immunization in infant and germ free mice. PLoS One 6:e27662

Laqueur GL, McDaniel EG, Mateumoto H (1967) Tumor induction in germfree rats with methylazoxymethanol (MAM) and synthetic MAM acetate. J Natl Cancer Inst 39:355–371

Leppi TJ (1964) A study of the uterine cervix of the mouse. Anat Rec 150:51–65. https://doi.org/10.1002/ar.1091500106

Levenson SM, Tenivant S (1963) Contributions of intestinal microflora to the nutrition of the host animal. Some metabolic and nutritional studies with germ-free animals. Fed Prec 22: 109–119

Lindenbaum J, Rund DG, Butler VP, Tse-Eng D, Saha JR (1981) Inactivation of digoxin by the gut flora: reversal by antibiotic therapy. N Engl J Med 305:789–794

Loesche WJ (1969) Accumulation of endogenous carbohydrate-containing compounds in the cecum of the germfree rat. Proc Soc Exp Biol Med 131(2):387–392. https://doi.org/10.3181/00379727-131-33885

Luckey TD (1963) Germfree Life and Gnotobiology. Academic Press, New York

Luckey TD (1968) Gnotobiology and aerospace systems. In: Miyakawa M, Luckey TD (eds) Advances in Germfree Research and Gnotobiology. CRC Press, Cleveland, OH, pp 317–353

Luczynski P, McVey Neufeld KA, Oriach CS, Clarke G, Dinan TG, Cryan JF (2016a) Growing up in a bubble: using germ-free animals to assess the influence of the gut microbiota on brain and behavior. Int J Neuropsychopharmacol 19(8):pyw020. https://doi.org/10.1093/ijnp/pyw020

Luczynski P, Whelan SO, O'Sullivan C et al (2016b) Adult microbiota-deficient mice have distinct dendritic morphological changes: differential effects in the amygdala and hippocampus. Eur J Neurosci 44(9):2654–2666. https://doi.org/10.1111/ejn.13291

Mack DR, Ahrne S, Hyde L, Wei S, Hollingsworth MA (2003) Extracellular MUC3 mucin secretion follows adherence of Lactobacillus strains to intestinal epithelial cells in vitro. Gut 52(6):827–833. https://doi.org/10.1136/gut.52.6.827

Mack DR, Michail S, Wei S, McDougall L, Hollingsworth MA (1999) Probiotics inhibit enteropathogenic E. coli adherence in vitro by inducing intestinal mucin gene expression. Am J Phys 276(4):G941–G950. https://doi.org/10.1152/ajpgi.1999.276.4.G941

Makulu DR, Wagner M (1967) Lysozyme activity in the serum, saliva and tears of germfree and conventional rats and mice. Proc Indiana Acad Sci 76:183–190

Mallapaty S (2017) Gnotobiotics: getting a grip on the microbiome boom. Lab Anim (NY) 46(10): 373–377. https://doi.org/10.1038/laban.1344

Maloy KJ, Powrie F. Intestinal homeostasis and its breakdown in inflammatory bowel disease. Nature. 2011;474(7351):298–306. Published 2011 Jun 15. doi:https://doi.org/10.1038/nature10208

McVey Neufeld KA, Mao YK, Bienenstock J, Foster JA, Kunze WA (2013) The microbiome is essential for normal gut intrinsic primary afferent neuron excitability in the mouse. Neurogastroenterol Motil 25(2):183–e88. https://doi.org/10.1111/nmo.12049

McVey Neufeld KA, Perez-Burgos A, Mao YK, Bienenstock J, Kunze WA (2015) The gut microbiome restores intrinsic and extrinsic nerve function in germ-free mice accompanied by changes in calbindin. Neurogastroenterol Motil 27(5):627–636. https://doi.org/10.1111/nmo.12534

Mendes R, Raaijmakers JM (2015) Cross-kingdom similarities in microbiome functions. ISME J 9(9):1905–1907. https://doi.org/10.1038/ismej.2015.7

Meropol SB, Chan KA, Chen Z et al (2008) Adverse events associated with prolonged antibiotic use. Pharmacoepidemiol Drug Saf 17(5):523–532. https://doi.org/10.1002/pds.1547

Mouse Genome Sequencing Consortium, Waterston RH, Lindblad-Toh K et al (2002) Initial sequencing and comparative analysis of the mouse genome. Nature 420(6915):520–562. https://doi.org/10.1038/nature01262

Nakanishi T, Tamai I (2015) Interaction of drug or food with drug transporters in intestine and liver. Curr Drug Metab 16(9):753–764. https://doi.org/10.2174/138920021609151201113537

Nardi RM, Vieira EC, Crocco-Afonso LC et al (1991) Bacteriological and immunological aspects of conventional and germfree mice infected with Salmonella typhimurium. Rev LatinoamMicrobiol 33(4):239–243

Nicholson JK, Holmes E, Kinross J et al (2012) Host-gut microbiota metabolic interactions. Science 336(6086):1262–1267. https://doi.org/10.1126/science.1223813

Nuttall G, Thierfelder H (1895) Tierisches Lebenohne Bakterienim Verdauungskanal. Hoppe Seyler's Zeitschrift PhysiolChem 21:109–112

O'Toole PW, Claesson MJ (2010) Gut microbiota: changes throughout the lifespan from infancy to elderly. Int Dairy J 20:281–291

Oh JZ, Ravindran R, Chassaing B, Carvalho FA, Maddur MS, Bower M, Hakimpour P, Gill KP, Nakaya HI, Yarovinsky F, Sartor RB, Gewirtz AT, Pulendran B (2014) TLR5-mediated sensing of gut microbiota is necessary for antibody responses to seasonal influenza vaccination. Immunity 41:478–492

Olson GB, Wostmann BS (1966) Cellular and humoral immune response of germfree mice stimulated with 7S HGG or Salmonella typhimurium. J Immunol 97(2):275–286

Orcutt RP, Gianni FJ, Judge RJ (1987) Development of an "altered Schaedler flora" for NCI gnotobiotic rodents. Microecol Ther 17:59

Pasteur L (1885) Observations relatives à lanote de M. Duclaux. Compte Rendu Académie des Sciences 100:68–69

Perez-Burgos A, Mao YK, Bienenstock J, Kunze WA (2014) The gut-brain axis rewired: adding a functional vagal nicotinic "sensory synapse". FASEB J 28(7):3064–3074. https://doi.org/10.1096/fj.13-245282

Phillips AW, Smith JE (1959) Germfree animal techniques and their applications. Adv Appl Microbiol 1:141–174. https://doi.org/10.1016/s0065-2164(08)70478-7

Pittermann W, Deerberg F (1975) Spontaneous tumors and lesions of the lung, kidney and gingiva in aged germfree rats. In: Fliedner TM, Heit H, Neithammer D, Pflieger H (eds) Clinical and Experimental Gnotobiotics. Gustav Fischer Verlag, New York

Pleasants JR (1959) Rearing germfree cesarean-born rats, mice, and rabbits through weaning. Ann N Y Acad Sci 78:116–126

Pleasants JR (1974) Gnotobiotics. In: Melby EC Jr, Altman NH (eds) Handbook of Laboratory Animal Science, vol vol. I. CRC Press, Cleveland, OH, pp 119–174

Pollard M (1971) Senescence in germfree rats. Gerontologia 17(5):333–338. https://doi.org/10.1159/000211839

Pollard M, Kajima M (1970) Lesions in aged germfree Wistar rats. Am J Pathol 61(1):25–36

Pollard M, Kajima M, Lorans G (1968) Tissue changes in germ-free rats with primary tumors. J Reticuloendothel Soc 5(2):147–160

Pollard M, Luckert PH, Meshorer A (1985) Xenogeneic bone marrow chimerism in germfree rats. Prog Clin Biol Res 181:447–450

Pollard M, Teah BA (1963) Spontaneous tumors in germ-free rats. J Natl Cancer Inst 31:457–465

Pope JL, Tomkovich S, Yang Y, Jobin C (2017) Microbiota as a mediator of cancer progression and therapy. Transl Res 179:139–154

Powley TL, Wang XY, Fox EA, Phillips RJ, Liu LW, Huizinga JD (2008) Ultrastructural evidence for communication between intramuscular vagal mechanoreceptors and interstitial cells of Cajal in the rat fundus. Neurogastroenterol Motil 20(1):69–79. https://doi.org/10.1111/j.1365-2982.2007.00990.x

Qin J, Li R, Raes J et al (2010) A human gut microbial gene catalogue established by metagenomic sequencing. Nature 464(7285):59–65. https://doi.org/10.1038/nature08821

Reddy BS, Pleasants JR, Wostmann BS (1969) Pancreatic enzymes in germfree and conventional rats fed chemically defined, water-soluble diet free from natural substrates. J Nutr 97(3): 327–334. https://doi.org/10.1093/jn/97.3.327

Reigstad CS, Salmonson CE, Rainey JF 3rd et al (2015) Gut microbes promote colonic serotonin production through an effect of short-chain fatty acids on enterochromaffin cells. FASEB J 29(4):1395–1403. https://doi.org/10.1096/fj.14-259598

Rekdal VM, Bess EN, Bisanz JE, Turnbaugh PJ, Balskus EP (2019) Discovery and inhibition of an interspecies gut bacterial pathway for Levodopa metabolism. Science 364:eaau6323

Reyniers JA (1957) The production and use of germ-free animals in experimental biology and medicine. Amer J Vet Res 18:678–687

Reyniers JA (1959a) Design and operation of apparatus for rearing germ-free animals. Ann New York Acad Sc 78:47

Reyniers JA (1959b) The pure-culture concept and gnotobiotics. Ann N Y Acad Sci 78:3–16

Reyniers JA, Sacksteder MR (1958) The use of germfree animals and techniques in the search for unknown etiological agents. Ann N Y Acad Sci 73(1):344–356

Reyniers JA, Wagner M, Luckey T D, and Gordon H A. Survey of germfree animals: the white Wyandotte Bantam and white Leghorn chicken, 1960 p. 7–159. In J. A. Reyniers (ed.), Lobund reports no. 3. University of Notre Dame Press, Notre Dame, Indiana

Ridaura VK, Faith JJ, Rey FE et al (2013) Gut microbiota from twins discordant for obesity modulate metabolism in mice. Science 341(6150):1241214. https://doi.org/10.1126/science.1241214

Round JL, Mazmanian SK (2009) The gut microbiota shapes intestinal immune responses during health and disease (published correction appears in Nat Rev Immunol. 2009 Aug;9(8):600). Nat Rev Immunol 9(5):313–323. https://doi.org/10.1038/nri2515

Schaedler RW, Dubos R, Costello R (1965) The development of the bacterial flora in the gastrointestinal tract of mice. J Exp Med 122(1):59–66

Sekirov I, Russell SL, Antunes LC, Finlay BB (2010) Gut microbiota in health and disease. Physiol Rev 90(3):859–904. https://doi.org/10.1152/physrev.00045.2009

Shanahan F (2002) Crohn's disease. Lancet 359(9300):62–69. https://doi.org/10.1016/S0140-6736(02)07284-7

Shimizu K, Muranaka Y, Fujimura R, Ishida H, Tazume S, Shimamura T (1998) Normalization of reproductive function in germfree mice following bacterial contamination. Exp Anim 47(3): 151–158. https://doi.org/10.1538/expanim.47.151

Smith K, McCoy KD, Macpherson AJ (2007) Use of axenic animals in studying the adaptation of mammals to their commensal intestinal microbiota. Semin Immunol 19:59–69

Srikanth CV, McCormick BA (2008) Interactions of the intestinal epithelium with the pathogen and the indigenous microbiota: a three-way crosstalk. Interdiscip Perspect Infect Dis 2008:626827. https://doi.org/10.1155/2008/626827

Sumi Y, Miyakawa M, Kanzaki M, Kotake Y (1977) Vitamin B-6 deficiency in germfree rats. J Nutr 107(9):1707–1714. https://doi.org/10.1093/jn/107.9.1707

Tazume S, Umehara K, Matsuzawa H, Aikawa H, Hashimoto K, Sasaki S (1991) Effects of germfree status and food restriction on longevity and growth of mice. Jikken Dobutsu 40(4): 517–522. https://doi.org/10.1538/expanim1978.40.4_517

Tripathi AK, Ray AK, Mishra SK. Molecular and pharmacological aspects of piperine as a potential molecule for disease prevention and management: evidence from clinical trials. Beni-Suef Univ J Basic Appl Sci. 2022;11(1):1–24

Turnbaugh PJ, Ley RE, Mahowald MA, Magrini V, Mardis ER, Gordon JI (2006) An obesity-associated gut microbiome with increased capacity for energy harvest. Nature 444(7122): 1027–1031. https://doi.org/10.1038/nature05414

Uchiyama R, Chassaing B, Zhang B, Gewirtz AT (2014) Antibiotic treatment suppresses rotavirus infection and enhances specific humoral immunity. J Infect Dis 210:171–182

Viaud S, Daillère R, Boneca IG et al (2015) Gut microbiome and anticancer immune response: really hot Sh*t! Cell Death Differ 22(2):199–214. https://doi.org/10.1038/cdd.2014.56

Viaud S, Saccheri F, Mignot G, Yamazaki T, Daillere R, Hannani D, Enot DP, Pfirschke C, Engblom C, Pittet MJ, Schlitzer A, Ginhoux F, Apetoh L, Chachaty E, Woerther PL, Eberl G et al (2013) The intestinal microbiota modulates the anticancer immune effects of cyclophosphamide. Science 342(6161):971–976

Walburg HE Jr (1973) Carcinogenesis in gnotobiotic rodents. In: Heneghan JB (ed) Germfree Research. Academic Press, New York, pp 115–122

Wang Y, Azevedo M, Saif LJ, Gentsch JR, Glass RI, Jiang B (2010) Inactivated rotavirus vaccine induces protective immunity in gnotobiotic piglets. Vaccine 28(33):5432–5436. https://doi.org/10.1016/j.vaccine.2010.06.006

Weisburger JH, Reddy BS, Narisawa T, Wynder EL (1975) Germ-free status and colon tumor induction by N-methyl-N'-nitro-N-nitrosoguanidine. Proc Soc Exp Biol Med 148(4): 1119–1121. https://doi.org/10.3181/00379727-148-38700

Williams SC (2014) Gnotobiotics. Proc Natl Acad Sci U S A 111(5):1661. https://doi.org/10.1073/pnas.1324049111

Wilson R, Bealmear M, Sobonya R (1965) Growth and regression of the germfree (axenic) thymus. Proc Soc Exp Biol Med 118:97–99. https://doi.org/10.3181/00379727-118-29766

Wostmann B, Bruckner-Kardoss E (1959) Development of cecal distention in germ-free baby rats. Am J Phys 197:1345–1346. https://doi.org/10.1152/ajplegacy.1959.197.6.1345

Wostmann BS, Knight PL, Keeley LL, Kan DF (1962) Metabolism and function of thiamine and naphthoquinones in germfree and conventional rats. In: Federation proceedings, pp 120–124

Wostmann BS, Larkin C, Moriarty A, Bruckner-Kardoss E (1983) Dietary intake, energy metabolism, and excretory losses of adult male germfree Wistar rats. Lab Anim Sci 33(1):46–50

Wostmann BS, Pleasants JR, Bealmear P, Kincade PW (1970) Serum proteins and lymphoid tissues in germ-free mice fed a chemically defined, water soluble, low molecular weight diet. Immunology 19(3):443–448

Wymore Brand M, Wannemuehler MJ, Phillips GJ et al (2015) The altered Schaedler Flora: continued applications of a defined murine microbial community. ILAR J 56(2):169–178. https://doi.org/10.1093/ilar/ilv012
Yano JM, Yu K, Donaldson GP, Shastri GG, Ann P, Ma L, Nagler CR, Ismagilov RF, Mazmanian SK, Hsiao EY (2015) Indigenous bacteria from the gut microbiota regulate host serotonin biosynthesis. Cell 161:264–276
Yugo DM, Heffron CL, Ryu J, et al. Infection dynamics of hepatitis E virus in wild-type and immunoglobulin heavy chain knockout $J_H^{-/-}$ gnotobiotic piglets. J Virol. 2018;92(21):e01208–e01218. Published 2018 Oct 12. doi:https://doi.org/10.1128/JVI.01208-18
Zachar Z, Savage DC (1979) Microbial interference and colonization of the murine gastrointestinal tract by Listeria monocytogenes. Infect Immun 23(1):168–174

Chapter 17
Gut Microbiome and Neurodegeneration: A Bioinformatics Approach

Swetanshu and Pratichi Singh

Abstract Dysfunction of the human microbiota is associated with various conditions ranging from antibiotic-resistant infections to inflammatory bowel disease. The evidence also proposes a bidirectional communication between the intestinal microbiota and the brain. This crosstalk may play a significant role in neurodegenerative diseases, including autism, multiple sclerosis, Parkinson's disease, anxiety, depression, Alzheimer's disease, etc. Lately, high throughput sequencing and bioinformatic tools deliver a powerful means to comprehend the role of the human microbiome in health and its potential as a target for therapeutic intercessions. This chapter discusses various techniques related to microbiome study such as shotgun sequencing technologies, 16S rRNA sequencing, metabolomics, metagenomics, metatranscriptomics, the computational challenges, and the approaches associated with these data. Furthermore, neural system pathways that connect the gut microbiota to the central nervous system are crucial to fully comprehending microbiomes.

Keywords Microbiota · Metagenomic · Metatranscriptomic · Shotgun sequencing · 16SrRNA sequencing

Abbreviation

CNS	Central nervous system
DNA	Deoxyribonucleic acid
FPR	False-positive rate
GI	Gastrointestinal tract
HMP	Human microbiome project
MS	Multiple sclerosis
NCBI	National Center for Biotechnology Information

Swetanshu · P. Singh (✉)
Department of Bio-Sciences, School of Basic and Applied Sciences, Galgotias University, Greater Noida, India
e-mail: pratichi.singh@galgotiasuniversity.edu.in

A. K. Tripathi, M. Kotak (eds.), *Gut Microbiome in Neurological Health and Disorders*, Nutritional Neurosciences,
https://doi.org/10.1007/978-981-19-4530-4_17

PD Parkinson's disease
rRNA Ribosomal ribonucleic acid

17.1 Introduction

An infant is unexposed to the world of microbes as long as it stays in his mother's womb. At the time of birth, billions of microbes are encountered, and among them, few become permanent residents in different parts of the body like the oral cavity, gut, skin, etc. Specifically, our gut is colonized by various archaea, bacteria, fungi, and viruses referred to as the human microbiome. Collectively, they contribute a few pounds to our body weight and have more genes than we have in our genome (Qin et al. 2010; Morgan and Huttenhower 2012). Studies also show that the total number of microbiomes present in the body is more than the entire human cell, maintaining several normal body functions like digestion, immunity, etc. (Qin et al. 2010 and Ackerman 2012).

The "Human Microbiome Project" (HMP), funded by the National Institute of Health (NIH), has been launched to characterize the different microbiomes for their detailed study (Peterson et al. 2009). The study derives that the microbes have numerous roles in emphasizing the normal body function (Aguiar-Pulido et al. 2016). The project successfully established a relationship between certain diseases and these microbes (HMP Consortium 2014). The result of HMP influenced researchers to launch the Metagenomics of the Human Intestinal Tract (MetaHIT) project. Initially, this project was funded by European Seventh Framework Program. The project was established to better understand the relationship between human health or disease and intestinal microbiota (Ehrlich 2010). Two more projects were launched, namely the American Gut Project and the Human Food Project, aiming to figure out a way to deliver a healthy microbiome in the gut via food (Goedert et al. 2014). The bioinformatic-based study in these various projects has provided us with very positive outcomes (Morgan and Huttenhower 2012).

17.2 Metagenomics

Metagenomics is an investigative study where the researcher studies the composition of the microbial community. The data gathered from the genetic material of a community is explored here. The next-generation sequencing technique is used for sequencing the DNA. 16S rDNA was found suitable for studying metataxonomics (Gonzalez and Knight 2012; Cole et al. 2014). Different databases have been established which hold the taxonomic profile of microbiomes. Various tools have also been developed for the detailed study of stored data. Problem-solving environments (PSEs) provide a working space where users can analyze. This includes a

Table 17.1 Bioinformatic tools for metagenomic datasets analyses

PIPELINE	FUNCTION
RIBOSOMAL DATABASE PROJECT	16S Databases
SILVA	
GREENGENES	
UCLUST	OTUs clustering
Cd-hit	
DADA2	Sub-OTU methods
UNOISE2	
DEBLUR	
Piphillin	Functional profiles prediction
PICRUSt	
Tax4Fun	
SOAPDenovo	Metagenome assembly
Genovo	
Velvet	
MetaVelvet	
BAMBUS2	
META-IBDA	
Ray-Meta	
PanPhlAn	
ViromeScan	
MetaGeneMark	Gene identification
FragGeneScan	
GLIMMER-MG	
IMG Database	Functional assignment
DBCAN	
MetaRef	
HUMANN	
UNIFrac	Communities' diversity comparison

remotely accessible tool, GALAXY, which allows a user to maximize the generality in the genomic analysis (Goecks et al. 2010; Kim et al. 2013). QIIME enables a user to analyze microbial DNA by using marker genes (Caporaso et al. 2010). MOTHER is another tool facilitating a similar function by using a marker gene (Schloss et al. 2009). PATHOSCOPE is used to identify bacterial strains from raw sequences and the generation of statistical reports (Hong et al. 2014).

The genes' functional annotation helped improve metagenomics (Meyer et al. 2008; Stark et al. 2010). Gene ontology (GO), Clusters of Orthologous Groups (COG), and Kyoto Encyclopedia of Genes and Genomes (KEGG) are some of the tools used for the annotation (Aguiar-Pulido et al. 2016). CAMERA, COMet, IMG/M, MEGAN, and METAREP are tools facilitating the generation of functional profiles of microbiomes (Aguiar-Pulido et al. 2016). Table 17.1 represents all the tools used to execute the metagenomic analysis.

17.2.1 Meta-transcriptomics

The focus is on studying the functional profile of microbiomes by analyzing the gene of microbes that are expressed (Moran 2009). For providing information about the function of any microbiome community, it examines the mRNA available in the sample. The meta transcriptome-based study is not performed regularly like other omics. The alignment tools of NCBI like BLAST, Bowtie, and BWA are preferably used. The result is annotated using different resources like COG, KEGG, GO, and UniProt, and at last, the downstream analysis is done.

17.2.2 Metabolomics

The identification and quantification of all the metabolites present in a sample of microbe are made. It helps in better understanding the homeostasis of a microbe with its surrounding environment. Moreover, there are some signature metabolites of all microbiota communities, variation in the production of the same will lead to changes in various metabolic processes. Therefore, this type of study is very beneficial in figuring out the metabolic pathways of a microbe (Krumsiek et al. 2015). The provided environmental condition highly influences the metabolic pathways of the microbe by the host (Human). This also reveals how the microbe interacts with the host and accommodates in his body (Manor et al. 2014). It describes internal metabolic processes as well as the impact of the external environment, allowing for a better understanding of biological processes by combining data from the other omics mentioned above.

Some of the databases associated with this are Human Metabolome Database, BioMagResBank, Madison-Qingdao Metabolomics Consortium Database, MassBank, Golm Metabolome Database, METLIN Metabolite Database, etc. (Aguiar-Pulido et al. 2016).

17.2.3 16S rRNA Sequencing

For fast recognition of the microbe in a specific environmental condition in the body, the genetic analysis of 16S ribosomal RNA (16S rRNA) is done. 16S rRNA gene is specific to bacteria with hyper-variable regions in its sequence, making it an ideal candidate for differentiating an organism from others and doing phylogenetic analysis. Green genes, Ribosomal Database Project, and Silva are some of the databases usually referred to for 16S rRNA sequences (Quast et al. 2013; Cole et al. 2014). The methodology is to align the 16S rRNAs with the sequence reads obtained during the experiment in a specific database so that the most similar and nearby species are assigned. Greengenes is a database designed to work on full-length 16S rRNA genes for taxonomic identification (DeSantis et al. 2006). A bioinformatics tool called Quantitative Insights into Microbial Ecology (QIIME) goes with the combinative approach for Operational Taxonomic Units (OTUs) where clustering is based on comprising the sequence using different algorithms, i.e., the use of pairwise alignment and sequence length (D'Argenio 2018). The OTUs have been a tool of great importance, but researchers have recently questioned their biological relevance. No doubt, the clustering of reads has effectively minimized the errors during sequencing. Still, it's been hypothesized that the method affects the original phylogenetic diversity as the discrimination between some similar taxonomic categories is not effectively done D'Argenio 2018). Therefore, other methods like sub-OTUs are coming into the light, for example, Divisive Amplicon Denoising Algorithm 2 (DADA2), UNOISE2, zero-radius OTUs (ZOTUs), etc. (Callahan et al. 2016; Edgar 2017).

17.2.3.1 Limitation of 16S rRNA Sequencing

1. Only be used for bacterial species identification.
2. The richness of species can sometimes be overestimated. This can be a result of sequencing errors.
3. It can only do taxonomic classification. The biological function is not included.
4. It lacks statistical analysis and modeling.

17.2.4 Shotgun Sequencing

To overcome the difficulties in 16S rRNA sequencing shotgun sequencing method was developed. The genomic content of the whole microbial community can be easily analyzed by this sequencing method. The approach to completing the job is quite similar to the way used to sequence a single bacterial genome (D'Argenio et al. 2014; D'Argenio et al. 2016), but we have an advantage here, i.e., here we obtain sequences of all the microbiota and even of the host. For the taxonomic arrangement of microbiota, it is necessary to have updated databases and highly efficient assembly tools like Bambus2, Genovo, Meta-IBDA, MetaVelvet, Ray-Meta, SOAPdenovo, etc. (D'Argenio 2018). We cannot conclude which assembly tool is best to use as assembling depends on both technical and biological factors that make one tool better in one case and the other in other cases (Quince et al. 2017). Even after having a bunch of bioinformatic tools for performing shotgun sequencing, we are still not able to complete it on viruses. Only a single tool is available for working with viruses called ViromeScan (Rampelli et al. 2016). Pangenome-based Phylogenomic Analysis (PanPhlAn) is a tool designed by Scholz et al. which is capable of isolating the novel strain and acquiring strain-level resolution from the metagenomic data (Scholz et al. 2016). The shotgun analysis is the unbiased approach to performing the sequencing. Shotgun allows the functional analysis, which is based on the gene prediction. The tool used are FragGeneScan, MetaGeneMark, Glimmer-MG, etc. (D'Argenio et.al. 2018). After identifying gene databases like dbCAN, HUMAnN, and IMG database, MetaRef is used to predict the specific function.

17.2.4.1 Advantages

1. It produces legit data for making a hypothesis about the composition of the microbial community. Moreover, it also predicts the function of the microbiota.
2. Here the total DNA of the community is sequenced independently after extraction producing a large number of DNA reads that can align at different genomic locations.
3. Using this sequencing approach answers the researcher about the composition and function of microbiota.
4. It does an unbiased study.
5. It is capable of differentiating the species based on gene content.

Table 17.2 depicts the comparison list of sequencing technologies used for the study.

17.3 Microbiota Signaling Pathway Influencing Diseases

In our body, there are multiple bidirectional pathways connecting the colonies of the gut's microbiota to the central nervous system (CNS). These pathways may be neural, hormonal, or any immune signaling cascade, as shown in Fig. 17.1. The gut microbiota influences the gastrointestinal (GI) tract cells to produce hormones and neurotransmitters, which may amend the brain to work differently (Clemmensen et al. 2017). It is found that they secrete some metabolite in the body that stimulates the vagus nerve causing the behavior change (Cox and Weiner 2018). CNS, on the other hand, has an effect on the gut microbiota, altering its function. It does so by modulating intestinal motility through the adrenergic nerve signaling system. By affecting a specific type of neurotransmitter that mediates the immune system, the CNS can also alter the activity and composition of microbiota (Collins et al. 2012).

Table 17.2 Comparison of sequencing technologies

Technology	Read length	Maximum insert size
ABI 3730	800 b	>1 Kb
454 FLX	+ 500–600 b	1200 b
Illumina GAIIx	76–101 b	500 b
454 FLX Titanium	300–400 b	800 b
Illumina HiSeq 2000	101–151 b	500 b
PacBio	1100 b	>1 Kb
Illumina MiSeq	36–151 b	500 b
IonTorrent	200 b	400 b

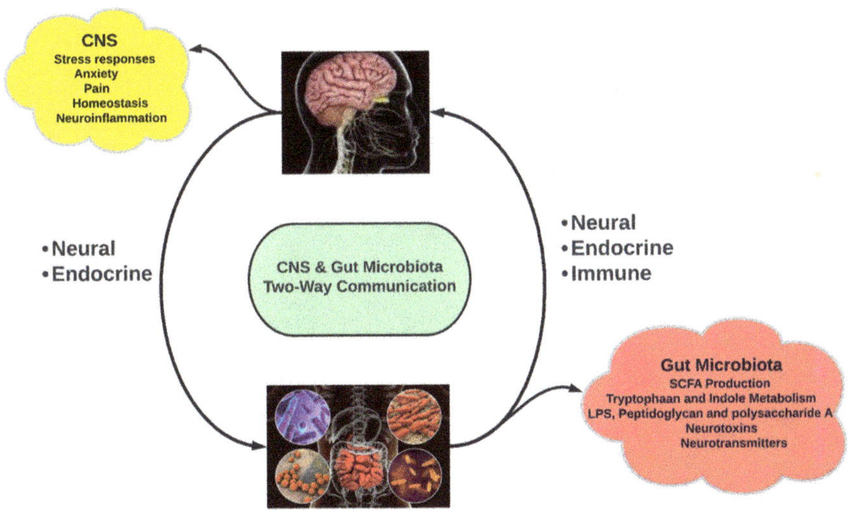

Fig. 17.1 Microbiota Influence on Neural Signaling

17.3.1 Afferent and Efferent Neural Signaling

The GI tract has major signaling pathways which connect the brain to microbiota. The autonomic nervous system has both afferent (GI to CNS) and efferent (CNS to GI) functions (Cox and Weiner 2018). These pathways bring change in motility, epithelial permeability, and secretion, modulating the physical environment and inhibiting the microbiota composition in the body. The signaling pathway recognizes the products of microbiota. *Lactobacillus rhamnosus* reduces depression and anxiety (Bravo et al. 2011). *Lactobacillus reuteri* targets the ion channel associated with sensory neurons, increasing the excitability altering action, potentially causing the change in perception of pain (Cox and Weiner 2018).

17.3.2 Neurotransmitters

The microbiota present in the GI tract secretes the neuroactive peptides like acetylcholine, dopamine, nor-adrenaline, gamma-aminobutyric acid, and 5-hydroxytryptamine (Collins et al. 2012). It is still unknown whether these produced metabolites are crossing the blood–brain barrier or if they act at the local level in an individual's gut (Foster and Neufeld 2013).

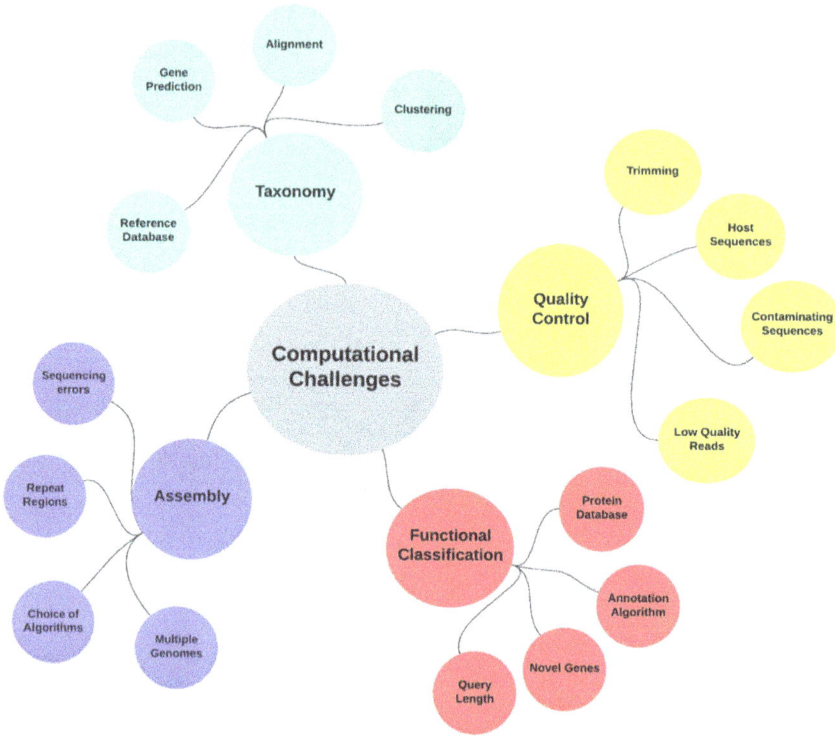

Fig. 17.2 A schematic overview outlining computational challenges allied with sequencing

17.3.3 Neurotoxins

Certain bacterial strain is accomplished with the ability to secrete neurotoxins like in tetanus and botulism, which results from either soft tissue injury or consumption of contaminated food which results from intoxication. Infant botulism is a severe neurological complication caused by the colonization of *Clostridium baratii, Clostridium botulinum,* or *Clostridium butyricum* in the GI tract separating toxins at a low level (Brown and Desai 2013; Fujinaga et al. 2013).

17.4 Microbiota Regulation of Endocrine Signaling

17.4.1 Hypothalamic–Pituitary–Adrenal Axis

The Hypothalamic–Pituitary–Adrenal (HPA) axis is a hormone signaling cascade that is associated with stress and is found to be influenced by the microbiota. The researcher experimentally proves this as they found that mice who don't have microbiota when exposed to stress show elevated HPA response, i.e., decreased cortically, increased hypothalamic corticotropin-releasing factor gene expression protein levels, decreased hippocampal brain-derived neurotrophic factor, and increased corticosterone and plasma adrenocorticotropic hormone. In contrast, these are physiologically normal in mice with microbiota (Cox and Weiner 2018). In studies, it has been found that *Bifidobacterium infantis* in mice lowered the stress. In contrast, the presence of an invasive strain of enteropathogenic *Escherichia coli* increased the stress response. These studies conclude that the microbiota can regulate and influence the physiological response either positively or negatively (Cox and Weiner 2018).

17.4.2 Peptide YY

A digestive hormone called Peptide YY has a chief role in regulating an individual's GI motility and appetite. This hormone has bidirectional regulation, i.e., the brain and microbiota (Wang and Kasper 2014). The hormone is chiefly produced by enteroendocrine cells in the GI tract when the Ffar3 (Gpr41) receptors, i.e., a G-Protein coupled receptor, sense fats, dietary protein, and microbe-derived short-chain fatty acids. This makes microbiota influence the production of Peptide YY, eventually affecting an individual's biological metabolism by bringing change in feeding pattern and GI tract leading to reduced food intake, slowed GI motility, etc. Suppose a low dose of penicillin is administered to an individual. In that case, the alteration in microbial composition occurs, causing a decreased Peptide YY level which may lead an individual to become obese due to increased food intake (Cox et al. 2014).

17.5 Neurodegenerative Disorders

17.5.1 Immune-Mediated Neurologic Diseases

An animal model-based study concluded that microbiota influence immune-mediated neurological diseases (Lee et al. 2011).

17.5.2 MS (Multiple Sclerosis)

Cognitive decline and motor dysfunction are caused by a demyelinating illness mediated by the immune system. According to recent studies, this condition is caused by a change in the microbiome composition. Researchers discovered that the population of *Methanobrevibacter smithii* and *Akkermansia muciniphila* had increased, whereas the population of *Butyricimonas* had decreased (Jangi et al. 2016). The therapy of MS also stabilises the microbial ecosystem, allowing the close link between the two. In a patient with proinflammatory immunological pathways, the population of Methanobrevibacter and Akkermansia increased (van Passel et al. 2011).

17.5.3 PD (Parkinson's Disease)

It is a neurodegenerative disorder that shows motor impairments and non-motor symptoms like anxiety, constipation, depression, sleep disorder, etc. The intestinal barrier of a patient with PD is found to be compromised. When someone is suffering from PD, the population of *Prevotella* gets decreased along with the *Faecalibacteriumprausnitzii,* which produces short-chain fatty acids.

17.5.4 Sequencing and Computational Challenges

Recently, there has been a remarkable development in bioinformatics, leading to new tools and methods (Levy and Myers 2016). Even after such advancement, there are still many difficulties in practicing these tools due to the unavailability of proper biological and metadata (Bharti and Grimm 2021). Many challenges are faced by the researcher while working, as shown in Fig. 2.

17.5.5 Challenges for Amplicon Sequencing Analysis

The difficulty arise in determining the error in sequencing from the actual nucleotides while performing amplicon sequencing. To resolve the error, the OUT approach, which uses tools like QIIME and Mothur, clusters the reads using a predetermined identity threshold (Westcott and Schloss 2015). UNOISE employs the denoising method on a sequence before the sequence fault is introduced, unlike the ASV method, which requires tools like DADA2 and MED (Tikhonov et al. 2015). According to many research, OTUs have a low resolution of results (Callahan et al. 2017).

17.5.6 Challenges of Metagenomic Sequencing Analysis

Since there are so many tools for metagenomic analysis, choosing the right one for the purpose might be challenging. Aside from that, maintaining quality is a difficult task. The elimination of contamination from raw reads and quality trimming both are part of quality control. Contamination removal is the process of detecting and removing contaminating host-associated sequences from reads. Quality trimming, on the other hand, is the filtration of raw reads to detect low-quality sequences.

17.5.7 Challenges in Short-Read Metagenomics

The key advantage of adopting the short-read sequence is that you can get a lot of reads in one go. The difficulty arises when assembling a sample with various genomes such as bacteria, viruses, fungus, archaea, and so on. Their sample contains errors in the sequencing. It may also contain intergenomic and intragenomic repetitions, as well as sequencing coverage in rare cases (Howe et al. 2014; Abram 2015). The problem also arises in predicting the gene because the gene has escape patterns that do not match any of the species-specific models available (Bharti and Grimm 2021).

Moreover, the false-positive rate (FPR) is another issue to be handled as sometimes genes are wrongly predicted. For taxonomic classification, many tools are available that match the given reads to the already known microbial genome in the databases. As the size of the databases is rapidly increasing, it becomes challenging to blast the sequences. Table 17.3 displays the list of some bioinformatics tools based on functional classification.

17.6 Conclusion

The human body is home to billions of microbes. These microbes stay in our body and negatively or positively regulate our body's metabolism. The research shows their influence on our body is more than we had ever thought. This is why a detailed study of these microbiotas is required. For doing so at the molecular level, researchers use various bioinformatics tools. There are plenty of tools available for specific jobs, and these tools are classified above. The Human Microbiome Project is among many other projects launched to study the microbiota community. Metagenomics, meta-transcriptomics, and metabolomics are the three most commonly used approaches for doing the job.

Moreover, 16S rRNA sequencing is the most common technique to identify microbes. Still, as this technique has many limitations, it is replaced by the short-gun sequencing technique, which has many advantages. The main drawback of the 16S rRNA technique is that it is limited to bacterial species and cannot be used for other types of microbes.

Further, studies also conclude that the microbiota has a significant influence on the human body in regulating behavior, feeding pattern, coping with stress and anxiety, etc. It affects neural functionality and immune responses. In different

Table 17.3 List of bioinformatics tools based on functional classification

Tools and approaches	Sub-type
NCBI RefSeq	Homology-based tools
UniProt	
SMART	
IMG/M	
MG-RAST	
Pfam	
TIGRFAM	
PROSITE	Motif- or pattern-based tools
PRINTS	
InterPro	
IMG/M	Context-based tools
SmashCommunity	
CAZy	Other functional prediction tools
PSORT	
CELLO	
Lipo	
DOLLOP	
SignalP	
ISsaga	
VFDB	
MvirDB	

diseases, they act differently, and the population of specific microbe either increases or decreases depending on the condition. Even after technological advancement, we still face some computational analysis issues like assembling, taxonomic classification, and controlling the quality.

17.7 Future Perspective

A detailed study on the microbe community present in our body should be done. While working, we find a strong relationship between these microbes and the body. The bidirectional regulation has been spotted here, concluding both body and microbes influence each other. Therefore, we hypothesized that there are many more relations between them that still need to be figured out. So, we can make a diagnosis of disease by studying these microbes. Moreover, we found that many difficulties in the whole computational process need to be addressed.

Conflict of Interest Author declares no conflict of interest.

References

Abram F (2015) Systems-based approaches to unravel multi-species microbial community functioning. Comput Struct Biotechnol J 13:24–32
Ackerman J (2012) The ultimate social network. Sci Am 306(6):36–43
Aguiar-Pulido V, Huang W, Suarez-Ulloa V, Cickovski T, Mathee K, Narasimhan G (2016) Metagenomics, metatranscriptomics, and metabolomics approaches for microbiome analysis: supplementary issue: bioinformatics methods and applications for big metagenomics data. Evol Bioinforma 12:EBO-S36436
Bharti R, Grimm DG (2021 Jan) Current challenges and best-practice protocols for microbiome analysis. Brief Bioinform 22(1):178–193
Bravo JA, Forsythe P, Chew MV et al (2011) Ingestion of Lactobacillus strain regulates emotional behavior and central GABA receptor expression in a mouse via the vagus nerve. Proc Natl Acad Sci U S A 108(38):16050–16055
Brown N, Desai S (2013) Infantile botulism: a case report and review. J Emerg Med 45(6):842–845
Callahan BJ, McMurdie PJ, Holmes SP (2017) Exact sequence variants should replace operational taxonomic units in marker-gene data analysis. ISME J 11:2639–2643
Callahan BJ, McMurdie PJ, Rosen MJ, Han AW, Johnson AJ, Holmes SP (2016) DADA2: high-resolution sample inference from Illumina amplicon data. Nat Methods 13:581–583
Caporaso JG, Kuczynski J, Stombaugh J et al (2010) QIIME allows analysis of highthroughput community sequencing data. Nat Methods 7:335–336
Clemmensen C, Müller TD, Woods SC et al (2017) Gut–brain crosstalk in metabolic control. Cell 168(5):758–774
Cole JR, Wang Q, Fish JA et al (2014) Ribosomal Database Project: data and tools for high throughput rRNA analysis. Nucleic Acids Res 42:D633–D642
Collins SM, Surette M, Bercik P (2012) The interplay between the intestinal microbiota and the brain. Nat Rev Microbiol 10(11):735–742
Cox LM, Weiner HL (2018 Jan) Microbiota signaling pathways that influence neurologic disease. Neurotherapeutics 15(1):135–145
Cox LM, Yamanishi S, Sohn J et al (2014) Altering the intestinal microbiota during a critical developmental window has lasting metabolic consequences. Cell 158(4):705–721
D'Argenio V (2018) Human microbiome acquisition and bioinformatic challenges in metagenomic studies. Int J Mol Sci 19(2):383
D'Argenio V, Notomista E, Petrillo M, Cantiello P, Cafaro V, Izzo V, Naso B, Cozzuto L, Durante L, Troncone L et al (2014) Complete sequencing of Novosphingobium sp. PP1Y reveals a biotechnologically meaningful metabolic pattern. BMC Genomics 15:384
D'Argenio V, Petrillo M, Pasanisi D, Pagliarulo C, Colicchio R, Talà A, de Biase MS, Zanfardino M, Scolamiero E, Pagliuca C et al (2016) The complete 12 Mb genome and transcriptome of Nonomuraeagerenzanensis with new insights into its duplicated "magic" RNA polymerase. Sci Rep 6:18
DeSantis TZ, Hugenholtz P, Larsen N, Rojas M, Brodie EL, Keller K, Huber T, Dalevi D, Hu P, Andersen GL (2006) Greengenes, a chimera-checked 16S rRNA gene database and workbench compatible with ARB. Appl Environ Microbiol 72:5069–5072
Edgar RC (2017) UNOISE2: Improved error-correction for Illumina 16S and ITS amplicon sequencing. BioRxiv
Ehrlich SD, MetaHIT Consortium (2010) Metagenomics of the intestinal microbiota: potential applications. Gastroenterol Clin Biol 34:S23–S28
Foster JA, Neufeld K-AM (2013) Gut–brain axis: how the microbiome influences anxiety and depression. Trends Neurosci 36(5):305–312
Fujinaga Y, Sugawara Y, Matsumura T (2013) Uptake of botulinum neurotoxin in the intestine. In: Rummel A, Binz T (eds) Botulinum neurotoxins. Springer Berlin Heidelberg, Berlin, Heidelberg, pp 45–59

Goecks J, Nekrutenko A, Taylor J (2010) Galaxy: a comprehensive approach for supporting accessible, reproducible, and transparent computational research in the life sciences. Genome Biol 11:R86

Goedert JJ, Hua X, Yu G, Shi J (2014) Diversity and composition of the adult fecal microbiome associated with history of cesarean birth or appendectomy: analysis of the American gut project. EBioMedicine 1(2):167–172

Gonzalez A, Knight R (2012) Advancing analytical algorithms and pipelines for billions of microbial sequences. Curr Opin Biotechnol 23:64–71

Hong C, Manimaran S, Shen Y et al (2014) PathoScope 2.0: a complete computational framework for strain identification in environmental or clinical sequencing samples. Microbiome 2(1):33

Howe AC, Jansson JK, Malfatti SA et al (2014) Tackling soil diversity with the assembly of large, complex metagenomes. Proc Natl Acad Sci U S A 111:4904–4909

Integrative HMP (iHMP) Research Network Consortium (2014) The integrative human microbiome project: dynamic analysis of microbiome-host omics profiles during periods of human health and disease. Cell Host Microbe 16(3):276–289

Jangi S, Gandhi R, Cox LM et al (2016) Alterations of the human gut microbiome in multiple sclerosis. Nat Commun 7:12015

Kim M, Lee K-H, Yoon S-W, Kim B-S, Chun J, Yi H (2013) Analytical tools and databases for metagenomics in the next-generation sequencing era. Genomics Inform 11(3):102–113

Krumsiek J, Mittelstrass K, Do KT et al (2015) Gender-specific pathway differences in the human serum metabolome. Metabolomics 11(6):1815–1833

Lee YK, Menezes JS, Umesaki Y, Mazmanian SK (2011) Proinflammatory T-cell responses to gut microbiota promote experimental autoimmune encephalomyelitis. Proc Natl Acad Sci U S A 108(Suppl. 1):4615–4622

Levy SE, Myers RM (2016) Advancements in next-generation sequencing. Annu Rev Genomics Hum Genet 17:95–115

Manor O, Levy R, Borenstein E (2014) Mapping the inner workings of the microbiome: genomic- and metagenomic-based study of metabolism and metabolic interactions in the human microbiome. Cell Metab 20(5):742–752

Meyer F, Paarmann D, D'Souza M et al (2008) The metagenomics RAST server – a public resource for the automatic phylogenetic and functional analysis of metagenomes. BMC Bioinformatics 9(1):386

Moran MA (2009) Metatranscriptomics: eavesdropping on complex microbial communities. Microbiome 4(7):329–334

Morgan XC, Huttenhower C (2012) Chapter 12: Human microbiome analysis. PLoS Comput Biol 8(12):e1002808

Peterson J, Garges S, Giovanni M et al (2009) The NIH Human Microbiome Project. Genome Res 19:2317–2323

Qin J, Li R, Raes J, Arumugam M, Burgdorf KS et al (2010) A human gut microbial gene catalogue established by metagenomic sequencing. Nature 464:59–65

Quast C, Pruesse E, Yilmaz P, Gerken J, Schweer T, Yarza P, Peplies J, Glöckner FO (2013) The SILVA ribosomal RNA gene database project: Improved data processing and web-based tools. Nucleic Acids Res 41:D590–D596

Quince C, Walker AW, Simpson JT, Loman NJ, Segata N (2017) Shotgun metagenomics, from sampling to analysis. Nat Biotechnol 35:833–844

Rampelli S, Soverini M, Turroni S, Quercia S, Biagi E, Brigidi P, Candela M (2016) ViromeScan: a new tool for metagenomic viral community profiling. BMC Genomics 17:165

Schloss PD, Westcott SL, Ryabin T et al (2009) Introducing Mothur: open-source, platform-independent, community-supported software for describing and comparing microbial communities. Appl Environ Microbiol 75(23):7537–7541

Scholz M, Ward DV, Pasolli E, Tolio T, Zolfo M, Asnicar F, Truong DT, Tett A, Morrow AL, Segata N (2016) Strain-level microbial epidemiology and population genomics from shotgun metagenomics. Nat Methods 13:435–438

Stark M, Berger SA, Stamatakis A, von Mering C (2010) Mltreemap-accurate maximum likelihood placement of environmental DNA sequences into taxonomic and functional reference phylogenies. BMC Genomics 11(1):461

Tikhonov M, Leach RW, Wingreen NS (2015) Interpreting 16S metagenomic data without clustering to achieve sub-OTU resolution. ISME J 9:68–80

van Passel MWJ, Kant R, Zoetendal EG et al (2011) The genome of Akkermansiamuciniphila, a dedicated intestinal mucin degrader, and its use in exploring intestinal metagenomes. PLoS One 6(3):e16876

Wang Y, Kasper LH (2014) The role of microbiome in central nervous system disorders. Brain Behav Immun 38(C):1–12

Westcott SL, Schloss PD (2015) De novo clustering methods outperform reference-based methods for assigning 16S rRNA gene sequences to operational taxonomic units. PeerJ 3:e1487